Cell Signalling

To Thomas and Annabel

Cell Signalling

THIRD EDITION

John T. Hancock

Reader in Molecular Biology
School of Life Sciences
Faculty of Health and Life Sciences
University of the West of England, Bristol

OXFORD
UNIVERSITY PRESS

OXFORD

UNIVERSITY PRESS

Great Clarendon Street, Oxford OX2 6DP

Oxford University Press is a department of the University of Oxford.
It furthers the University's objective of excellence in research, scholarship,
and education by publishing worldwide in

Oxford New York

Auckland Cape Town Dar es Salaam Hong Kong Karachi
Kuala Lumpur Madrid Melbourne Mexico City Nairobi
New Delhi Shanghai Taipei Toronto

With offices in

Argentina Austria Brazil Chile Czech Republic France Greece
Guatemala Hungary Italy Japan Poland Portugal Singapore
South Korea Switzerland Thailand Turkey Ukraine Vietnam

Oxford is a registered trade mark of Oxford University Press
in the UK and in certain other countries

Published in the United States
by Oxford University Press Inc., New York

British Library Cataloguing in Publication Data

Data available

Library of Congress Cataloging in Publication Data

Data available

Typeset by MPS Limited, A Macmillan Company
Printed in Great Britain on acid free paper by
Ashford Colour Press Ltd, Gosport, Hampshire

ISBN 978–0–19–923210–9

5 7 9 10 8 6

■ PREFACE

This is the third edition of *Cell Signalling*, and hopefully this text will continue to be useful for those that wish to find out more about the subject. The original book was written with the intention that it would be useful for degree students in their second and final years of undergraduate studies. It was also hoped that the book would be of interest to those on a medical degree, or those embarking on postgraduate studies, where they were perhaps new to this area or need to refresh earlier studies. It is therefore hoped that now it is the third edition that this vein has continued. However, in re-writing the text a second time, many new diagrams have again been added, with some of the ones in the second edition being re-drawn for clarity. Colour is still included to highlight important aspects of the figures. The text has once again been altered to attempt to make it more readable and accessible, and this is still at the detriment of many details that some may find are missing. Many more references have been added, but unfortunately for those who actually do the experimental work, it is impossible to quote all the primary references, and reviews have been prioritized to enable the reader to work their way through the vast amount of literature on the topic.

The field of cell signalling continues to be very large and to expand at what seems like an exponential rate, with a great deal of new literature published in the journals every month. Here the aim was to describe the main components used by cells in their communications, with discussion on the ways in which they interact. This is not an approach taken by many books which cover this area of biochemistry, but it is becoming apparent that pathways are not just single lines of signal transducers leading simply from the original signal to the cellular response, but an intricate inter-play of many components, with one pathway having profound effects on another. Therefore, the discussion of such components in relative isolation allows the knowledge gained about one particular component to be used when considering many pathways, not just the one in which it was first discovered. The aim was also to have a very generic approach, as the ubiquity of some of these systems means that an understanding of the mechanisms is relevant to all organisms throughout nature and not just mammals, or indeed the animal kingdom.

It has proved impossible to do justice to every area in this field for this edition, and the overall emphasis still reflects my interests so some people may feel that I have overlooked their particular area of research. To those that feel ignored I once again apologize.

The book has now been partitioned, too. Part 1 is an overview of signalling and the principles involved. Part 2 deals with the components that may be involved in more depth. Part 3 is a selection of some pathways and cellular events that help to illustrate the role and activity of signalling in cells.

The final part is once again some final thoughts and ideas that the reader may reflect on.

Chapter 1 contains further reading, which includes several excellent chapters published in major textbooks covering cell signalling, as well as other books published on this topic, and as such would be a good place to start further reading.

The book starts with an overview chapter, encompassing many of the ideas and discussions that could be applicable to all areas of cell signalling. Chapter 2 is new and highlights how signalling components are grouped together in pathways. Chapter 3 concentrates on the history and an overview of some of the main techniques that could be used to study cell signalling. Chapters 4 and 5 concentrate on the use of extracellular signals and their perception, whereas Chapters 6–10 concentrate on the intracellular components thought to be of most importance. Chapters 11 and 12 take more specific examples of signalling pathways and try to show how these components might come together to complete pathways. Chapter 13 is a new chapter on some, but by no means all, of the signalling components that may be involved in the control of gene expression and development, whereas Chapter 14 focusses on cell death. The final chapter tries to brings some ideas and thoughts together, and summarizes some of the discussion from the rest of the book.

I am as always indebted to all the scientists who did the work and published their papers so I could obtain the information needed for this book. I wish to particularly thank those who deposited structural data, which was used for several figures in the book. These structures were obtained from the RCSB Protein Data Bank (www.rcsb.org/pdb/), to whom I would like to say a big thank you.

As with the original, the content of the book was greatly shaped by many conversations, and e-mails, with colleagues. I am indebted to many of my colleagues at the University of the West of England (UWE), Bristol, including Heather Macdonald for sending me papers, and collaborators elsewhere, such as Tihana Teklic and Hrvoje Lepedus in Osijek, Croatia. I must also thank the OUP editors, particularly Jonathan Crowe and Dewi Jackson. Jon who not only persuaded me to do the 3rd edn, but also made very useful comments along the way. Meanwhile, Dewi kept me on track and helped greatly. Lastly I would like to again thank my wife Sally-Ann, and Thomas and Annabel for their never-ending help and, not least, encouragement.

■ CONTENTS

▪ LIST OF FIGURES

■ LIST OF TABLES

■ LIST OF ABBREVIATIONS

2-D	two-dimensional
7TM	seven-transmembrane receptors
AA	arachidonic acid
ABA	abscisic acid
ADP	adenosine diphosphate
ADAM	a disintegrin and metalloproteinase
AF-1	activation function 1
AF-2	activation function 2
AHL	N-acetyl-L-homoserine lactone
AIF	apoptosis-inducing factor
AP-1	activator protein 1
Apaf-1	apoptotic protease-activating factor 1
April	a proliferation-inducing ligand
βARK	β-adrenergic receptor kinase
AMP	adenosine monophosphate
AMPK	5'-AMP-activated protein kinase
AMPKK	5'-AMP-activated protein kinase kinase
AOS	active oxygen species
APC	adenomatous polyposis coli
ARF	ADP-ribosylation factor
ATP	adenosine triphosphate
BMP	bone morphogenetic protein
Boss	Bride of sevenless
bp	base pair
cADPr	cyclic ADP-ribose
$[Ca^{2+}]_i$	intracellular calcium ion concentration
cAMP	cyclic adenosine monophosphate (adenosine 3',5'-cyclic monophosphate)
cAPK	cAMP-dependent protein kinase (otherwise known as PKA)
CAP	Cbl activating protein (see Chapter 9)
CAP	catabolic gene activator protein
CAPP	ceramide-activated protein phosphatase
CARD	caspase-activation and -recruitment domain
caspase	cysteinyl-aspartate-specific protease
CaVs	voltage-gated calcium ion-selective channels
CBP	CRB-binding protein
CCE	capacitative Ca^{2+} entry

cCMP	cyclic cytidine monophosphate (cytidine 3',5'-cyclic monophosphate)
CDK	cyclin-dependent kinase
cDNA	complementary deoxyribonucleic acid
CGD	chronic granulomatous disease
cGI-PDE	cGMP inhibited phosphodiesterases
cGMP	cyclic guanosine monophosphate (guanosine 3',5'-cyclic monophosphate)
cGPK	cGMP-dependent protein kinase
Ci	Cubitis interruptus
CIF	Ca^{2+} influx factor
cIMP	cyclic inosine monophosphate
CK	cytokinin
CK1	casein kinase 1
CO	carbon monoxide
Cos2	Costal-2
CRADD	caspase and RIP adaptor with death domain
CRD	cysteine-rich domain
CRE	cAMP-response element
CREB	CRE-binding protein
CRLR	calcitonin receptor-like receptor
CRP	cAMP receptor protein
CSL	CBF1, Suppressor of hairless, Lag-1
DAF-2DA	diaminofluorescein diacetate
DAG	diacylglycerol
DBD	DNA-binding domain
DED	dead effector domain
DGK	diacylglycerol kinase
DISC	death-inducing signalling complex
DNA	deoxyribonucleic acid
DR	death receptor
Dsh	Dishevelled
DUBs	de-ubiquitylating enzymes
EDRF	endothelium-derived relaxing factor
EET	epoxyeicosatrienoic acids
EGF	epidermal growth factor
E. coli	Escherichia coli
ER	endoplasmic reticulum
ERE	estrogen-response element
ERK	extracellular signal-regulated kinase
ESI	electrospray ionization mass spectroscopy
FAD	flavin adenine dinucleotide

FADD	Fas-associated protein with death domain
FGF	fibroblast growth factor
FLIM	fluorescence lifetime imaging microscopy
FLIP	FLICE-like inhibitory protein
FMN	flavin mononucleotide
FRET	fluorescence resonance energy transfer
FSH	follicle stimulating hormone
Fu	Fused
Fz	Frizzled
GA	gibberellin
GABA	γ-aminobutyric acid
GAP	GTPase activating protein
GAPDH	glyceraldehyde-3-phosphate dehydrogenase
G-CSF	granulocyte colony-stimulating factor
GDP	guanosine diphosphate
GEF	guanine nucleotide exchange factor
GFP	green fluorescent protein
GM-CSF	granulocyte-macrophage colony-stimulating factor
GMP	guanosine monophosphate
GNRP	guanine nucleotide releasing protein
GPCR	G protein-coupled receptors
GPx	glutathione peroxidise
GRB2	growth factor receptor-bound protein 2
GRK	G protein-coupled receptor kinases
GSH	glutathione (reduced)
GSK3	glycogen synthase kinase 3
GSNO	S-nitroso-glutathione
GSSG	glutathione (oxidized)
GTP	guanosine triphosphate
GTPγ	guanosine 5' –[γ-thiol] triphosphate
H_2S	hydrogen sulphide
HCR	haem-controlled repressor
HETE	hydroxyeicosatetraenoic acids
HGF	hepatocyte growth factor
Hh	Hedgehog
HR	hypersensitive response
HPETE	hydroperoxyeicosatetraenoic acids
HRE	hormone-response element
HSC	hedgehog signaling complex
Hsp	heat-shock protein
IAA	indole-3-acetic acid
IAP	inhibitor of apoptosis protein

ICC model	$InsP_3$-Ca^{2+} cross-coupling model
ICE	interleukin-1β-converting enzyme
I_{CRAC}	Ca^{2+}-release-activated-current
I_{DAC}	depletion-activated-current
IFN	interferon
IGF	insulin-like growth factor
IKK	IκB kinase
IL	interleukin
$InsP_3$	inositol 1,4,5-trisphosphate (IP_3)
$InsP_3R$	inositol 1,4,5-trisphosphate receptor
$InsP_4$	inositol 1,3,4,5-tetrakisphosphate (IP_4)
$InsP_6$	inositol hexaphosphate (IP_6)
IP_3	inositol 1,4,5-trisphosphate
IP_4	inositol 1,3,4,5-tetrakisphosphate
IP_6	inositol hexaphosphate
IRAK	IL-1 receptor associated kinase
IRBIT	$InsP_3R$-binding protein released with $InsP_3$
IRE	insulin-responsive element
IRS1	insulin receptor substrate 1
ISPK	insulin-sensitive protein kinase or insulin-stimulated protein kinase
JA	jasmonic acid
JAK	Janus kinases
JIPS	jasmonate-induced proteins
LBD	ligand binding domain
LEF	lymphoid enhancer factor
LH	luteinizing hormone
LHC	light-harvesting chorophyll a/b complex
L-NAA	L-N$^\omega$ aminoarginine
L-NAME	L-N$^\omega$ arginine-methyl ester
L-NMA	L-N$^\omega$ methylarginine
LPA	lysophosphatidic acid
LPC	lysophosphatidylcholine
LPS	lipopolysaccharide
LRP	low density lipoprotein receptor-related protein
LTBP	latent TGFβ-binding protein
LTN	lymphotactin
Mae	modulator of the activity of Ets
MALDI-TOF	matrix assisted laser desorption ionization-time of flight
MAMPs	microorganism-associated molecular patterns
MAPK	mitogen-activated protein kinase
MAPKK	mitogen-activated protein kinase kinase

MAPKKK	mitogen-activated protein kinase kinase kinase
MCP	monocyte chemoattractant protein
MEK	MAP/ERF kinase
MEKK	MEK kinase
MiCa	mitochondrial Ca^{2+} channel
MIG	monokine induced by IFN-γ
MIP	macrophage inflammatory protein
MLCK	mysosin light chain kinase
mRNA	messenger ribonucleic acid
MSP	macrophage-stimulating protein
$NAADP^+$	nicotinate adenine-dinucleotide phosphate
NAD^+	nicotinamide adenine dinucleotide (oxidized)
NADH	nicotinamide adenine dinucleotide (reduced)
$NADP^+$	nicotinamide adenine dinucleotide phosphate (oxidized)
NADPH	nicotinamide adenine dinucleotide phosphate (reduced)
NAP	neutrophil activating protein
NBC1	Na^+/HCO_3^-cotransporter 1
NF-κB	nuclear factor κB
NICD	intracellular domain of Notch
NMR	nuclear magnetic resonance
NO	nitric oxide
NOS	nitric oxide synthase
PA	phosphatidic acid
PAGE	polyacrylamide gel electrophoresis
PAMPs	pathogen-associated molecular patterns
PAP	phosphatidate phosphohydrolase
PC	phosphatidylcholine
PCD	programmed cell death
P-choline	phosphorylcholine
PCR	polymerase chain reaction
PDE	cyclic nucleotide phosphodiesterase
PDGF	platelet-derived growth factor
PDK1	3-phosphoinositide-dependent kinase
PE	phosphatidylethanolamine
PEPCK	phosphoenolpyruvate carboxylase
PF	platelet factor
PGHS	prostaglandin G/H synthase or cyclooxygenase
PH domain	pleckstrin homology domain
P_i	inorganic phosphate
PI	phosphatidylinositol
PIP_2	phosphatidylinositol 4,5-bisphosphate

PKA	cAMP-dependent protein kinase
PKB	protein kinase B
PKC	protein kinase C
PLA$_2$	phospholipase A$_2$
PLC	phospholipase C
PLD	phospholipase D
PM	plasma membrane
PMA	phorbol 12-myristate 13-acetate
PMCA	plasma membrane Ca^{2+} ATPase
PP1	protein phosphatase 1
PPi	inorganic pyrophosphate
PRRs	pattern-recognition receptors
Ptc	Patched
PtdIns	phosphatidylinositol (PI)
PtdIns 3-kinase	phosphatidylinositol 3-kinase
PtdIns 4-kinase	phosphatidylinositol 4-kinase
PtdIns4P	phosphatidylinositol 4-phosphate
PtdInsP$_2$	phosphatidylinositol 4,5-bisphosphate (PIP$_2$)
PtdInsP$_3$	phosphatidylinositol 3,4,5-trisphosphate
PTP	protein tyrosine phosphatase
RAMP	receptor-activating-modifying protein
Rantes	regulated on activation, normal T-cell expressed and secreted
RGS	regulators of G protein signalling
RING	really interesting new gene (domain)
RNA	ribonucleic acid
RNAi	RNA interference
RNS	reactive nitrogen species
ROS	reactive oxygen species
RTK	receptor tyrosine kinase
RT-PCR	real-time polymerase chain reaction
RyR	ryanodine receptor
SAGE	serial analysis of gene expression
S. cerevisiae	*Saccharomyces cerevisiae*
SDS	sodium dodecyl sulphate
SERCA	smooth endoplasmic reticulum calcium ATPase
SH2 domain	src homology domain 2
SH3 domain	src homology domain 3
siRNAs	small interfering RNAs
SM	sphingomyelin
SMDF	sensory and motor neuron derived factor
Smo	Smoothened

SOCs	store-operated channels
SOD	superoxide dismutase
SPC	sphingosylphosphorylcholine
Sos	Son of sevenless
SR	sarcoplasmic reticulum
STAT	signal transducers and activators of transcription
SuFu	Suppressor of fused
TAK1	TGFβ-activated kinase 1
TCF	T-cell factor
TGF	transforming growth factor
TIR	Toll-interleukin 1 receptor domain
TLR	Toll-like receptor
TNF	tumour necrosis factor
TNF-R	tumour necrosis factor receptor
TPA	12-O-tetradecanoyl-phorbol-12-acetate
Trail	tumour necrosis factor-related apoptosis-inducing ligand
Trance	tumour necrosis factor-related activation induced cytokine
TRIF	TIR-containing adaptor protein inducing IFN-β
TRP ion channels	transient receptor potential ion channels
TSP-1	thrombospondin
VSP	vegetative storage protein
XO	xanthine oxidase
XOR	xanthine oxidoreductase

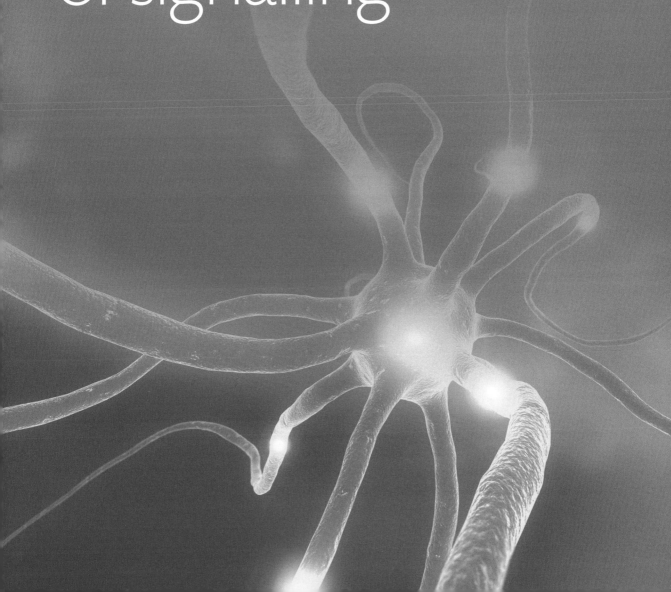

Part 1

An overview
of signalling

Aspects of cellular signalling

1

Cell signalling has become a vital and integral part of modern biology, and has an innate complexity. It controls the inner workings of organisms, allowing them to respond, adapt and survive. However, the basic workings of cell signalling events are not vastly diverse across different organisms, but rather the regulatory needs of organisms' cells are similar. Principles and mechanisms can be seen to be repeated across the kingdoms of species. The molecules that constitute the signals, the ways they travel to their targets, and the ways that they are perceived are similar, whether the organism is a plant or an animal. The details may be different, and there are many unique aspects that can be seen to break the rules, but in Chapter 1 some of the more overlying aspects are discussed, which should help in the understanding of the chapters to follow.

1.1 Introduction

Cell signalling has arguably become one of the most important aspects of modern biochemistry and cell biology. An understanding of aspects of cell signalling is vital to a wide range of biologists, from those who are investigating the causes of cancer, to those who are concerned about the impact of environmental pollutants on the ecosystem.

All cells, whether they live as individuals or in a multicellular organism, are bombarded by signals in many forms in a continual manner. It is the ability of organisms, or individual cells within an organism, to sense and respond to their environment that is crucial to their survival. All cells must have the ability to detect the presence of extracellular molecules and conditions, and must also be able to instigate a range of intracellular responses. Such systems have to be carefully orchestrated and controlled. To enable living organisms to do this, a complex and interwoven range of signalling pathways has evolved. Signalling systems of a single cell organism are complex enough, but a multicellular organism has to coordinate the functioning of cells that may be adjacent, or a huge distance apart, possibly in the order of metres. Such cells may also have extremely specific and different functions.

The main principles and components behind the signalling mechanisms are essentially the same across the diverse range of organisms, from bacteria, fungi, plants and animals. It is the fine detail of the components that gives the specificity or unique functionality that is required for a particular cell. In today's research, the knowledge and understanding of a signalling system in one tissue or species is often used to speed the discovery of an analogous system in a completely different tissue or species. For example, systems discovered and characterized in mammals are now sought in plants, with the tools used for the original research being adopted for this new use. Much of the early work on organism development was carried out on the fruit fly, but is applicable to human biology. This can greatly enhance the rate of new discoveries, but, to add a note of caution, may also throw up some anomalies. Even between what may be regarded as very similar species, for example mice and man, the specific differences may be very important. Having said that, it certainly holds that the underlying principles and mechanisms are similar across a broad range of organisms, and knowledge of one is often transferrable to another, for example humans.

■ The use of animal models is further discussed in Chapter 13, along with references discussing their relevance.

Cell signalling is not only important for the understanding of the functioning of a normal cell, but also is of vital importance to understand the growth and activity of an aberrant cell, or that of a cell that is combating adverse conditions. The discovery of oncogenes, genes that cause the uncontrolled growth of cells which may lead to cancerous growths, was heralded as a major breakthrough in the understanding of cancer in humans, and the discovery of cytokines held great hopes to be a cure for a variety of diseases. It is the role and impact of such gene products and blood-borne molecules on the cell signalling pathways that ensued in cells which made them so important, and highlights the importance of an understanding of cell signalling mechanisms to modern biology.

The literature in the field of cell signalling is still growing exponentially, with several journals dedicated to the publication of research in this field. Other journals specifically highlight papers on cell signalling either in the journals themselves or on web pages and E-mail alerts. In early March 2009, entering "cell signalling" and the American spelling "cell signaling" into PubMed returned 361,604 entries. This is because, following elucidation of the functioning of a protein, it is often the control of its activity and the control of its synthesis to which the researcher turns to get a better understanding of how the protein fits into the overall biochemistry of the cell. The literature is expanding at such an alarming rate, as more researchers realize the importance of the subject to their own interests. For instance, entering "cell signaling disease" into PubMed returned 32,281 entries, a number which is no doubt higher on the day that you read this book. It is hard for any single text to do justice to the subject. Here, the aims of the book are to discuss the main components found in cell signalling pathways, rather than describe many specific pathways that might contain common elements. The details in the text are not directly referenced, but further reading is given at the end of each chapter to enable the reader to refer to the original research or to read review articles on specific areas.

> **PubMed** is a publicly available database of scientific literature, a service of the U.S. National Library of Medicine and the National Institutes of Health. It can be accessed at: http://www.ncbi.nlm.nih.gov/sites/entrez/

1.2   The main principles of cell signalling

The main underlying event for many cell signalling mechanisms is the arrival at the cell of something that requires the cell to respond. This might be the arrival of light photons, as in the case of cells in the eyes of humans and other animals. Likewise, the arrival may be a chemical released from another organism, or from another cell in the same organism. Most, but not all, signals arrive at, and are perceived at, the outer borders of the cell, commonly the plasma membrane. Therefore, a scenario such as shown in **Figure 1.1** is common.

Several events follow the arrival of a signal:

- Perception of the signal, usually by dedicated proteins referred to as receptors.
- Transmission of the signal by the receptor into the cell.
- Passing on of the "message" to a series of cell signalling components, often referred to as a cell signalling cascade.
- The arrival of the "message" at the final destination in the cell.
- A response by the cell, an action carried out so that the outcome is appropriate to the original signal, or stress.

The response may be anywhere in the cell, and the cascade will be designed to carry the "message" to a defined place. Some cascades end in the cytoplasm, for example in the control of glycogen metabolism, whereas others might end in the nucleus, as is seen with the control of gene expression.

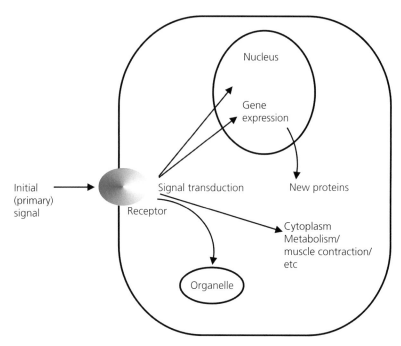

Figure 1.1 The main event in cell signalling may be the arrival of a signal at the cell surface. A cascade of events commonly follows, carrying the "message" to its final destination, perhaps the cytoplasm or the nucleus.

Several points should of course also be borne in mind. A single signal arriving at a cell's surface may lead to more than one outcome, and a single signal may lead to the activation and use of more than one signalling cascade, for example hormones that signal through phospholipase C initiate both inositol signalling and kinase signalling (see Chapter 8). These cascades might share components with other signalling cascades, too. Secondly, an extremely small amount of the signal might be perceived by the cell, but a rather large response might be needed. For example, a few molecules of hormone arriving at the cell surface might require a large number of individual enzyme proteins to be activated, and therefore an element of amplification is required by the system. Thirdly, cells do not have the luxury of accepting signals neatly one at a time, but they are likely to be bombarded by them. A myriad of hormones, cytokines etc., are present in human blood, and a cell may wish to respond to a selection of these simultaneously. However, several signals may lead to the activation or inactivation of the same signalling component in the cell, and the idea of crosstalk, where one signalling transduction pathway is influenced by another, has been introduced to aid in the elucidation of such complications. Also, a cell does not respond to all signals present, but rather a defined array, tailored for that cell or type of cell. For example, in a leaf some cells such as the guard cells of the stomata might need to respond to hormones that signal that the plant is short of water, but the rest of the leaf might not need to respond. Similarly, some cells of the mammalian immune system might need to respond to a pathogen attack, but not all the cells.

Therefore, in the discussion of cell signalling such principles and factors need to be borne in mind, and hopefully the more in-depth discussion within this book will help to unravel how cell signalling manages both the subtlety and the complexity that is required for cells to survive and thrive.

1.3 What makes a good signal?

Signals reported in the literature come in a wide range of various shapes and sizes, some of which appear to be almost bizarre, but they generally have a set of common characteristics. Probably the most important characteristic is that a signal must have specificity. A signal must be unique enough to relay a defined signal, and to only be detected by the molecular machinery designed for that detection. If made and perceived incorrectly, the signal will have failed to relay its specific message.

To be an effective signal, a molecule usually is relatively small and therefore able to travel from the site of manufacture to the target site reasonably easily. Extracellular signals can take advantage of the vascular system of the organism to assist in their travels but inside the cell the signal molecule often appears to rely on diffusion, although the cellular cytoskeleton is often thought to play a part. Therefore, intracellular signals need to be small enough to diffuse rapidly and often need to have the capacity to diffuse through membranes, either by the aid of a carrier protein or through their hydrophobic nature. However, as with all rules, there are exceptions, and some signals are very large proteins, or lipids that appear to stay embedded in membranes and thus not able to travel very far.

A further important characteristic is that a signal should be able to be made, mobilized or altered relatively quickly. There are two main ways of rapidly creating a signal. Firstly, an enzyme can be employed to manufacture the signal as and when required, usually from a substrate that is ubiquitous, or at least easily available, perhaps a vital molecule such as ATP. Secondly, a pre-made signal can be sequestered, and only released to be perceived at the required time. Such mechanisms are shown in Figure 1.2.

An example of the first mechanism is exemplified by the production of cyclic AMP (cAMP). It is produced enzymatically from ATP inside the cell, by a dedicated enzyme that is under the control of other cell signalling components. Therefore, once initiated by the arrival of a hormone at the cell surface, the signalling cascade can ensure the activation of the enzyme and cAMP is rapidly produced to initiate the next stage of the signalling cascade. The second mechanism is less common in intracellular signalling, although it is used for the control of calcium ion concentrations. As calcium ions cannot be created or destroyed, they are sequestered until needed. However, it is a common mechanism for the rapid "creation" of an extracellular signal.

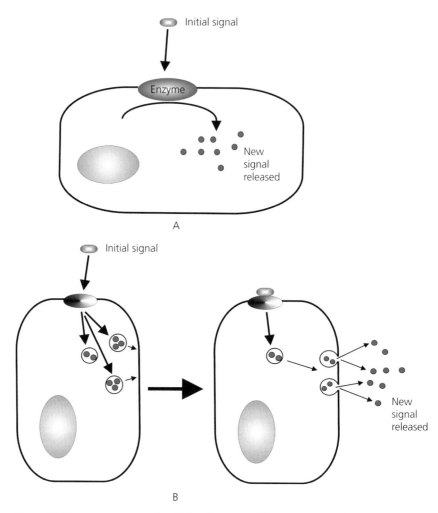

Figure 1.2 Two common ways in which cells may rapidly muster a signal. A. They may either be made as and when required, the manufacturing need itself under the control of signals. B. A signal may be pre-made, and held and sequestered out of the way until needed. Signalling then triggers the release of the signal to allow it to move to its site of recognition.

Hormones, such as insulin, are often made and held in intracellular vesicles, and on being triggered, the appropriate signalling cascade will induce the movement of the vesicles to the cell surface and the release of the hormones into the extracellular space, perhaps the vascular system of the organism.

Although the cell often has to have a fast response and the signal has to be relayed through the cell as quickly as possible, the responses are rarely required to last very long. Therefore, once it has been produced and had its effect, a signal must have the capacity to be turned off. Often the cessation of a signal also needs to be rapid. If a person is suddenly panicked, a rush of adrenaline stimulates metabolism and aids in what has been referred to as the "fight or flight" response. However, the person will wish to relax after the event, and so the "message" that adrenaline carried will need to be turned off. Similarly, if a cell is instructed to produce a certain gene product having received an extracellular message, it will not be required to produce that product forever more, until the cell dies. The stimulation of gene expression needs to be reversed to allow the cell to return to its non-stimulated state. How is such a cessation of signals achieved?

Often the signals are physically destroyed, perhaps enzymatically. Cyclic adenosine monophosphate is de-cyclized to AMP for example (see Chapter 6). Alternately, a signal may be re-sequestered, or a receptor internalized into the cell to effectively cease its signalling. Although cells have developed an array of enzymes involved in turning off signals, such as phosphodiesterases and phosphatases, often it is this part of the pathway which is more poorly understood, and in many cases ignored by researchers in the field. For example, many studies will show that a kinase is required to stimulate a certain cellular activity but often the reversal of the stimulation is brushed aside, or the molecular mechanism is assumed. However, it is almost certainly the balance between the on and off signals that is in fact important, and so studies of the off mechanisms are as important as those of the activation processes.

1.4 Different ways in which cells signal to each other

Cells may signal to each other in a variety of different ways. Often this is classified depending on the distance between the signalling cell and the target cell.

If the cells are touching, the signalling may simply occur through pores in the membranes, such as gap junctions (in animals) or plasmodesmata (in plants), or may be due to a membrane bound ligand being identified by a receptor in the membrane of a neighbouring cell. If the cells are further apart they may communicate via the release of molecules, which are then detected by the target cell or via the transmission of an electrical signal.

> **A ligand** The term ligand is used throughout this book as a generic term for any molecule that binds to specific sites on a protein, such as a hormone binding to its receptor. The word ligand comes from the Latin, *ligare*, which means to bind.

Electrical and synaptic

A fast and efficient method of signalling over long distances is via changes in the electrical potential across the plasma membrane of a cell, such as in the cells known as neurones of the nervous system in animals. Such a potential is propagated along the length of the cell, which is commonly extraordinarily long, and so the signal can therefore travel considerable distances in a body very quickly.

Neurones usually contain four regions: the axon for conduction of the signal; a cell body containing the nucleus and normal cellular functions; dendrites that receive chemical signals from other neurones or other cells; and axon termini, from where the signal is passed on to the next neurone or target tissue (**Figure 1.3**). Neurones communicate through the use of synapses. These are areas of close contact through which the signal can be propagated. Synapses generally fall into two groups, either transmitting the electrical signal directly through gap junctions or chemically by the release of neurotransmitters, which themselves are detected by the target cell (see also **Section 4.6**).

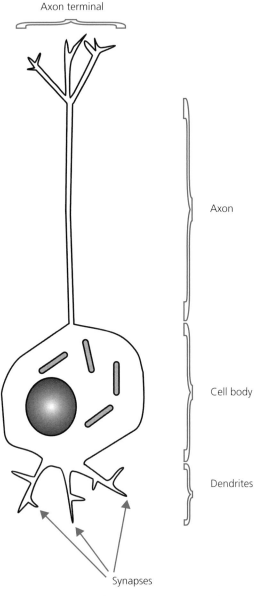

Figure 1.3 A schematic representation of a neurone, showing the four main areas: axon terminal, axon, cell body and dendrites. The axon section may be extremely long, relatively far longer than shown here, reaching out into the outer parts of an organism.

Although electrical signalling was always thought to be unique to animals, plants also use a primitive type of electrical impulse in response, for example, to wounding. Over relatively long distances the signals are carried by the vascular system, but short signalling distances involving the plasmodesmata have also been seen.

Endocrine

Many signalling systems involve the release of a signalling molecule from one cell, the movement of that molecule to a target cell, and its perception there. Such a scheme is shown in **Figure 1.4A**. Endocrine signalling exemplifies this. Here, cells release signalling molecules, such as hormones, which can travel vast distances in the organism, usually having an effect in a different tissue. In animals, the blood stream supplies the route to carry these messengers, but it

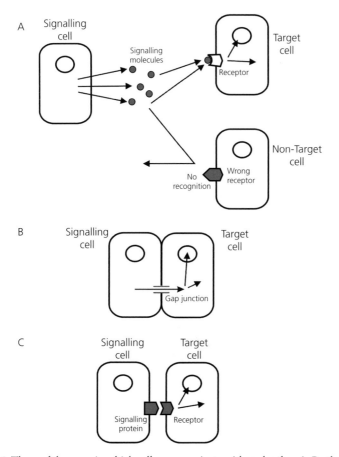

Figure 1.4 Three of the ways in which cells communicate with each other. A. By the release of molecules by one cell, which are detected by a receptor on another cell. Here, the presence of the wrong receptor on the cell surface will mean that the ligand is not perceived. B. By the direct transfer of small molecules through a gap junction. C. By the detection of a membrane protein on the surface of one cell by a receptor on a second cell.

is relatively slow and unspecific. All the tissues will be bathed in a supply of the signalling molecule and specificity comes from the target cell's ability to detect it. Plants also use endocrine signalling, hormones once again being carried by the vascular system of the organism. Even unicellular organisms have been shown to rely on a type of hormonal signalling, for example amoeba rely on diffusion in the surrounding medium to carry signals to the next organism.

Paracrine

As seen with endocrine signalling, a common way for cells to communicate is by the release of molecules, which diffuse to and are detected by a second cell. The difference here is that in paracrine signalling the released molecules are detected and have their effect in the local area of the signalling cell. Diffusion of the signalling molecules is very limited because they are either rapidly destroyed by extracellular enzymes or alternatively they are rapidly immobilized on or taken up by neighbouring cells. An example of such signalling is the release of neurotransmitters from neurones, which have their effect in the neighbouring neurone or in the neighbouring cells such as a muscle cell. A further example would be cells involved in the immune response, where a variety of white blood cells release cytokines and chemokines to modulate the activity of other white blood cells, and perhaps other local cells, in response to an insult to the organism, such as pathogen attack.

Autocrine

In a similar way to paracrine, autocrine describes a local effect of diffusable signalling molecules, but here the term autocrine is used specifically to describe the mechanism in which the released signalling molecules act on the same cell that released them. This type of signalling is common in growth hormones and sometimes with molecules that fall into the eicosanoid group. It is often found in cells that are in a developmental or differentiating stage and it may be seen as an emphasizing feature. Once a cell has been instructed to follow a certain developmental route, an autocrine signal may ensure that the set route is followed by releasing a signal to emphasize that message. Autocrine signalling is also commonly found in tumour cells, where they produce growth hormones stimulating the uncontrolled growth and proliferation.

Autocrine This is signalling to self; that is, a cell signalling to itself.

Direct cell–cell signalling

Cells that are in physical contact with each other are often very active in signalling to each other. They can achieve this either by the recognition of molecules on each others' surfaces or through direct communication via specialized areas of the cell surface.

Receptor–ligand signalling

A common way that cells communicate is through the recognition of surface markers (see **Figure 1.4B**). A good illustration of this is the communication of cells in the eyes of the fly *Drosophila* during their development. The fly's compound eyes contain approximately 800 separate photoresponsive units, or ommatidia, which each consist of 22 cells. Eight of these are responsible for photoreception and are called retinula or R cells. During development of ommatidia, the R8 cells express on their surface a protein known as "Bride of sevenless", or Boss. This is effectively a fixed ligand on the surface of these cells. The presence of the Boss protein is detected by another protein (actually a receptor tyrosine kinase: see **Chapter 6** for details) known as "Sevenless", or Sev, which is on the surface of the neighbouring R7 cell. The binding of Sev receptor to the Boss ligand protein triggers the R7 cells to develop, and enables the fly to detect ultraviolet light. This system was discovered due to mutations in the Sev protein where no R7 cells were signalled to develop, hence the naming of the protein as Sevenless. Such flies were identified by their inability to detect ultraviolet light and such animals have been extremely useful in the elucidation of signalling pathways.

Son of Sevenless

The story of the signalling of the sevenless mutants continued, as expected, and other downstream components were identified. One of these was a protein that interacts with G proteins, and was called Son of Sevenless, Sos. The Sos protein has now been found in many other signalling pathways, and despite the fact that many have nothing to do with the original system the name has stuck. Sos is an extremely important protein in humans. Mutations of Sos have been implicated in hereditary gingival fibromatosis.

Other proteins involved in this type of cell to cell interaction that have been identified in *Drosophila* include Notch (as discussed in **Chapter 13**), Armadillo and the product of the *dlg* gene, and such proteins may help to control cell proliferation.

Gap junctions and plasmodesmata

Animal tissues cells are often seen to have an area where the two plasma membranes of adjoining cells are apparently held at a fixed but very small distance apart. These are areas known as gap junctions. The plasma membranes in this area contain proteins that form tubes, and when the tubes in the two membranes come into perfect alignment a complete pipe is formed, which allows the passage of small molecules directly from one cytoplasm to the other. The passage of molecules of less than 1200 Da can therefore take place directly from one cell to the next. For example, small signalling molecules such as cAMP or Ca^{2+} ions can move through gap junctions (see **Figure 1.4C**), proteins which are much larger will not be able to move from one cell to the next in this way.

Dalton A unit of mass used commonly in science. One dalton (1 Da) is equivalent to the mass of a hydrogen atom, and was named after John Dalton (1766–1844). In biological sciences it is often used as the kilodalton (kDa), which is equal to 1000 Da.

The tubular structures, known as connexons, in the membranes are composed of a group of six identical proteins called connexins, which are arranged in a ring. The connexons protrude slightly from the membrane and, therefore, the two plasma membranes are held apart by a distance equal to the length of the protrusions from each connexon added together (see **Figure 1.5**). Each gap junction of the cells may contain several hundreds of connexons, which can be revealed by freeze fracture techniques.

Freeze fracture techniques This technique effectively allows the two parts of the bilayers of the membrane to be peeled back to reveal proteins within the membrane. Other techniques with fluorescent tagging will also allow the identification of proteins such as connexons in membranes.

The protein structure of the connexins has been predicted from the amino acid sequences. At least 11 isoforms coded for by separate genes have been identified, and therefore conserved features have been used to determine the likely topology of the polypeptides. It has been predicted that each protein contains four membrane spanning helices, with a particularly conserved helix sequence being used to line the hole made when the six subunits come together.

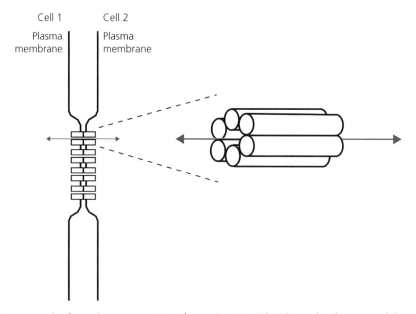

Figure 1.5 A schematic representation of a gap junction. This shows the alignment of the connexin proteins forming connexons.

Isoforms

Isoforms are different forms of a protein. Therefore, such polypeptides would be assumed to have the same function, have similar interactions and similar structures. However, isoforms often differ, sometimes quite subtly, in their kinetics of action, in their control and in their presence in different tissues. Many proteins in signalling seem to exist in a variety of isoforms, suggesting that the subtlety in their differences is very important.

The expression of the isoforms differs between tissues, as does the permeability of the gap junctions, suggesting that the movement between cells is carefully managed. Though a single cell can contain more than one isoform the 12 connexins of a single junction are always of the same type.

As might be expected, the passage through the gap junctions is not a free for all but is in fact very well controlled. Gap junctions are rapidly closed if the Ca^{2+} or H^+ concentration rises for example. The connexins undergo a reversible conformational change, probably working in a similar manner to that of an aperture on a camera lens. Closure of the gap junctions is of particular importance if the neighbouring cell dies, so stopping the leakage of material from the still living cell, or the import of any death signals. The separation of the intracellular part of one cell from the cytosol of another in which there is limited or no control of the input of molecules, would be vital to a cell's survival. This may be of particular pertinence during development of tissues, when certain cells undergo cell suicide, whereas others are required to survive and develop.

Mutations on genes that encode connexins have been implicated in a variety of human diseases. Amongst these are the formation of cataracts and skin disorders, deafness and cardiovascular problems.

Many, but not all, plant cells also have direct cell to cell connections, but owing to the presence of a cell wall they differ from gap junctions. The connections in plants are called plasmodesmata (**Figure 1.6**). The cell walls contain holes lined by the plasma membrane of the cell, such that the plasma membrane of the cells joined by plasmodesmata can be considered to be continuous. This lined hole is approximately 20–40 nm in diameter and also allows the cytosol to be continuous with the neighbouring cell. Through the hole is a membranous tube called the desmotubule, which is derived from the endoplasmic reticula of the cells. Therefore, a cell's cytosol, plasma membrane and endoplasmic reticulum are shared with the neighbouring cell.

Once again, this does not mean the free passage of all molecules between the cells. The cut-off is thought to be approximately 800 Da, allowing the passage of small signalling molecules but not macromolecules such as proteins, although interestingly the transfer of some proteins and even viruses has been reported. Passage through plasmodesmata is also under tight regulation, but the processes controlling this are poorly understood.

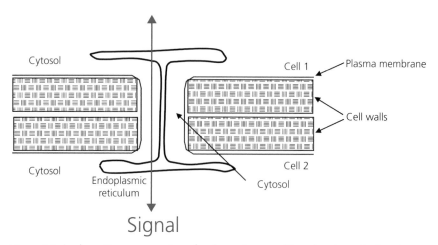

Signal

Figure 1.6 A schematic representation of a plasmodesmata. Note that the endoplasmic reticulum and cytosol are continuous across the two cells.

1.5 Amplification and physical architectures

The mechanisms of signal transduction from one cell to the next and subsequently into the interior of the target cell often need to involve formation of chains of signalling molecules, each passing on the message to the next molecule in the line. An extracellular signalling molecule, or first messenger, perceived by a cell often leads to the production of small and transient signalling molecules on the inside of the cell often referred to as second messengers (see *Second messengers* box). Such intracellular messengers would then activate or alter the activity of the next component of the transduction pathway, for example, a kinase. Historically, a good example of a "second messenger" is said to be cAMP, as discussed in **Chapter 6**.

Second messengers

The term second messengers was originally coined to describe the signalling molecules found to be produced by cells in response to the perception of the "first" message, or extracellular ligand. However, they are often not the second component in the signalling pathway and the term is now confusing and misleading, and should be avoided.

One very important feature of the production of intracellular messengers and the existence of these cell signalling cascades is the capacity to allow amplification of the original signal, so that the end result is effective in the target cell. Binding of a single hormone molecule to a receptor on the cell surface will not generally cause activation of a single enzyme molecule. Rather, a situation could be envisaged as shown in **Figure 1.7**.

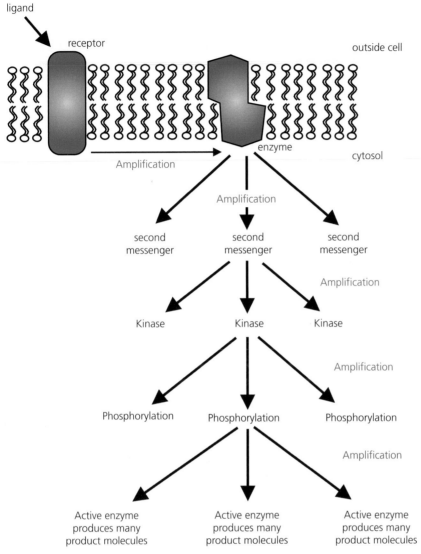

Figure 1.7 Amplification: an example of how amplification could take place in a signal transduction cascade.

Here, binding of ligand will cause the receptor to switch on many molecules of enzyme on the plasma membrane. Such transduction may involve G proteins acting on enzymes such as adenylyl cyclase. Ligand binding to a single recep-tor leads to the potential activation of many G proteins, which in turn can potentially activate several adenylyl enzymes each. The activation of each one of these adenylyl cyclase polypeptides leads to the production of many cAMP molecules, causing further amplification of the signal. These cAMP molecules will, in turn, activate many kinases, which will phosphorylate a great many proteins, causing yet further amplification of the signal. Therefore, it can be seen that the binding of one hormone molecule to the exterior surface of the cell can potentially lead to the activation of hundreds or thousands of enzyme molecules inside the cell, each of which will catalyze many rounds of a reaction. The final signal, causing the desired cell event, will therefore not be reliant on the presence of a large concentration of extracellular signals or the vast amount of ligand binding. Relatively small amounts of ligand can have profound intracellular effects if the cellular machinery is in place to allow its perception and this amplification.

However, as discussed above, signal cascades are usually not straightforward pathways with a defined route, going straight from beginning to end. Most signalling molecules can have an effect on several different other signalling systems in the cell, so that the stimulation of a single element of one pathway may lead to the further stimulation of several other signalling branches.

1.6 Coordination of signalling

One of the major mysteries in the field of cell signalling is the question of how a cell coordinates all the signals that it has to deal with? As discussed above, cells are inundated with a myriad of signals all the time, some of which might require the same response, and others which might require a completely opposite response. Researchers talk about integration of signalling pathways and "crosstalk", but there are few examples where there is a clear understanding of how a cell handles the multitude of signals that it has to deal with and how the cell manages different signal pathways simultaneously. It is quite possible for example, for a cell to detect two or more signals on the outside simultaneously, via its receptors. These receptors will then activate the relevant signalling pathways inside the cell, but quite often these distinct signalling pathways will have common elements. For example, two signals may both use the modulation of intracellular calcium ion concentrations as part of their signalling cascade. How does the cell determine if the alteration of Ca^{2+} concentration is due to turning on of receptor 1 or receptor 2? Or indeed, is interaction of the signalling pathways the key? Does the cell rely on the subtle changes made to one signal by a second signalling pathway and does this give the cell the fine control required? Often, the end result of the activation of receptors may in fact be different or indeed opposite in the cell, and yet both pathways when studied in isolation may again appear to have common components.

An example of such a dilemma is seen when studying the signalling induced in insect gut muscles, where two receptors that apparently have differing effects on the muscle cells both induce the production of diacylglycerol (DAG) and presumably the activation of protein kinase C (PKC). How does the cell know what the production of DAG is supposed to mean if it is forming part of two cascades that have differing results? If a common element is used, how is the specificity of any single signal maintained? And yet it appears that it is.

Although we are far from obtaining a full understanding of cell signalling, it is now proposed that there is a greater compartmentalization of cells than classically thought, and this might help to explain why common signals have different effects. In the past, the cytoplasm for example has been thought of as a homogenous organelle, in which one part is the same as any other. How could it not be, as it is just a liquid phase of the cell? Diffusion across the cytoplasm is not as free as imagined, and the cytoplasm is more viscous than most think. Studies with Ca^{2+} ions and other small molecules have shown that "hot-spots" exist within the cytoplasm. There are areas of high calcium ion concentration, and areas with virtually no calcium. Therefore, the calcium signalling can only be perceived by the cell in certain areas of the cytoplasm. How these hot-spots are formed and maintained clearly requires further studies, and will open a way forward for a more full understanding of crosstalk in signalling. However, the idea of hot-spots is not confined to calcium signalling, and the idea has been discussed in other signalling events such as those involving reactive oxygen species and nitric oxide, and especially with cAMP signalling.

To help explain the apparent lack of diffusion across cells that takes place in some signalling mechanisms, research has uncovered proteins that have a role as a structural scaffold inside cells (discussed below), which restrict the movement of signalling components through the cytoplasm. In light of such research, a new perception of what a cell actually looks like on the inside is needed before the answers to many of these problems are solved.

Almost opposed to this view of the signalling web of events, along with the proteins of defined signalling function, other polypeptides have also been postulated to be involved in the signalling systems of cells, but to transfer any message per se. Such proteins are thought to have a scaffolding role, holding the relevant components of a cascade in a particular physical environment, allowing efficient transfer of the signal, but in doing so restricting their interaction with other structurally and functionally similar components in a related signalling cascade. One such protein is the STE5 polypeptide of yeast, whereas some mitogen activated protein kinases in mammals, for example MEKK1, have been seen to have such roles too, as well as active phosphorylation. Other examples include the protein Axin, which is involved in the Wnt pathway (see **Chapter 13**). However, such a restriction of movement of components would put a question mark over the potential amplification or divergence of the signal that could take place. If proteins are held and can only further interact with a limited range of other proteins, perhaps only one, then neither amplification nor divergence is possible at this point in the cascade. It is likely that cells have evolved a pay-off whereby amplification is compromised for a vital need to retain specificity of a signal transduction cascade. Examples here would include the MAP kinase pathways, where a message can be relayed through the cell, perhaps with restricted divergence of the signal, but with a defined target, often the expression of genes in the nucleus in this case.

It can also be seen that signals have the potential for both amplification and divergence. For example, a kinase may lead to the phosphorylation of several proteins that have very different effects in the cell, including the activation of other kinases. Likewise, the activation of phospholipase C leads to the release of molecules that ultimately turn on both protein kinase C and calcium signalling pathways, each having their own defined effects in the cell. Conversely, several pathways may lead to the stimulation of the same enzyme, showing convergence of pathways. An example here would be the breakdown of glycogen, which can be stimulated by the arrival of a hormone at the cell surface, leading to intracellular signalling through cAMP, or separately, by the release of intracellular stores of calcium.

Therefore, it is misleading to think of cell signalling in simple terms of pathways, each distinct, functioning independently and simultaneously. Rather, signalling is a complex web of events (see **Figure 1.8**), with vast amounts of divergence, amplification and in many cases convergence. A cell will be bombarded by many signals, and will only react to those it is equipped to recognize. Pathways will need to share common components, and often signals might lead to opposing action on those common components. This complexity has yet to be understood in many cases, but by investigating the potential interactions of all signalling elements in cells, the potential complexity of the signalling web can be determined and in some cases modelled to predict outcomes.

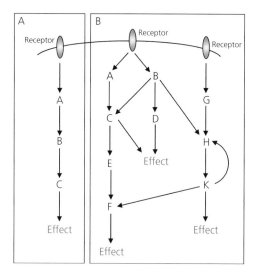

Figure 1.8 Cell signalling cascades can be viewed as simple pathways, or a more complex web of events. A. The receptor invokes a response down a single simple pathway leading to an effect. B. The activation of one receptor may lead to the involvement of components in more than one pathway, and even modulate the pathway leading from another receptor, and there might be several final effects.

1.7 **Domains and modules**

> **Consensus sequence**
>
> If the sequences for several related proteins (or DNA/RNA sequences) are aligned, and a commonality revealed, then the sequence that represents the common sequence which can be found in these related proteins is referred to as a consensus sequence. The consensus sequence can then be used to hunt for similar regions in newly discovered proteins or proteins of unknown function.

The details of signalling pathways of cells look incredibly complex but quite often components contain areas of similarity or homology that are repeated. Different kinases, for example, might have sequences that can be recognized as the catalytically active regions of the polypeptide and, although not identical in different proteins, a consensus for the sequence in that area can be drawn up. On to this basic structure, other areas may be attached, such as amino acid sequences that are responsible for regulating the activity. Again, this second area may share homology with regions in other proteins that are controlled in the same or similar ways, but in proteins that contain a different catalytic region and therefore have a completely different function. It can be seen that these polypeptide sequences and regions are like functional modules that can be "glued" together in an incredibly varied arrangement (**Figure 1.9**).

Evolution appears to have led to a good structure and design within proteins, and then adapted and modified it for other uses, rather than starting the design of a new enzyme or protein from scratch. A good example is found if proteins that contain a redox function are looked at closely. Many cells have the same redox pathways, but often they are subtly different, allowing these apparently similar proteins to have a wide range of functions, including proton transfer in energy production, the detoxification of xenobiotics and, of particular relevance here, the production of free radicals such as nitric oxide and superoxide (see **Chapter 10**).

Now that such domains are recognized to be similar across a range of proteins, consensus sequences for their primary structure can be determined, and with the use of bioinformatics such domains can be hunted for in new gene sequences or in whole genomes. For example, it is not uncommon to read papers where domains such as an EF-hand (see **Chapter 9**) have been found in a gene sequence, and the authors therefore suggest that the resulting protein will be controlled by the presence of calcium ions. Such domains can be identified if the peptide sequence is known by the use of secondary databases, such as Prosite. However, caution should be exercized here. Just because the domain sequence is present, it does not automatically mean that it is functional, but such bioinformatic data is very constructive to inform future biochemical studies, where activation profiles and functionality can be determined experimentally.

> **Redox (reduction/oxidation) proteins** Such proteins are involved in the transfer of electrons and as such can exist in both reduced and oxidized states.

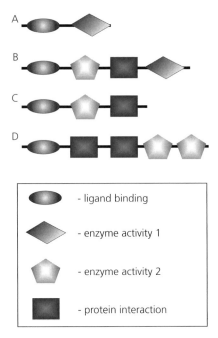

Figure 1.9 Protein domains. Often proteins can be seen as being made of domains, each with a separate function. During evolution, such domains have been put together in different orders, and different amounts, to create families of related proteins, but each with a specific function and activity.

Bioinformatics

Bioinformatics is the use of computers in biological sciences, and it is very common for computers to be needed to scan DNA sequences or carry out complex calculation so that the structure of a protein can be predicted. To aid such studies, there is a considerable resource of databases and information available through the internet and one of the best places to start for protein characterization is http://www.expasy.org

As discussed above, some domains are repeated and used in many proteins, and common domains that are seen in proteins involved in cell signalling are those involved in protein/protein interactions. Two very important polypeptide domains found in many proteins that have such a role in cell signalling are the Src-homology domains: the so-called SH2 and SH3 domains. Both are involved in the interactions of proteins, acting like a "protein Velcro" to hold them together, but the specificity of the two types of sequence differs.

SH2 domains are, in general, approximately 100 amino acids in length and have a binding affinity for phosphotyrosine amino acids, that is, tyrosine residues of a protein that have been phosphorylated. The domain has the structure of a deep pocket lined with positively charged amino acids, which interact with the phosphate group of the phosphotyrosine. Binding is very much reduced if dephosphorylation has occurred. A second region of importance of SH2 domains has also been used to categorize them into two groups. The first group also contains a second binding pocket, which is partly responsible for bestowing the specificity on the binding as it recognizes a single amino acid of the target polypeptide. The second SH2 group contains a region that is indented and shows binding to a group of mainly hydrophobic amino acids, a little way removed from the target phosphotyrosine residue.

SH3 domains, on the other hand, bind to a sequence on the target polypeptide containing several proline residues, which are often held in a left-handed helical secondary structure, known as a polyproline helix type II.

X-ray diffraction

X-ray diffraction is a method for studying the structures of biological materials such as proteins. It was the method that allowed the solving of the double helix structure of DNA by Watson and Crick. See also Chapter 3.

It is quite common for a single protein to contain both of these domains, and even multiple copies of them. They are particularly common where proteins that are not normally associated have to come together leading to activation of one of the proteins. An excellent example of this is provided by adaptor proteins, such as growth factor receptor-bound protein 2 (GRB2). The structure of this protein has been studied and solved at 3.1 angstrom resolution: a representation is shown in Figure 1.10. GRB2 is a 25 kDa polypeptide that contains one SH2 domain and two SH3 domains, and which can be in association with guanine nucleotide releasing proteins, for example Sos. These proteins are often involved in signalling from receptor tyrosine kinases, such as the epidermal growth factor receptor, which are autophosphorylated on binding to their ligand. The formation of the phosphotyrosine residues on the receptor then allows the binding of these adaptor proteins through their SH2 domains. A conformational change will take place, and in so doing cause the activation of their associated guanine nucleotide releasing proteins through the use of the SH3 domains, resulting in the stimulation of a G protein and the activation of the rest of the cascade (see Chapter 7 for details). Other adaptor proteins include Shc, Crk and Nck, although to date these are not as well characterized as GRB2.

Another example of the involvement of SH2 domains is on the STAT proteins. These are phosphorylated by Janus kinases (JAKs), which become associated with activated cytokine receptors (see Chapter 5). The STAT proteins are phosphorylated on tyrosine, and because they also contain SH2 domains they dimerize. The dimers so formed then go on to cause the enhancement of transcription. At least three STAT proteins have been identified to date. STAT1α (91 kDa) and STAT1α (84 kDa) arise from the same gene via alternate splicing, whereas STAT2 is slightly larger at 113 kDa.

> **Topology** the topology of a protein refers to the way the amino acids are arranged in three dimensions, and therefore its tertiary structure.

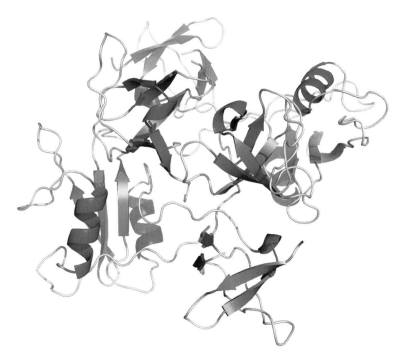

Figure 1.10 Structure of GRB2, solved using X-ray diffraction. Structure was obtained from the RCSB Protein Data Bank (http://www.rcsb.org/pdb) PDB ID: 1gri (Maignan, S., Guilloteau, J.P., Fromage, N., Arnoux, B., Becquart, J. and Ducruix, A. (1995) Crystal structure of the mammalian Grb2 adaptor. *Science*, 268, 291–293).

Other proteins may not contain such domains themselves, but on becoming phosphorylated gain the topology to interact with such domains resident on other proteins. These proteins, often referred to as relay proteins, are also seen in signalling cascades from receptor tyrosine kinases. A good example is insulin receptor substrate 1 (IRS-1). This protein is phosphorylated on multiple tyrosines, so creating sites that can interact with the adaptor protein GRB2 via its SH2 domain. This particular pathway is discussed in more detail in Chapter 11.

SH2 and SH3 domains are not the only domains allowing protein/protein interactions. Others include the pleckstrin homology (PH) domain, which was first identified in the platelet protein pleckstrin. Such domains have now been found in over a hundred human proteins, including Sos and some calcium-independent protein kinase C isoforms. Within proteins, the PH domain may be anywhere along the sequence. At the sequence level, PH domains appear to be a rather diverse group, with limited amino acid homology (10–30% only), but at the structural level they are more conserved. They are usually 100 amino acids long, folded into two anti-parallel β-sheets with an α-helix along the top, the latter being the amino acids of the C-terminal end of the domain. As for their function, they have been associated with the interactions of proteins with phosphoinositide lipids, for example as seen with protein kinase B, with interactions of proteins with G proteins and in anchoring proteins to membranes.

It is very likely that further binding domains will be found as further cloning and sequencing work allows consensus sequences to be deduced.

14-3-3 proteins

A family of polypeptides that have a regulatory role, and act like scaffolds, are the 14-3-3 proteins. Their rather strange name was derived from a survey of brain proteins, but they have now been found to be important in a plethora of pathways, across the kingdoms. They exist in multiple forms in organisms, for example the *Arabidopsis* genome encodes 15 versions of 14-3-3, and are involved in a wide range of signalling pathways. Target proteins for the 14-3-3 polypeptides include kinases, phosphatases, ATPase and enzymes such as nitrate reductase.

14-3-3 proteins act by their ability to bind to phosphorylated proteins. In some instances it is thought that the phosphorylation of a protein may not be sufficient in itself to control that protein. However, interaction of the phosphoprotein with a regulatory peptide, such as 14-3-3, may then enable a full level of control to be reached.

The sequences of the 14-3-3 polypeptides are highly conserved between species, suggesting that the structures are also similar. The central core of the structure in particular is thought to be conserved, whereas the N- and C-termini seem to be diverged in sequence and also fail to resolve in crystal structures. This suggests that the core has a common function, but that variations in protein interactions and role may be conferred on the protein by the sequences at each end. The 14-3-3s exist as dimers, and in three dimensions the central core forms a double barrelled W-shaped clamp. This is created by anti-parallel helices of a dimer, where each of the polypeptides forms a channel for interaction with a phosphorylated protein. The sequence to which 14-3-3 proteins binds has also been studied in many proteins, and the consensus R/KxxpS/TxxP has been derived. Therefore, potentially new interactions can be postulated if such a sequence is found in a protein previously not associated with the 14-3-3s. However, caution has to be exercized, as not all consensus sequences are necessarily used in a protein, and not all interacting proteins will contain the consensus.

As the 14-3-3 proteins are dimers, and each member of the dimer has a protein interaction site, it is possible that the 14-3-3 acts as a bridge across two phosphoproteins, so bringing those proteins into close proximity. However, 14-3-3 has been shown to bind to non-phosphorylated proteins too, suggesting a wider role for this family of proteins.

> **Use of protein crystals**
>
> To determine the structure of proteins, often it is attempted to obtain protein crystals, and these are examined using X-ray diffraction.

1.8 **Oncogenes**

The discovery of oncogenes and their functions was heralded as a great advance in our fight against cancer and in the understanding of why tumour cells form. The first hint of what was going on was supplied as far back as 1911 by Peyton Rous with his work on chicken tumours and his discovery of tumour viruses. Even today, many of the oncogenes that have been characterized were originally isolated from retroviruses.

However, the discovery of oncogenes really highlights the importance of an understanding of cell signalling as it is now clear that most oncogenes code for proteins that have an influence on cellular signalling mechanisms. They do not all influence a single part signalling mechanism but can be found spread through the plethora of signalling components. An oncogene product may well disrupt the way a cell perceives an extracellular signal, or indeed the way in which that signal is transmitted through the cell. Therefore, because of their diverse nature, they have been grouped into four classes. Class 1 includes oncogenes that code for growth factors, such as the *sis* gene, which codes for platelet derived growth factor. The oncogenes that encode growth factor receptors make up class 2, and include receptors for epidermal growth factor, the *erbB* gene, and nerve growth factor gene, *trk*. The oncogenes that code for intracellular signalling components make up the third and fourth classes, the class into which they fit being determined by their intracellular location. Class 3, includes the G protein gene *ras*, as well as the kinases genes *src* and *raf*, and are all cytosolic proteins. The nuclear proteins that make up class 4 include transcription factors, such as *jun* and *fos*, and steroid receptors such as the *erbA* product.

So what is it that makes genes oncogenes? Often they are versions of normal genes but actually encode defective components of a signalling pathway, which in many instances are in a permanently activated state. For example the *erbB* gene codes for a receptor that has been truncated at both ends and signals to the cell the presence of epidermal growth factor even in its absence. Likewise, the *ras* oncogene traps the G protein in a GTP bound state, where GTP hydrolysis cannot take place and the G protein remains permanently active, permanently relaying an "on" signal. Therefore, although many oncogenes are coded for on viral genomes, many are actually coded for by our own genomes too. In the normal state, where the gene products are not defective, the genes are known as proto-oncogenes. The product of the *ras* gene as we shall see later (Chapter 7) is an instrumental protein involved in critical signalling pathways. It is only when the protein is defective that the gene encoding it becomes classed as an oncogene.

■ Summary

- The potential for an organism or individual cell to signal to its neighbouring organism or cell and to detect and respond to such signals is crucial for its survival.

- Signals used in biological systems appear to be very diverse, ranging from a simple change in the concentration of an intracellular ion such as Ca^{2+}, to complicated compounds.

- The production of the signal usually needs to be reasonably rapid, particularly of intracellular signals and it must be able to be conveyed from its site of production to its site of action.

- A signal must be readily reversible.

- Very crucially, a signal has to contain specificity, relaying a defined message that can be translated into the correct cellular response.

- Signals between cells can be via:

  - electrical potential changes, involving synaptic signalling;

  - the release of compounds that are detected either over relatively large distances or even by the cell producing the signal, that is paracrine, endocrine or autocrine;

  - the passage of small chemicals through pores in the cell, such as gap junctions and plasmodesmata;

  - the direct physical interaction between cells, often mediated by protein–protein interactions.

- Signalling pathways can diverge to cause a multitude of cellular changes in response to a single signal, or they can converge, for example, where two or more signals effectively control the same metabolic pathway.

- In nearly all signalling pathways, a great deal of amplification takes place, allowing a small number of signalling molecules to precipitate a large effect.

- Signalling is best viewed as a complex signalling web, with many interactions between pathways and individual signalling components.

- The molecular study of the protein components often has revealed consensus patterns within the polypeptides and many of the proteins can be considered to be made of functional modules.

- Many signalling components have been identified as products of proto-oncogenes, highlighting the importance of these signalling components in the control of cellular proliferation and differentiation, and the study of the functioning of these gene products has helped enormously in the elucidation of many pathways.

→ Further reading

Alberts, B., Johnson, A., Walter, P., Lewis J., Raff, M. and Roberts, K. (2008) *Molecular Biology of the Cell* (5th edn). Garland Science, New York. ISBN 978-0815341055.

Barritt, G.J. (1994) *Communications Within Animal Cells*. Oxford Science Publications, Oxford, UK.

Berridge, M.J. (2005) Unlocking the secrets of cell signalling. *Annual Review of Physiology*. 67, 1–21.

Bowles D.J. (ed.) (1994) *Molecular Botany: Signals and the Environment*. Portland Press, London UK.

Bradshaw, R.A. and Dennis, E.A. (eds) (2003) *Handbook of Cell Signalling* (3 volume set). Academic Press, San Diego, California, USA. ISBN 0121245462 [a very specialist book on many aspects of the field].

Cotter, T. (ed.) (2003) *Programmed Cell Death*. Essays in Biochemistry, 39. Portland Press, UK. ISBN 1855781484.

Heldin, C.-H. and Purton, M. (eds) (1996) *Signal Transduction*. Nelson Thornes, Cheltenham, UK. ISBN 0748740740 [a collection of essays on topics in the field].

Helmreich, E.J.M. (2001) *The Biochemistry of Cell Signalling*. Oxford University Press, Oxford, UK. ISBN 0-19-850820-4 [A good text but with an over-emphasis on structure and mechanism].

Gap junctions

Wei, C.J., Xu, X. and Lo, C.W. (2004) Connexins and cell signaling in development and disease. *Annual Review*

Kornbluth, S. and Pines, J. (2003) Cell division, growth and death. *Current Opinion Cell Biology*, 15, 645–648 [and articles within this issue].

Kumar, S. and Bentley, P.J. (eds) (2003) *On Growth, Form and Computers*. Elsevier, London, UK. ISBN 0124287654 [Chapter 3, pp. 64–81, in particular on *The Principles of Cell Signalling*].

Lodish, H., Berk, A., Kaiser, C.A., Krieger, M., Scott, M.P., Bretscher, A., Ploegh, H. and Matsudaira, P. (2008) *Molecular Cell Biology* (6th edn). W.H. Freeman, New York. ISBN 9781429203142 [Chapters 15, 16, 21 and 22 in particular].

Marks, F., Klingmüller, U. and Müller-Decker, K. (2009) *Cellular Signal Processing: An Introduction to the Molecular Mechanisms of Signal Transduction*. Garland Science, New York.

Plant Cell (May 2002) 14 supplement, S1-S417 [An issue dedicated to signalling in plant cells, including aspects of plant development].

Science (May 2002) 296, 1632–1657 [An excellent collection of articles on a range of aspects of cell signalling].

Stryer, L. (1995) *Biochemistry* (4th edn.). W.H. Freeman. New York [An excellent general biochemistry text, but particularly good for the signalling in the eye]. N.B. The new edition is Berg *et al*. ISBN 978-0716746843.

Cell Development Biology, 20, 811–838.

Amplification and physical architectures

Herskowitz, I. (1995) MAP kinase pathways in yeast: For mating and more. *Cell*, 80, 187–197 [discussion of scaffolding proteins].

Oparka, K.J. (2004) Getting the message across: how do plant cells exchange macromolecular complexes? *Trends in Plant Science*, 9, 33–41.

Šamaj, J., Baluška, F. and Hirt, H. (2004) From signal to cell polarity: mitogen-activated protein kinases as sensors and effectors of cytoskeleton dynamicity. *Journal Experimental Botany*, 55, 189–198 [includes scaffold proteins in discussion].

Coordination of signalling

Dumont, J.E., Pécasse, F. and Maenhaut, C. (2001) Crosstalk and specificity in signalling. Are we crosstalking ourselves into general confusion? *Cellular Signalling*, 13, 457–463.

Dumont, J.E., Dremier, S., Pirson, I., and Maenhaut, C. (2002) Cross Signaling, cell specificity, and physiology. *American Journal of Physiology Cell Physiology*, 283, C2–C28.

Taylor, J.E. and McAinsh, M.R. (2004) Signalling crosstalk in plants: emerging issues. *Journal Experimental Botany*, 55, 147–149 [and articles within this special issue (No 395) entitled *Crosstalk in Plant Signal Transduction*].

Domains and modules

Alex, L.A. and Simon, M.L. (1994) Protein histidine kinases and signal transduction in prokaryotes and eukaryotes. *Trends in Genetics*, 10, 133–138 [illustration of protein modules].

Cohen, G.B., Ren, R. and Baltimore, D. (1995) Modular binding domains in signal transduction proteins. *Cell*, 80, 237–248.

Feller, S.M., Ren, R., Hanufusa, H. and Baltimore, D. (1994) SH2 and SH3 domains as molecular adhesives: the interactions of Crk and Abl. *Trends in Biochemical Science*, 19, 453–458.

Koch, C.A., Anderson, D., Moran, M.F., Ellis, C., Moran, M.F. and Pawson, T. (1991) SH2 and SH3 domains: elements that control interactions of cytoplasmic signalling proteins. *Science*, 252, 668–674.

Lemmon, M.A. and Ferguson, K.M. (1998) Pleckstrin homology domains. *Current Topics Microbiology Immunology*, 228, 39–74.

Rebecchi, M.J. and Scarlata, S. (1998) Pleckstrin homology domains: a common fold with diverse functions. *Annual Review of Biophysics and Biomolecular Structure*, 27, 503–528.

Ren, R., Mayer, B.J., Cichetti, P. and Baltimore, D. (1993) Identification of a 10 amino acid proline rich SH3 binding site. *Science*, 259, 1157–1161.

14-3-3 proteins

Fu, H., Subramanian, R.R. and Masters, S.C. (2000) 14-3-3 proteins: structure, function, and regulation. *Annual Review Pharmacology and Toxicology*, 40, 617–647.

Sehnke, P.C., DeLille, J.M. and Feri, R.J. (2002, Supplement) Consummating signal transduction: the role of 14-3-3 proteins in the completion of signal-induced transitions in protein activity. *The Plant Cell*, S339–S354 [an excellent review of 14-3-3 proteins, and not just from plants].

Oncogenes

Cantley, L.C., Auger, K.R., Carpenter, C., Duckworth, A., Graziani, A., Kapeller, R. and Soltoff, S. (1991) Oncogenes and signal transduction. *Cell*, 64, 281–302.

Rak, J.W. (2003) *Oncogene-directed therapies*. Humana Press, Totowa, New Jersey. ISBN 089603982X.

Watson, J.D., Gilman, M., Witkowski, J. and Zoller, M. (1992) *Recombinant DNA*. 2nd edn. Scientific American Books, USA [Chapter 18 in particular].

Weinstein, I.B. and Joe, A. (2008) Oncogene addiction. *Cancer Research*, 68, 3077–3080.

Weinstein, I.B. and Joe, A. (2006) Mechanisms of disease: oncogene addiction—a rationale for molecular targeting in cancer therapy. *Nature Clinical Practice Oncology*, 3, 448–457.

Pathways are the key to signalling

2

The key to any signalling is to get a message from one place to another, and cellular signalling is no exception. However, as with many other situations where signalling is required, there are barriers to be crossed and distances to be travelled, and therefore in biological signalling the relay of the message usually involves many components and many mechanisms. In this chapter over-simplified examples are used to illustrate this point, but these examples are based on real ones found later on in the book. Clearly, there are many other examples that could be used, but these well-characterized examples are used to illustrate pertinent points that can be used to understand more complicated pathways encountered later.

2.1 Introduction

Signalling involves the production of a message, the relay of that message to a new place and then the response to that message. Signalling in an organism, between organisms, or within a cell is like this too. Often a "message" is given off in the form of a hormone, and then this moves to a new cell where it is perceived and a response is seen. However, although it can clearly be seen in such an instance that the hormone (message) can be moved easily in the extra-cellular medium, perhaps the bloodstream or phloem, it is not easy to see how this can work inside a cell. There are examples of mass movement of materials in some cells, in a process known as cytoplasmic streaming, and in such cases it would be possible for signalling molecules to be swept along inside the cell, moving from one place to another with surprising efficiency; however, this would not be the norm for intracellular events. It would lead to problems such as the lack of specific action, or would require the production of far more signal than really required. In extracellular hormone signalling much of the hormone is ignored by many cells as it sweeps past in the bloodstream - is this a situation that could be tolerated inside a cell? The cell would need to make vast amounts of material that may never be used, and in most situations cells are not able to be frivolous, but need to conserve and recycle to survive. Signalling, therefore, needs to be as efficient and streamlined as it can be, but still maintain the urgency, specificity and reversibility that are required, and this was discussed in Chapter 1.

Signalling in cells has evolved in such a way that it involves many components, and these often interact in a sequential way. An analogy here may be the messages that are seen to be sent in submarines in many films. The captain may be looking through the periscope, and has an urgent message that should result in the submarine as a whole having a response, perhaps diving to the depths of the ocean to evade a depth-charge attack. The captain turns to his first officer and tells him what needs to be done. The first officer shouts the message to the next seaman, who shouts the message down the boat to the next and so on, until the person with their hand on the dive controls gets told what to do. In a cell the receptor on the surface will "see" the hormone, relay the message to the next component, which then relays that message to the next component and so on, until the effect is seen on the metabolic enzyme or cellular component that needs to cause the effect on the cell. This has many advantages. Firstly, it enables a single captain to tell many men what is happening, and so the message is amplified. Secondly, it allows more than one response, perhaps other seamen need to make urgent actions before the submarine tilts and dives, like batten the hatches. Thirdly, as long as the right seamen get the message it is very efficient and quite specific. The message could get scrambled, but each person in the chain is able to hear and understand, and knows what to do. So it can be seen in cells that the same approach is used. Signalling components lie in a pathway, and this is often called a "signal transduction pathway". In the simplest sense it can be linear, but as discussed in Chapter 1, this is rarely the case. Using some simplified examples the idea of signal transduction cascades will be illustrated below.

Signalling pathways can use a variety of "players" or components. They do not all need to be the same type of chemical, or indeed be related to each other in any way. All that needs to happen is for the message to be conveyed. In some wars pigeons were used to relay messages; it did not matter that they could not speak, they still could carry a message. So in cells it can be found that very disparate compounds and components will interact and pass on the signal. All that matters is that they "talk" to each other, that the specificity and direction of movement of the signal is maintained, and that it is as efficient as needs be for the cell. All types of cellular material can be drafted in to be involved, and evolution has enabled a vast array of components to be used. These include peptides, large proteins, large complexes of proteins, nucleotides, lipids and their derivatives, small but reactive compounds and even simple ions. In the subsequent chapters, the use of these will be discussed in more detail, with real examples used in Part 3 to show how they may come together to form real signal transduction pathways. Here, a simplified overview is given, with reference to how to find the detail in the later parts of the book if required.

2.2 Simplified examples of signalling pathways

In a signalling pathway one of the first events is the arrival of a signal, perhaps a hormone such as adrenaline, at the cell surface. The first thing that needs to happen is for this signalling component to be recognized, and this can happen if the cell contains the correct receptor, which has a unique binding site for the hormone. Many cells will not need to respond, so the receptor is not expressed, or the cell may not need to respond at the moment in time as the receptor has been internalized as discussed in Chapter 1, or the receptor for some other reason has been inactivated. However, assuming that the cell needs to respond, the receptor will bind to the hormone, and then relay the message inside the cell. Here, one of the greatest challenges of cell signalling is overcome. The cell has an external barrier, the plasma membrane and the "message" has to be able to cross it. This is the job of the receptor, and by binding to the hormone the receptor can alter its conformation, and as the receptor spans the membrane the new conformation of the receptor can be recognized inside the cell. The message has arrived.

Receptors are usually proteins. They commonly span the membrane at least once, and sometimes more, spanning seven times seems to be a common feature. However, some receptors are already inside the cell, as will be discussed in Section 5.2. But the common feature is that once bound to their specific ligand they are able to interact with the next component in the pathway.

In the example in Figure 2.1, on arrival of the ligand the receptor interacts with a protein. This may be a protein that is already in contact with the receptor, and recognizes the change that has taken place on that receptor, or it may be a protein that is recruited as the receptor changes. A typical example of an interaction that is changed would be a G protein interacting with the receptor, which then goes on to relay the message at the appropriate time (see Chapter 7).

Often in a signal transduction pathway there is the production of what would commonly be referred to as a "second messenger". These are small diffusible molecules, and were often found to have their concentration changed in signalling in cells, in response to a first (hormone) message, and hence the name. To produce a signal of such small molecules requires either their release from stores at an appropriate time, or their immediate production from a common substrate by the action of an enzyme. Here, in our example, the protein intermediary activates a plasma membrane bound enzyme, which produces

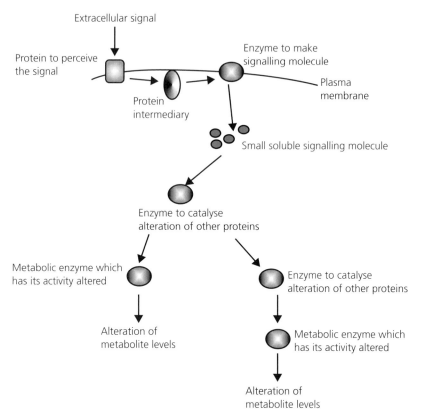

Figure 2.1 A simplified cell signalling pathway. The arrival of an extracellular signal leads to the initiation of a series of sequential events, which includes the generation of small soluble signalling molecules (shown in blue).

the small soluble signal. This has two major effects. Firstly, although there was scope for amplification before, a massive amplification is possible here. A single enzyme polypeptide can generate hundreds or thousands of small molecules, especially if the substrate is ubiquitous and available. Secondly, up to this point the signalling was restricted to the vicinity of the plasma membrane, but now there is the presence of a small and readily diffusible molecule, so the signalling can rapidly move inside the depths of the cell. An excellent example here would be the production of a cyclic nucleotide by an enzyme, often using ATP as a substrate. This is discussed further in **Chapter 7**.

Once produced, the small diffusible molecule can move, but to be acted on it once again needs to be recognized as a signal, and this requires a component with receptor-like characteristics. There are, in fact, proteins that will recognize these molecular messengers, and on binding to them, just as a ligand binds to its receptor, there will be a structural change in the protein undertaking the recognition, and the protein will now be able to be recognized, or able to function differently. In this case, the protein in **Figure 2.1** will be activated and its function is to alter other proteins. An excellent example, and one which is ubiquitous, is the alteration of the downstream proteins by the addition of phosphate groups, in a process known as phosphorylation. This is discussed in **Chapter 6**.

Activation of a protein that is able to alter other proteins has several advantages. Firstly, as with the production of a small signalling molecule above, as the protein here is also an enzyme, there is scope for huge amounts of amplification. A single enzyme can alter many downstream proteins. Secondly, there is scope for divergence. There may be many different proteins that can be modified. Some of these may be activated by such a modification, but others may be inhibited when modified. In the example given in **Figure 2.1**, one protein modified is a metabolic enzyme, which means that there could be a direct effect on a metabolic pathway, either increasing or decreasing its rate, depending on the exact effect of the modification to that enzyme. However, in this example there is a second protein also modified, and this is another protein that can go on to alter further proteins downstream. Often such "proteins which catalyze protein alteration" do come one after the other in a pathway, and a further example is discussed below.

To sum up for this example (**Figure 2.1**), there is a protein receptor that responds to the arrival of the extracellular signal, the receptor passing its message to a protein intermediary, then an enzyme, which makes a small soluble signal, which then actives another enzyme, which alters the activity of yet further enzymes, which themselves may alter the activity of further enzymes, which in this example alters the metabolism that the extracellular signal wished to act upon. This is a simplified version of the control of glycogen metabolism, which is discussed in more detail in **Chapter 7** and shown in **Figure 7.1**.

Figure 2.2 shows a similar example. Again, an extracellular signal causes the initiation of the pathway by binding to the receptor, and a protein intermediary causes the activation of an enzyme that produces further signalling. However, in this case, this enzyme produces two signals. One is a small soluble signal, whereas the other remains at the plasma membrane. Therefore, there is clear divergence of the signalling, with now two potential pathways leading from one initial signal. These two, now separate, pathways may go on to cause the same effect in the cell, but not necessarily. They may lead to very different effects. In this example, the small soluble signal produced leads to the alteration of another signalling pathway, which may be under the control of other signals too. Meanwhile, the hydrophobic signal generated leads to the activation of an enzyme that can alter other proteins, and a new cascade of signalling is possible. Once again amplification is possible here, but the key is that there are still specific interactions and events that are leading to the overall response by the cell. It is not random, but still tightly controlled and orchestrated. **Figure 2.2** is a simplified depiction of the inositol and diacylglycerol pathways, which are discussed further in **Chapter 8**.

However, this is really too simple. Questions need to be asked. For example, why does the protein intermediary only interact with one other protein, i.e. the enzyme? Why does the small signalling molecule only get recognized by one protein and send on its message? The answer is that they do not in

Downstream

Often in pathways terms such as downstream and upstream are used. The former means later on in the pathway, whereas upstream means coming before. Such terms are also used in genetics when discussing gene sequences.

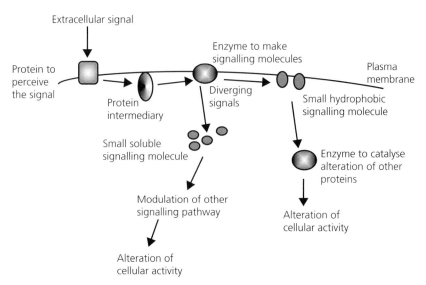

Figure 2.2 A simplified signal transduction pathway that involves divergence, leading to two distinct downstream pathways. The route shown on the left leads to the alteration of the activity of another signalling pathway, perhaps that involving calcium ions (not shown).

many cases. If a protein recognizes the small signalling molecule because it has the appropriate binding site, why should other proteins not also have that binding site? As discussed in **Chapter 1**, many proteins have a module-like (domain) structure, and one domain may be the one for binding this small soluble signal. Therefore, many different proteins may contain this domain, and therefore may be affected by the rise in the presence of this signal, and all these different proteins may have different effects in the cell. So a more real picture may be as depicted in **Figure 2.3**. Each star is a point in the pathway where divergence is theoretically possible, and, in reality, many points of divergence are indeed used. The protein intermediary known as the G protein, as discussed in **Chapter 7**, does have more than one "output" as it is a complex that breaks and both parts produced can have an effect. Small signalling molecules such as those based on inositol, as discussed in **Chapter 8**, have the capacity to have many effects, some related to each other, but often disparate. The conclusion that needs to be drawn from this is that even the complex pathways discussed in subsequent chapters are in reality far more complex than discussed. Computer modelling and a more holistic approach (systems biology) will be required to fully understand all the potential interactions, points of divergence and convergence of pathways.

The last example simplified and discussed here is that which is seen with growth hormone perception, and leads to alterations of gene expression. Such effects on the activity of the nucleus in cells are key to the events that control organism development and this will be discussed further in **Chapter 13**.

In the example shown in **Figure 2.4** the pathway does not need or use the generation of a small soluble signalling molecule, so there is no "second messenger".

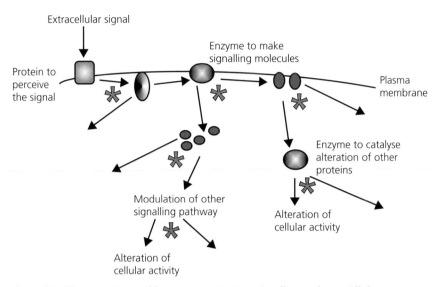

Figure 2.3 Divergence is possible at many points in a signalling pathway. All the steps marked with the blue stars could potentially allow alteration of two or more downstream events.

Convergence

Although the notion of divergence is highlighted here, signal transduction pathways may also converge, that is they meet at a common control point to bring about the overall effect. A good example here is again glycogen metabolism, which can be controlled by two major pathways, i.e. those controlled by cyclic AMP and calcium ion concentrations. This allows multiple controls over a single important event.

Here, the extracellular signal activates a receptor, which will then allow the interaction of the receptor with a protein intermediary. This looks similar to before, but in the cases above, when the detail is examined, it is found that the protein with the intermediary function actually has an activity of its own (it is a G protein and therefore as discussed in **Chapter 7** it is a GTPase enzyme). However, here in **Figure 2.4**, this intermediary has no activity as such, but allows the other proteins to come together. It is really a bit like a Lego brick, allowing one protein to "stick" to, or complex, with another. However, it needs to recognize the receptor only when required, and then undergo a conformational change and pass that "message" on through a new interaction with a protein downstream. Good examples of this sort of action are the adaptor proteins, which will be discussed again later in **Chapters 6 and 7**, and see for example **Figure 7.12**.

Intermediary proteins, such as the adaptor, cause the activation of the next proteins, which activate the proteins downstream, and, as can be seen in **Figure 2.4**, there is a cascade of proteins being activated. These activations are specific, and, although seemingly the same in the diagram, each will have

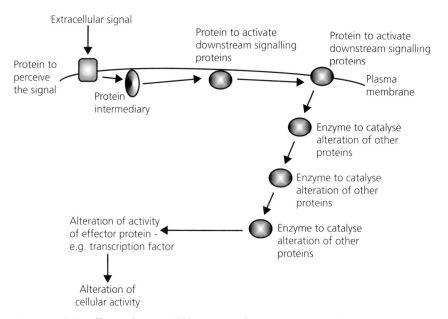

Extracellular signal

Protein to activate downstream signalling proteins

Protein to activate downstream signalling proteins

Protein to perceive the signal

Protein intermediary

Plasma membrane

Enzyme to catalyse alteration of other proteins

Enzyme to catalyse alteration of other proteins

Alteration of activity of effector protein - e.g. transcription factor

Enzyme to catalyse alteration of other proteins

Alteration of cellular activity

Figure 2.4 A signalling pathway could be a series of proteins in sequential interactions. They may act in different ways, but the "message" may be passed from one protein to another in a series, perhaps leading to alterations deep within the cell, such as in the nucleus.

their own unique actions and effects. However, although depicted as a linear cascade, many of the steps will have the capacity to have large amounts of divergence, as well as amplification of the signal.

In this particular example (**Figure 2.4**) the end result is in the nucleus, and therefore one of the proteins needs to be altered in such a way as for the cell to recognize it as being needed to be located in the nucleus, whereas before it was resident in the cytoplasm. Therefore, a conformational change of the protein, or modification, will be perceived by the mechanisms of the cells that drive nuclear import of that protein into the nucleus, where it can have its effect. Alteration of the protein will usually be twofold. Firstly, to flag it as a protein that needs to be imported to the nucleus, and, secondly, to activate it, so it carries out the correct functioning once in the nucleus. Later, the modification that was so key for the protein's nuclear import may be reversed, and the protein will once again be inactivated and can be re-exported ready for another round of signalling.

It can be seen therefore that components in signalling do not act alone. They each have a defined role in a pathway. With the present state of knowledge it is not always clear exactly why each step is required, but cells have evolved over millions of years, and it is very likely that steps in a pathway are only there for a reason, perhaps yet to be elucidated. But it is clear that such pathways allow for the amplification of the signal, for divergence and convergence of pathways, and for the movement of the message, or messages, from one place to another within the cell. Therefore, having the components in pathways seems to be key to successful cell signalling.

■ Summary

- The initiation of signalling is often the arrival of a "message" at the cell surface.

- The arriving "message" needs to be recognized and responded to, if the cell is to act on this message.

- Often the signal needs to be transmitted into a cell.

- Signalling inside the cell requires the use of a sequential group of components arranged in a pathway.

- Such arrangement of components is called a "signal transduction pathway".

- Many different biological materials may constitute a pathway, including proteins, lipids and nucleotides.

- Many pathways enable large amounts of amplification of the signal.

- Many pathways have the opportunity for divergence of the pathways to cause multiple effects in the cell.

- Convergence of pathways is also often seen, allowing multiple control over single events.

- Not all pathways require the use of small diffusible compounds, but could be a cascade of protein activations.

- Pathways in reality are often far more complex than depicted.

- Computer modelling and a systems approach will no doubt unravel the complexity of pathways in the future.

→ Further reading

Janes, K.A. and Yaffe, M.B. (2006) Data-driven modelling of signal-transduction networks. *Nature Reviews Molecular Cell Biology*, 7, 820–828.

Shimmen, T. (2007) The sliding theory of cytoplasmic streaming: fifty years of progress. *Journal of Plant Science*, **120**, 31–43.

Stewart, M. (2007) Molecular mechanism of the nuclear protein import cycle. *Nature Reviews Molecular Cell Biology*, **8**, 195–208.

3

A look at some of the history and techniques of cell signalling

Cell signalling has been studied for over a century, and still seems to be increasing at an alarming, if not exponential, rate. A brief history is included here, and, although many of the names and events are covered briefly, unfortunately not all landmarks can be discussed. Hopefully, however, it will give the reader a taste of the immense amount of work that this book tries to cover, and the immense effort that has been put in over the years by thousands of researchers to bring cell signalling to the point it has reached today.

With similar mechanisms and chemistry being used by a wide variety of organisms, tissues and cells, it is of no surprise that the techniques used for their study are also similar. A technique used to elucidate the signalling cascades of a mouse liver can often be used to study a plant leaf. Some of the common techniques are discussed here, with references at the end of the chapter where readers can find more details if they wish to delve more deeply. Often techniques are named or mentioned with little detail of what is entailed in carrying them out, but an overview of the methods available should be useful to guide further reading. However, throughout the rest of the chapters, techniques and relevant methodologies will also be mentioned to illustrate how signalling components have been discovered or characterized.

3.1 A brief history

The history of cell signalling goes back a surprisingly long way, and certainly does not just span the molecular age of biological sciences. Observations were made that led to the idea of molecular communication in organisms and cells, but it was only as techniques developed that a more full understanding of what was happening could be obtained. Of course it would be naïve and foolish to believe that elucidation of cell signalling is complete, and it will probably be a long time before that goal is achieved. Here, a brief overview of the history of cell signalling will be given (see **Figure 3.1**), but it is impossible to do justice to the subject here, and many researchers will note the absence of some events. Some points of historical interest not discussed here are mentioned in later chapters. It is worth pointing the reader to an excellent collection of key papers, edited by Burgoyne and Peterson (see *Further reading*), but, of course, future work will undoubtedly lead to further landmarks, and technology continues to move on considerably.

A quick search of a literature database using any key-words associated with the subject of cell signalling is enough to frighten the most enthusiastic researcher, and it will reveal the vast extent of work in the area and the

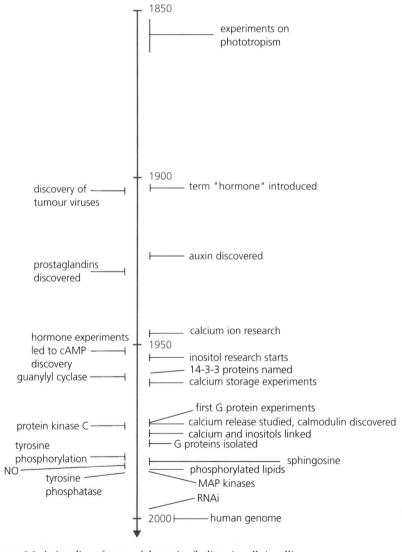

Figure 3.1 A time-line of some of the major findings in cell signalling.

immense interest there is across the whole of biological sciences. At the time of writing, putting "cell signalling" into a PubMed search lists approximately 25,000 papers with the alternate spelling of "cell signaling" yielding a further 307,000 papers. However, this interest is not new and has spanned the period of biochemical research (see **Figure 3.1**). Even in the middle of the nineteenth century, Darwin was experimenting with the phototropism of coleoptiles, and suggested the presence of a substance that was transported around plants having its effect at a site distant from that of its production and secretion. This substance was discovered by Went in 1928 and was later identified as indoleacetic acid, or, to give it its common name, auxin. Today, auxin is still the subject of intense research around the world.

The term hormone, really in an animal context, was introduced from the Greek meaning "to excite" or "to arouse", by Starling as long ago as 1905. Lipid soluble hormones such as prostaglandins, however, had to wait 30 years to be discovered. This family of compounds was originally named after the organ that was supposed to be their site of production, but this has subsequently proved to be misleading. Even today, the number and variety of extracellular signals discovered still seems to be increasing, especially for any researcher in the cytokine/chemokine field.

Although extracellular events were beginning to be unravelled in the early part of the twentieth century, some of the first clues as to what was happening inside cells came in the late 1940s and 1950s. In 1947, experiments in which small quantities of Ca^{2+} ions were injected directly into the cell showed that an increase in intracellular calcium led to skeletal contractions, but it was not until the early 1960s with the work of the likes of Ebashi and Lipmann that the way calcium ions were stored in cells was beginning to become clear. It took work on another system, that of inositol lipids, and the proposals by the likes of Michell, to elucidate the release of Ca^{2+} from intracellular stores, whereas proteins that are under the control of Ca^{2+}, such as calmodulin came to light through the work of Cheung and others in 1970. Complications were introduced to the field of calcium signalling in the 1980s by the work of researchers such as Woods, Cuthbertson and Cobbold, when it was discovered that calcium ion concentrations could be measured as transient spikes. Other complications, such as hot-spots in the cytoplasm, are now being investigated too.

In the 1950s, Earl W. Sunderland and Theodore Rall, working at the Western Reserve University, found that the addition of epinephrine and glucagon to cells, led to the binding of these hormones to cell receptors, and experimentation could show the subsequent production of cAMP. They named the enzyme responsible for this increase in intracellular cAMP concentration as adenyl cyclase, but it is now known as adenylyl cyclase or adenylate cyclase. However, it was not until 1989 that Krupinski and colleagues cloned the first mammalian enzyme. Adenosine monophosphate and guanosine monophosphate were isolated from rat urine in 1963, whereas guanylyl cyclase was identified in the late 1960s. Investigations of the role of adenylyl cyclase and cAMP were further helped by the isolation of adenylyl cyclase deficient cells, known as cyc⁻.

■ Earl W. Sutherland, received the Nobel Prize for Physiology or Medicine in 1971 for work on cAMP.

These cells were first reported in 1975 by Gordon Tomkins, Henry Bourne and Philip Coffinio working at the University of California.

The involvement of a separate signalling pathway, that of the inositol phosphate pathway was also starting to unravel in the 1950s. The role of phospholipids such as phosphoinositides, was first suggested in 1953 when Hokin and Hokin showed that the addition of acetylcholine to cells was found to lead to the incorporation of P^{32} phosphate into what was regarded as a minor lipid, phosphatidylinositol. This pathway was shown to involve the muscarinic receptors. The pathway leading to the biosynthesis of these lipids in the endoplasmic reticulum was reported by Agranoff and Paulus and their respective co-workers in the late 1950s. However, a link was made to another signalling pathway in 1975, when Michell suggested that messengers which stimulated the breakdown of phosphatidylinositol also caused an increase in cytosolic Ca^{2+}. A new pathway involving the membrane inositol lipids was more recently suggested when, in 1988, Whitman *et al.* first identified phosphatidylinositol 3-phosphate in transformed lymphocytes. This is now an exciting new area of lipid signalling, which is being shown to be implicated in the signal transduction of several systems, including responses to insulin and the onset of apoptosis. Similarly, evidence in 1986 that sphingosine inhibited protein kinase C has led to elucidation of the sphingomyelin cycle.

Protein phosphorylation is a common theme of many signal pathways, and it was over 30 years ago that the regulation of glycogen breakdown by phosphorylation of glycogen phosphorylase was suggested. In this system, the protein kinase was itself regulated by the intracellular cAMP concentration (PKA, see **Chapter 7**), but 1970 saw the first reports of a kinase that was more sensitive to cGMP than cAMP, cGMP-dependent protein kinase. Another, extremely important kinase, protein kinase C (PKC), was identified during the 1970s, through the work of Takai, Kishimoto, Nishizuka and others. However, until 1980, it was thought that only serine and threonine were targets for the phosphorylation of proteins, but it is now clear that one of the major phosphorylation events is the addition of the phosphoryl group to tyrosine residues. It was not until 1988 that Tonks obtained the first partial sequence of a tyrosine phosphatase however, while a year earlier the first MAP kinase was identified through the work of Ray and Sturgill. It has been suggested that the human genome might contain as many as 2000 kinase genes and up to 1000 phosphatase genes, and other genomes are likely to have an equally high number of genes encoding proteins involved in phosphorylation events.

The history of the elucidation of the G protein story takes us back to the 1970s. Gilman, working at the University of Virginia, identified G proteins as crucial intermediates in signal transduction in the latter part of the decade, and in 1980 Paul Sternweis and John Northup, working in Gilman's laboratory, purified the G protein G_s. It was not long afterwards that Lubert Stryer and Mark Bitensky independently discovered the existence of a second G protein, this time, using rod cells from the eye. This G protein was transducin or G_t.

■ Alfred Gilman and Martin Rodbell shared the Nobel Prize in Physiology or Medicine in 1994 for their work on G proteins.

■ Louis Ignarro along with Robert Furchott and Ferid Murad shared the 1998 Nobel Prize in Physiology or Medicine for work on nitric oxide.

■ Andrew Fire and Craig Mello were awarded the Nobel Prize in 2006 for their early work on RNAi.

Today, it would be folly to claim that all the signalling pathways have been discovered, even if we are not sure of the function of many of them. For example, it was not until 1987 that Moncada, amongst others, suggested that endothelium-derived relaxing factor (EDRF) was in fact nitric oxide (NO), opening up a whole new field of research, some of the founders of which later received the Nobel prize. Very recently, the idea of the importance of the reduction and oxidation states of cysteine residues has been discussed, with the term nano-switch, or nano-transducer, being coined. However, with the advances in molecular genetic technology, the rapid rate of cloning of new genes and the human genome project, it will not be long before all potential new members of existing pathways will eventually come to light, but the challenge to understand how all these pathways interact will remain. Perhaps that challenge will be overcome by the use of new technologies, such as RNA interference (RNAi). This relatively new and clever way of reducing the expression of specific genes was discovered in 1998, and helped to explain the results presented in several previous studies.

Future work in many aspects of cell signalling is now moving to more holistic studies, involving investigations into the changes of expression of all the genes in the genome (transcriptomics) leading to changes in the protein content of the cell (proteomics), and how all these changes impinge on each other and are coordinated, in what has been dubbed as "systems biology". As well as immediate changes, such as the breakdown of stored carbohydrates, giving a short term response in the cell, longer term changes are important to understand and perhaps manipulate in the future. Long term effects of changes in gene expression, perhaps through changes in the levels and activities of transcription factors, can lead to adaptive responses, and with the media constantly discussing global climate change, longer term adaptive responses of organisms and cells will need to be fully understood.

3.2 A brief look at some techniques

The understanding of the molecular events of cell signalling, like all areas of biochemistry, has been facilitated by the techniques available, and it is probably true that the science of cell signalling has been led by the technology and the popular trends at the time. However, the study of cell signalling requires a holistic approach, where the physiological events that ensue from stimulation still need to be studied, but where the molecular events inside the cell also need to be unravelled.

In a proposed study by a research group that suspects "Factor X" is involved in the signalling induced by a particular ligand during a specific physiological event, what are the general questions that usually need to be addressed? They are probably along the lines of:

• On addition of the ligand, does the concentration of "Factor X" change?

• If "Factor X" has activity, does this change during the response?

- If "Factor X" is inhibited, or removed altogether, is there still a response to the ligand?
- If the level of "Factor X" is increased in some way, can the response be induced?
- If the level of "Factor X" is manipulated does the activity or levels of other signalling molecules change?
- Does the manipulation of "Factor X" lead to changes in gene expression, and so potential changes in the proteome of the cell?
- If the physiological response is triggered by more than one ligand, is "Factor X" involved in many or all of these pathways?

and often most importantly:

- Does the manipulation of "Factor X" have an application, such as in the development of a new drug?

Therefore, any proposed use of research techniques needs to try to address these, and many other questions, and try to at least lead to an answer, if only in part.

Early studies were based on the observable events with whole cells, or even whole tissues or organisms, such as does the stem of a plant bend towards the light? It is fair to say that this sort of study is still instrumental today. If a researcher wishes to know what molecular event makes a stomatal aperture on a leaf close, they still need to watch to see if it happens, and often such observations need specialist equipment such as microscopes. Today, molecular biology makes a major impact, and even allows us to discover new prospective signalling molecules without even knowing their function, or even if they really exist at all! Having said that, even with the advent of high-powered molecular techniques, physically looking at what happens is still important. For example, if a cell signalling protein has its gene expression stopped, as in a "knock-out" plant, the growth and development of that plant still needs to be characterized and documented. Of course, the same can be said for animals, where the characterization of the growth, development and longevity of the knock-out animal is crucial to know.

Biochemistry and labelling studies

Confocal microscope

A confocal microscope commonly has a scanning laser as a light source instead of white light, and the focusing nature of the lenses allows the user to take images at different depths through the sample and therefore to "optically section" a sample. By layering these images it allows the researcher to build up a three-dimensional picture. Therefore, it is a powerful technique allowing the researcher to "look" inside a cell or tissue. Confocal microscopy will be further discussed in Chapter 10.

Despite the plethora of new technologies in recent years, there is still a place here for more traditional biochemical approaches. For enzymes involved in signalling it is still important to determine their catalytic characteristics. For example, what is the preferred substrate, or substrates? What is the product or products? And very importantly, what are the kinetics of the reaction, and how do they change during a signalling response? The kinetics of reaction are often the focus to determine the exact action of a new drug too, especially to check that it does not inhibit other related proteins. For proteins that are known to bind other things, such as for receptors or calcium binding proteins, it is important to determine the binding constants, and especially to establish whether significant binding takes place under physiological concentrations of the ligand under physiological conditions.

Immunological techniques have been a great benefit to the discovery and the study of the distribution of receptors, and, coupled with modern fluorescence techniques, the pattern of receptors on a cellular surface may be visualized in a three-dimensional image using confocal microscopic technology (discussed further below and in **Chapter 10**). Antibodies raised against particular epitopes on a protein can be labelled, and assuming that the antibody only binds to the correct proteins against which it was raised, then the presence of the labelled antibody will reveal the presence of the protein of interest. Antibodies have also been invaluable in the elucidation of the interactions between polypeptides, or to knock out a particular active site, so stopping a signal transduction pathway and watching the effect on the cell or cellular function such as transcription. If an antibody binds to a site on a peptide used for interaction with another protein, then that interaction can no longer take place. Antibodies in which the epitope is part of the active site of a protein often, but not always, totally, or more often partially block the activity of the protein. Therefore, assuming once again that the antibody is specific, this will determine if that protein's activity is required for a particular response to take place.

However, antibodies are not the only way of identifying proteins and their activity. Photoreactive radioactive analogues of ligands have been used in the study, for example, of LPA receptors. In another example, radioactive tagging of TGF-ß, adopting the use of ^{125}I-TGF-ß, was able to identify the presence of three polypeptides (55 kDa, 85 kDa and 280 kDa), so finding a family of TGF-ß receptors.

The use of probes

The combination of fluorescence and confocal microscopy has also been widely adopted in the absence of antibodies, with some companies able to supply hundreds of probes to investigate the functioning of cell signalling components. Here, a normally non-fluorescent probe, but one that gains fluorescence when in interaction with its target molecule, is added to the cell, and then the subsequent fluorescence seen is proportional for the amount of the signalling molecule present. Such an approach is probably most commonly used to study the levels of calcium in cells, but it can also be used to measure nitric oxide production from cells and a plethora of other events. However, some caution is required here. Often the probes are not as specific as first reported, and this is especially true for those used in nitric oxide research. Often researchers hope to use such probes in tissues, but penetration of the probe into that tissue can be difficult, and the lack of a signal from the probe may simply be because it was not there. Furthermore, even if the probe is present, the penetration of light to excite the fluorescence may be a problem, or the emission light is quenched by chromophores in the tissue and so once again no signal is seen.

A further advance in this area is the use of Fluorescence Resonance Energy Transfer (FRET) measured with what is referred to as Fluorescence Lifetime Imaging Microscopy (FLIM). These technologies allow the tracing of the catalytic activity of fluorescently tagged proteins inside still living cells, and can also be used to determine the state of activation of proteins in fixed cells, for example in pathological samples.

Pharmacological tools

Pharmacological studies have made a big impression on our understanding of signalling events. Here, the study of signalling pathways involves the addition of stimulators or inhibitors. For example, the question of what happens inside the cell on addition of a particular receptor ligand may be asked. Once a cellular event that is induced has been demonstrated, the method by which it can be inhibited is quite often soon addressed. Other inhibitors, targeted against potential components in a cascade are also used, and by building up an inhibition profile, the possible cascade can be elucidated. This might include kinase or phosphatase inhibitors to modulate particular phosphorylation events, or inhibitors of specific enzymes, such as phopholipases or cyclases.

However, caution is required here. Many of the inhibitors used, when discovered, were thought to be quite specific, but often subsequent studies show that their specificity can be doubted and that they have effects on more than one pathway. Interpretation of the results then becomes more difficult, and the search for better inhibitors continues. Having said this, the basis of many medicines used today is the inhibition of cell signalling components, especially receptors, and often the ultimate aim of finding and characterizing new inhibitors is not so much to help researchers elucidate pathways as to develop a new drug for the pharmaceutical market-place.

Structure and protein interactions

Once a component of a pathway has been identified, the search for an understanding of its exact mechanism of action usually requires its purification, and further characterization. If a protein is soluble, or can easily be over-expressed, then isolation of the peptide may be straightforward. However, if the peptide is of low abundance, or membrane bound, or is part of a complex, then isolation to homogeneity may be very difficult. Often the gene sequence is known, which will easily yield a translation to give the primary structure of the peptide. This should be straightforward, but it is becoming increasingly apparent that alternative splicing of gene products, and post-translational modifications of proteins make life more complicated in some instances. If the full primary structure of the polypeptide is known, the secondary and tertiary structures can be predicted by computer analysis, for example with tools available on the internet (see www.expasy.org). Alternatively, pure protein analysis can be carried out by time-consuming nuclear magnetic resonance (NMR) or X-ray diffraction studies. This will give data about the real structure of the protein, perhaps in different states of regulation.

X-ray diffraction

X-ray diffraction is a method for solving protein structures. It can be used to solve the structures of other biological materials too, such as DNA, which is how this molecule was found to have a double helical structure. The resolution of the "image" is given in Angstroms, where the smaller the number the more detailed the structure is said to be. Several structures in this book have been created using this method, for example, see Figure 1.10.

Often, once several components of a system have been isolated, they can be recombined in a cell free reconstitution system where the concentrations and conditions can be carefully controlled and altered. For example, proteins can be added to membrane fractions to get a better understanding of which proteins from the cytoplasm may be activating a membrane associated activity.

If interacting partners have not been identified then techniques such as the yeast two-hybrid system may be used. Once interacting partners have been identified then other techniques can be used to verify or further characterize the interactions. One such technique is surface plasmon resonance, a technique in which the reflected angle of a laser light is dependent on the amount and the mass of an interacting molecule that might interact with an immobilized membrane fraction. Alternatively, if two particular polypeptides are thought to undergo an interaction, the potential amino acid sequences involved in those interactions can be synthetically produced as short peptides. If these are made and added to an assay, they may bind to a prospective docking site on one polypeptide, thus flooding it with "false" peptides, and therefore potentially upsetting the propagation of the signal or activity.

Other methods for characterizing protein–protein interactions include the use of the split-GFP method. In this technique, two sections of a fluorescent protein are expressed separately but as part of potential interacting signalling proteins. If the two signalling proteins do interact, the fluorescent probe is reformed and can then be visualized, showing that the proteins have indeed come together, and also where in the cell they have come together.

Molecular genetic techniques

The greatest advances made in finding new signalling components and in the study of their interactions have come through the use of molecular genetic techniques. One of the most powerful is the ability to knock-out, or knock-down the expression of a protein, and ask the question "Does this make a difference?".

Knock-out, or knock-down

Knock-out expression is to stop the expression of a protein altogether. However, some technologies allow the expression levels of a proteins to be suppressed, and this is referred to as knock-down. This is useful particularly if knock-out expression is impossible. If a protein is of vital importance to a cell removing its expression altogether may be lethal, and therefore the cells, or organism will not survive. In that case, low levels of expression may be viable, but the phenotype seen may not always reflect the level of protein, the cell might not need very much expression to be seen to be normal.

Knock-out expression often involves the disruption of the gene, with perhaps an insertion, so that the protein of interest is not expressed at all. Without the protein present, its role in a signalling pathway should be easy to determine. Alternatively, anti-sense technology can be used to specifically knock-out, or knock-down, the expression of elements in a pathway. Although the use of anti-sense oligomers has yielded many results, the smart method of choice now is RNAi. Here, specific double stranded RNA is utilized. This is introduced into a cell where it is processed into small interfering RNAs (siRNAs), and these subsequently cause the degradation of homologous endogenous mRNA. Therefore, this method can specifically reduce the expression of a particular gene, enabling the researcher to answer the questions: What would happen if the cell could not express a specific protein fully, for example a kinase? What is the effect of a reduced amount of that protein on the signalling cascade being studied? Is that protein essential for the signalling pathways used by this particular ligand?

DNA clones of signalling components are often useful to have, especially if their over-expression is sought. An antibody, or an oligonucleotide designed from the back translation of a N-terminal protein sequence, can be used to pull out a clone from a DNA library. This would prevent the need for much of the protein sequencing undertaken previously. Oligonucleotide sequences based on the consensus sequences of several related polypeptides or cDNA clones can be used to find the genes for more members of a polypeptide family, where the functions have not even been discovered. Once full length DNA sequences have been cloned, they can be ligated into plasmids and over-expressed in cells, so enhancing or reducing the pathway in the cell. For example, the over-expression of a defective kinase (a so-called dominant-negative mutation) in a cell will mean that the defective kinase preferentially receives the message from the signal pathway, but is unable to perform any kinase activity, so effectively stopping the pathway at that point. Similarly, the over-expression of a specific phosphatase will mean that a specific kinase activity may be negated. Furthermore, full length sequences can be used to screen data bases to identify related genes, perhaps genes that fall in a family in the genome. Or they can be used to identify similar genes in different organisms. If up-stream genomic sequence is then sought, the promoter that drives the gene expression can be studied, perhaps giving an insight into the regulation of the expression of that particular signalling component.

Molecular techniques can yield extremely valuable information about the location of a protein in a cell too. By adding an extra sequence that encodes a protein domain which can be "seen" to a gene sequence that encodes a signalling protein to be studied, the expressed new composite protein may be tracked in the cell. Often green fluorescent protein (GFP) is used for such studies. The gene encoding it can be added at the end of the gene encoding the protein of interest, and as the GFP is naturally fluorescent it can be seen using a confocal microscope. Therefore, with optical sectioning and scanning via confocal microscope, a three-dimensional image of the location of the protein can be obtained. This can then be superimposed on to, for example, a bright field image of the same sample to add clarity to the data. Therefore, if it is suspected that a protein only signals in the nucleus, it can be discovered if this is true. If the protein only appears in the cytoplasm, but has an apparent nuclear function, perhaps the research team need to re-think the possible function of that protein in the cell.

Green fluorescent protein

Although GFP was the original protein used for such purposes as identifying the cellular location of other proteins, there are now many varieties with different fluorescent properties, such as yellow fluorescence (YFP), and many others. They now have a wide range of uses too, from measuring the redox environment inside cells to helping to study protein–protein interactions. They are further mentioned in Chapter 10. Roger Tsien along with Martin Chalfie and Osamu Shimomura were awarded the 2008 Nobel Prize in Chemistry for their work on GFP.

Microarrays and proteomics

Current technologies that are helping to elucidate the signalling events of cells include those that allow the study of holistic events in cells: microarray analysis and proteomics. Microarray analysis allows the expression profiles of hundreds or thousands of genes to be determined all at the same time, perhaps for all those in a genome. For certain species, particular microbes, model mammals such as mice, model plants such as *Arabidopsis thaliana*, and humans, the whole genomes have now been sequenced, so that representative example sequences from all the genes can be placed on a microarray and studied simultaneously. Therefore, if a ligand is added to a cell, all the genes that are expressed, or turned off, in response to that ligand can be determined. If signalling has gone awry in cells, such as in cancer, the expression profiles can be compared to those exhibited by normal cells, to unravel the molecular events that might be involved. It is clearly a powerful technique, but, like all experiments, the results have to be assessed carefully, and studies to confirm the findings using Northern analysis or real-time polymerase chain reaction (RT-PCR) need to be undertaken, giving a relatively independent and quantifiable measure of gene expression. Secondly, just because a gene is expressed does not mean that the protein will be made and be functional in the cell, and therefore proteomic techniques are required to study the protein profiles of cells. This often involves separation of all the proteins of interest with two-dimensional (2-D) gel electrophoresis, and then subsequent mass spectrometry and complex bioinformatic analysis, but is again a powerful technology, and one that is becoming more commonly used.

As with all techniques, the original methods for holistic investigations of both gene and expression, and protein profiles have been superseded. Microarrays and 2-D gels are still used, but other methods are becoming more common such as in solution studies for proteins, whereas bespoke arrays are often used if target genes are already known, or alternate methods such as serial analysis of gene expression (SAGE) are employed.

Computer networks

Signalling pathways are so complex and interwoven that the logic of computer networks is being used to try to understand the complexity of the interaction of the signalling pathways. Much research studies single pathways and events, but it is becoming more important to take a holistic view of the events in cells. It is likely that combinations of certain pathways and their relative contributions to the overall signal will be important for the overall outcome for the cell, and therefore modelling such combinations of pathways and their relative levels of activation will require computers and software development. Such studies are now becoming encompassed under what is being referred to as "systems biology". Here, all the workings of a cell are viewed as a single system, and the challenge is to look at it as a whole, rather than isolated separate components. Once this can be done, any perturbation to the system can be studied. For example in a cancer cell, if it is suspected that a particular gene is defective, the potential end result of that defect could be predicted if the workings of the cell are modelled. Or, if a drug is being developed to inhibit a particular signalling component, the overall effect on the cell can be envisaged so that potential side-effects or lack of a correct effect can be determined early in the drug development. However, taking this approach for cell signalling is a complex one, but is becoming more realistic as more about signalling is being unravelled. Perhaps one day a "virtual cell" will be available so that effects of drugs can be studied *in silico* before laboratory experiments are needed.

■ Summary

- The history of cell signalling goes back further than is perhaps appreciated, and many Nobel Prizes have been won along the way.

- In all cell signalling studies there are some fundamental questions that need to be answered, such as is a signalling component really involved?

- A wide range of technologies can be adopted for signalling studies, including the use of antibodies, fluorescent probes, confocal microscopy as well as more traditional biochemical approaches.

- The development of modern molecular biology techniques has allowed the further understanding of how many of these proteins function, with the use of gene sequence analysis, over-expression, knock-out and knock-down studies. Such approaches may even predict the existence of isoforms of proteins yet to be discovered.

- With the advent of post-genomics technologies, such as microarray analysis and proteomics, cell signalling is entering a new and exciting stage, where holistic studies will inform the future direction of signalling research.

- The ambition of "systems biology" is to determine how all the signalling and activities of a cell work in unison. Once this has been achieved, perturbations of the "system", for example drug treatment or disease, could be more comprehensively studied.

→ Further reading

A brief history

Agranoff, B.W., Bradley, R.M. and Brady, R.O. (1958) The enzymatic synthesis of inositol phosphatide. *Journal of Biological Chemistry*, 233, 1077–1083.

Ashman, D.F., Lipton, R., Melicow, M.M. and Price, T.D. (1963) Isolation of adenosine 3', 5'-monophosphate and guanosine 3', 5'-monophosphate from rat urine. *Biochemical Biophysical Research Communications*, 11, 330–334.

Burgoyne, R.D. and Petersen, O.H. (eds) (1997) *Landmarks in Intracellular Signalling*. Portland Press, London, UK. ISBN 1-85578-101-8.

Cheung, W.Y. (1970) Cyclic 3'-5'-nucleotide phosphodiesterase: demonstration of an activator. *Biochemical Biophysical Research Communications*, 38, 533–538.

Ebashi, S. and Lipmann, F. (1962) Adenosine triphosphate-linked concentration of calcium ions in aparticulate fraction of rabbit muscle. *Journal of Biological Chemistry*, 14, 389–400.

Fire, A., Xu, S.Q., Montgomery, M.K., Kostas, S.A., Driver, S.E. and Mello, C.C. (1998) Potent and specific genetic interference by double-stranded RNA in *Caenorhabditis elegans*. *Nature*, 391, 806–811 [The discovery of RNAi].

Hokin, M.R. and Hokin, L.E. (1953) Enzyme secretion and the incorporation of P^{32} into phospholipids of pancreatic slices. *Journal of Biological Chemistry*, 203, 967–977.

Hunnum, Y.A., Loomis, C.R., Merrill, A.H. and Bell, R.M. (1986) Shingosine inhibition of protein kinase C and activity of phorbol dibutyrate binding *in vitro* and in human platelets. *Journal of Biological Chemistry*, 261, 12604–12209.

Hunter, T. and Sefton, B.M. (1980) Transforming gene product of Rous sarcoma virus phosphorylates tyrosine. *Proceedings of the National Academy of Science, USA*, 77, 1311–1315.

Krupinski, J., Coussen, F., Bakalyar, H.A., Tang, W-.J., Feinstein, P.G., Orth, K., Slaughter, C., Reed, R.R. and Gilman, A.G. (1989) Adenylyl cyclase amino acid sequence: possible channel-like or transporter-like structure. *Science*, 244, 1558–1564.

Kuo, J.F. and Greengard, P. (1970) Cyclic nucleotide-dependent protein kinases. *Journal of Biological Chemistry*, 245, 2493–2498.

Lefkowitz, R.J. (2004) Historical review: A brief history and personal retrospective of seven-transmembrane receptors. *Trends in Pharmacological Sciences*, 25, 413–422.

Moore, B.W. and Perez, V.J. (1968) Specific acidic proteins of the nervous system. In *Physiological and Biochemical Aspects of Nervous Integration*. F. Carlson (ed.) Prentice Hall, Englewood Cliffs, N.J. [early report of 14-3-3 proteins].

Northup, J.K., Sternweis, P.C., Smigel, M.D., Schleifer, L.S., Ross, E.M. and Gilman, A.G. (1980) Purification of the regulatory component of adenylate cyclase. *Proceedings of the National Academy of Science, USA*, 77, 6516–6520.

Palmer, R.M.J., Ferrige, A.G. and Moncada, S. (1987) Nitric oxide release accounts for the biological activity of endothelium-derived relaxing factor. *Nature*, 327, 524–526.

Paulus, H. and Kennedy, E.P. (1960) The enzymatic synthesis of inositol monophosphate. *Journal of Biological Chemistry*, 235, 1303–1311.

Rall, T.W. and Sutherland, E.W. (1958) Formation of a cyclic adenine ribonucleotide by tissue particles. *Journal of Biological Chemistry*, 232, 1065–1076.

Rall, T.W., Sutherland, E.W. and Berthet, J. (1957) The relationship of epinephrine and glucagon to liver phosphorylase. *Journal of Biological Chemistry*, 224, 463–475.

Ray, L.B. and Sturgill,T.W. (1987) Rapid stimulation by insulin of a serine/threonine kinase in 3T3-L1 adipocytes that phosphorylates microtubule-associated protein 2 *in vitro*. *Proceedings of the National Academy of Science, USA*, 84, 1502–1506.

Sattin, A., Rall, T.W. and Zanella, J. (1975) Regulation of cyclic adenosine 3', 5'-monophosphate levels in guinea-pig cerebral cortex by interaction of adrenergic and adenosine receptor activity. *Journal of Pharmacology and Experimental Therapeutics*, 192, 22–32.

Sutherland, E.W. and Rall, T.W. (1957) The properties of an adenine ribonucleotide produced with cellular particles, ATP, Mg^{++}, and epinephrine or glucagon. *Journal of the American Chemical Society*, 79, 3608.

Sutherland, E.W. and Rall, T.W. (1958) Fractionation and characterisation of a cyclic adenine ribonucleotide formed by tissue particles. *Journal of Biological Chemistry*, 232, 1077–1091.

Takai, Y., Kishimoto, A., Kikkawa, U., Mori, T. and Nishizuka, Y. (1979) Unsaturated diacylglycerol as a possible messenger for the activation of calcium-activated, phospholipid-dependent protein kinase system. *Biochemical Biophysical Research Communications*, 91, 1218–1224.

Tonks, N.K., Diltz, C.D. and Fischer, E.H. (1988) Characterisation of the major protein-tyrosine-phosphatases of human placenta. *Journal of Biological Chemistry*, 263, 6731–6737.

Whitman, M., Downes, C.P. Keeler, M. Keller, T. and Cantley, L. (1988) Type 1 phosphotidylinositol kinase makes a novel inositol phospholipid, phosphatidylinositol-3-phosphate. *Nature*, 332, 644–646.

Woods, N.M., Cuthbertson, K.S.R. and Cobbold, P.H. (1986) Repetitive transient rises in cytoplasmic free calcium in hormone-stimulated hepatocytes. *Nature*, 319, 600–602.

A brief look at some techniques

Arenz, C. and Schepers, U. (2003) RNA interference: from an ancient mechanism to a state of the art therapeutic application. *Naturwissenschaften*, 90, 345–359 [an excellent review of RNAi].

Alberts, B., Bray, D., Lewis, J. Raff, M., Roberts, K. and Watson, J.D. (1994) *Molecular Biology of the Cell*, 3rd edn, Garland Press, New York, USA [in particular pp. 778–782 for discussion on computer networks: removed from the later edition].

Barnard, E., McFerran, N.V., Trudgett, A., Nelson, J. and Timson, D.J. (2008) Detection and localisation of protein-protein interactions in *Saccharomyces cerevisiae* using a split-GFP method. *Fungal Genetic Biology*, 45, 597–604.

Berg, T. (2008) Small-molecule inhibitors of protein-protein interactions. *Current Opinions in Drug Discovery Development*, 11, 666–674.

Causton, H.C., Quackenbush, J. and Brazma, A. (2003) *Microarray Gene Expression Data Analysis: A Beginner's Guide*. Blackwell Publishers, Oxford, UK. ISBN 1405106824.

Desikan, R., Hagenbeek, D., Neill, S.J. and Rock, C.D. (1999) Flow cytometry and surface plasmon resonance analyses demonstrate that the monoclonal antibody JIM19 interacts with a rice cell surface component involved in abscisic acid signalling in protoplasts. *FEBS Letters*, 456, 257–262 [an example of a paper which uses surface plasmon resonance to study cell signalling events].

Desikan, R., A.-H.-Mackerness S., Hancock, J.T. and Neill, S.J. (2001) Regulation of the Arabidopsis transcriptome by oxidative stress. *Plant Physiology*, 127, 159–172 [an example of a microarray study, looking at cellular responses to a signal].

Hannon, G.J. (ed.) (2003) *RNAi; A Guide to Gene Silencing*. Cold Spring Harbor Laboratory Press, ISBN 0879696419.

Hidaka, H. and Kobayashi, R. (1994) *Essays in Biochemistry*, 28, Portland Press, UK, pp. 73–97 [discussion on kinase inhibitors].

Ivakhno, S. (2007) From functional genomics to systems biology. *FEBS Journal*, 274, 2439–2448.

Kendall, D.A. and Hill, S.J. (eds) (1995) *Signal Transduction Protocols*. Humana Press, New Jersey [a particularly useful collection of protocols for cell signalling research].

Liebler, D.C. (ed.) (2001) *Introduction to Proteomics: Tools for the New Biology*. Humana Press, New Jersey. ISBN 0896039927.

Murray, D., Doran, P., MacMathuna, P. and Moss, A.C. (2007) *In silico gene expression analysis – an overview*. *Molecular Cancer*, 6, 50.

Matsumoto, B. (ed.) (2003) *Cell Biological Applications of Confocal Microscopy (Methods in Cell Biology)*. Academic Press, London, UK. ISBN 012480277X.

Ng, T., Squire, A., Hansra, G., Bornancin, F., Prevostel, C., Hanby, A., Harris, W., Barnes, D., Schmidt, S., Mellor, H., Bastiaens, P.I. and Parker, P.J. (1999) Imaging protein kinase C-activation in cells. *Science*, 283, 2085-2089 [a study using Fluorescence Resonance Energy Transfer (FRET) in conjunction with Fluorescence Lifetime Imaging Microscopy (FLIM)].

Palzkill, T. (2002) *Proteomics*. Kluwer Academic Publishers, Norwell, Massachusetts, USA. ISBN 0792375653.

Polverari, A., Molessini, B., Pezzotti, M. Buonaurio, R., Marte, M. and Delledonne, M. (2003) Nitric oxide-mediated transcriptional changes in *Arabidopsis thaliana*. *Molecular Plant-Microbe Interactions*, 12, 1094-1105 [an example of a paper in which the expression of a large number of genes is studied, following the addition of a cellular signal to cells, and it is not surprising that many of the genes encode proteins involved in cell signalling].

Ratushny, V. and Golemis, E. (2008) Resolving the network of cell signaling pathways using the evolving yeast two-hybrid system. *Biotechniques*, 44, 655–662.

Rutter, G.A., White, M.R.H. and Tavare, J.M. (1995) Involvement of MAP kinase in insulin signalling revealed by noninvasive imaging of luciferase gene-expression in single living cells. *Current Biology*, 5, 890–899.

Schena, M. (2002) *Microarray Analysis*. Wiley-Liss, Hoboken, New Jersey. ISBN 0471414433.

Stepanenko, O.V., Verkhusha, V.V., Kuznetsova, I.M., Uversky, V.N. and Turoverov, K.K. (2008) Fluorescent proteins as biomarkers and biosensors: throwing color lights on molecular and cellular processes. *Current Protein Peptide Science*, 9, 338–369.

Tavaré, J.M., Fletcher, L.M. and Welsh, G.I. (2001) Review: Using green fluorescent protein to study intracellular signalling. *Journal Endocrinology*, 170, 297–306.

Terrian, D.M. (2002) *Cancer Cell Signalling: Methods and Protocols*. Humana Press, Totowa, New Jersey, ISBN 1588290751.

Walker, John M. (series ed.) *Methods in Molecular Biology*, Humana Press ISSN: 1064-3745. [This is an excellent up-to-date series of books of which several are applicable to cell signalling research. For example: Willars, G.B., Challiss, R.A.J. (eds) (2004) *Receptor Signal Transduction Protocols*, Vol. 259, ISBN 978-1-58829-329-9; Stockand, J.D., Shapiro, M.S. (eds) (2006) *Ion Channels: Methods and Protocols*, Vol. 337, ISBN 978-1-58829-576-7; Paddock, S.W. (ed.), in press, *Confocal Microscopy: Methods and Protocols*, ISBN 978-1-58829-351-0].

Point of note

When doing a computer-based search, care is needed with spelling as many proteins, etc., have alternate names or spellings. Cell signalling itself can be spelt two ways, "signalling" and "signaling", so often more than one search will be needed.

Part 2

Components that comprise signalling pathways

Extracellular signals: hormones, cytokines and growth factors

4

One of the basic tenets of cell signalling is that two cells need to communicate with each other. They may be a long way apart, or they may be neighbours, but still they often need to signal to each other. Many types of molecules are used as signals, and often the way they are classified appears to be a little arbitrary, but here the main groups are discussed. Animal hormones are commonly used as intercellular signals, and plants use hormones too.

The group of peptides known as the cytokines is used by animals for short-range signalling, and new members of this group of signals seem to be discovered on a regular basis. Some of this group were thought to be elixirs with which to cure many diseases, but their full potential has yet to be uncovered.

On the other hand, long-range signalling between organisms involves the pheromones, and these too are discussed here.

It is by the uncovering of the multitude of intercellular signals used in nature that an understanding of the messages sent between cells can be gained, giving an insight into how cells respond and survive. Often it is the dysfunction of these messages that leads to disease, and a greater understanding may also lead to ways to modulate such messages. Such manipulation may allow organisms to survive in new conditions; for example, perhaps new plants can be developed that require less water, or new drugs can be developed to alleviate the symptoms of a disease, or even cure the disease.

4.1 Introduction

One of the most significant ways in which cells communicate with each other is by the release and detection of extracellular signalling molecules, such as hormones, cytokines and growth factors. Such molecules often can be released some considerable distance from their point of action, and in general have an effect over a relatively long time scale with an unspecific transport to the site of action. For example, a hormone might be released and carried by the blood stream where it is transported to all parts of the body and washes around many types of cells, some of which will detect its presence, and many of which will not. Many cells will be unaffected by such a release of hormone, the specificity of effect being determined by the presence of specific receptor molecules on the surface of the detecting cells (see Chapter 5 for detailed discussion).

Many of these extracellular signalling molecules are involved and have been implicated in disease states, a defect either being the cause of the disease, or the presence of the molecule propagating the manifestation of the disease. One of the most well studied and understood is the role of insulin in diabetes. This multifactorial disease may involve aberrations in the synthesis of the hormone, so relaying no message to the target cells, or may involve a lack of detection by the target cells. Insulin and the signalling pathways it induces are further discussed in Chapter 11.

The role of hormones has been most well studied in animals, although such extracellular signals are widely used in the plant kingdom and, in fact, as discussed in Chapter 3, a plant hormone, or at least its effects, was studied by Darwin well over a hundred years ago. Hormones have also been found to be important for unicellular eukaryotes, where the release of hormone-like molecules has been seen in the transfer of messages between individual organisms, in a similar manner to that used by mammals between tissues.

Although the distinction between hormones, cytokines and growth factors appears to be a somewhat loose one, a rough division is given here and they are considered separately.

4.2 Hormones

Under the broad name of hormones, a term taken from the Greek meaning "to arouse" or "to excite", are several diverse types of molecule. Although the exact definition of a hormone appears to be somewhat vague, here the term is used to describe substances that are released into the extracellular medium by the cells of one tissue, to be carried to a new site of action, such as a different tissue, where they provoke a specific response. They can be split into several broad classes: small water soluble molecules; peptide hormones; lipophilic molecules that are detected by surface receptors; and lipophilic molecules that are detected by intracellular receptors. However, what is common to all hormones, is the fact that they have specificity, being released only when required, and detected only by cells that need to respond to them.

Small water soluble molecules

These hormones share common characteristics in that they are all water soluble and cannot cross the plasma membrane of the cell, usually because they are too large or carry a charge at physiological pH and as such their detection requires the presence of specific receptors on the cell surface of the responding cell (see **Chapter 5**). They are usually released into the vascular system of the organism, to be carried freely to their site of perception. Usually, relatively small organic molecules and hormones fall into this category, including histamine and epinephrine (adrenaline) (see **Figure 4.1**).

Histamine is produced in the mast cells and is responsible for the control of blood vessel dilation. Often topical drugs, for example, for the treatment of bee stings, will contain anti-histamines, which are used to counteract the release of histamine during an inflammatory response. Epinephrine, or as it used to be known, adrenaline, was classically known as the hormone released at times of panic, and has been referred to as the "fright, fight and flight" hormone. It is produced by the adrenal medulla and causes an increase in blood pressure and pulse rate, contraction of smooth muscles, an increase in glycogen breakdown in muscles and the liver, and an increase in lipid hydrolysis in adipose tissue, all factors that would ready the body for a situation where a greater energy requirement is envisaged, such as in fight and flight. In most cases, the effects of epinephrine are fast, and, importantly, quickly reversed. An organism would not wish to remain in a state of panic for too long.

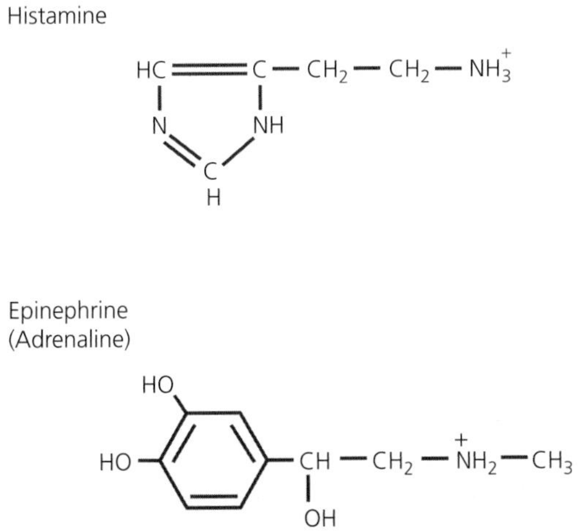

Figure 4.1 The molecular structures of histamine and epinephrine: examples of small water soluble hormones.

Peptide hormones

Like histamine and epinephrine, the peptide hormones are water soluble and are carried to their site of action by the vascular system. Usually their release is rapid, as they are often stored within vacuoles in the cell, ready to be evicted from the cell by exocytosis. Such vacuoles will remain dormant in the cell until such time as the cell is stimulated, involving a signal transduction pathway (often involving a rise in calcium ion concentrations), at which time the vacuoles will be moved to the plasma membrane. Fusion of the vacuole membrane with the plasma membrane will cause the hormone to be released to the extracellular fluid (see **Figure 1.2B**). The breakdown of peptide hormones is also rapid, usually by proteases in the blood and tissues. The two that have probably been most extensively studied are glucagon and insulin, which in fact have antagonistic effects.

Insulin is produced by the β cells of the pancreas as a single pre-proinsulin polypeptide, which is converted to a single proinsulin polypeptide as illustrated in **Figure 4.2**. This folds into its correct secondary and tertiary structure aided and stabilized by the presence of disulphide bridges. Once correctly folded, active insulin is produced by proteolytic cleavage in two places causing removal of a substantial part of the middle of the polypeptide. The result of this is that insulin contains two polypeptide chains, an A chain of 21 amino acids and a B chain of 30 amino acids held together by disulphide bonds and electrostatic forces.

Once released, insulin stimulates the uptake of glucose by muscle and fat cells, increases lipid synthesis in the adipose tissues and causes a general

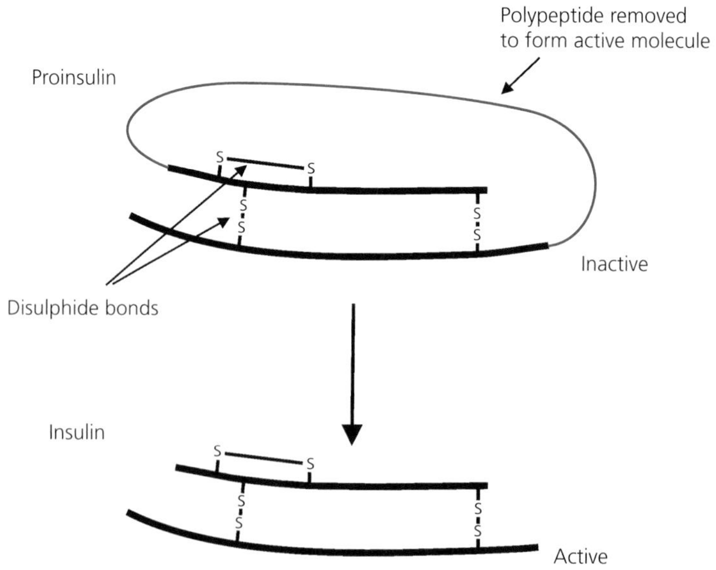

■ The structure of insulin was identified by Dorothy Hodgkins. Born 1910, she studied the structures of several biomolecules, including penicillin and vitamin B_{12}. She was awarded the Nobel Prize for Chemistry in 1964.

Figure 4.2 The production of insulin from proinsulin by the removal of the linking polypeptide (shown by the blue line).

increase in cell proliferation and protein synthesis. Detection of insulin is due to the presence on the surface of target cells of a multipolypeptide receptor (actually a tetramer), which contains intrinsic tyrosine kinase activity, and its activation leads to a cascade of events including the activation of phosphatidylinositol 3-kinase, activation of G proteins and the stimulation of a MAP kinase pathway. Amongst its effects on glucose homeostasis, insulin can also lead to the stimulation of transcription factors and the increase in selected gene expression, as discussed further in **Chapter 11**.

Glucagon, like insulin and all peptide hormones, is also made from a precursor polypeptide. Active glucagon contains a single polypeptide chain of 29 amino acids. It is released from the α cells of the pancreas and leads to glycogen breakdown and lipid hydrolysis, allowing an increase in glycolysis and respiratory rates.

Other hormones in this class include follicle stimulating hormone (FSH) and luteinizing hormone (LH). Both are like insulin in having two polypeptide chains. However, they are much bigger, FSH having an α chain of 92 amino acids along with a β chain of 118 amino acids. Both hormones are produced by the anterior pituitary. FSH stimulates the growth of ovarian follicles and oocytes, and also increases the production of oestrogen. LH controls the maturation of oocytes and increases the release of oestrogen and progesterone.

Hormones often influence the release or action of other hormones, and here, the release of LH is under the control of another peptide hormone, this time a single polypeptide hormone called LH-releasing hormone. This is produced by the hypothalamus and neurons. This type of interaction of different hormones and other extracellular signals, controlling each other's synthesis and secretion is not unusual, and, in fact, interactions of this nature are exemplified by the action of cytokines and chemokines.

Lipophilic molecules that are detected by cell surface receptors

As well as detection of the water soluble hormones above, some receptors on the surface of cells also detect the presence of a group of hydrophobic, or lipophilic, molecules that act like hormones. The main group of signalling molecules here are prostaglandins.

Prostaglandins are synthesized from arachidonic acid, a 20-carbon fatty acid. Arachidonic acid is often found covalently attached to the glycerol backbone of phospholipids, which constitute the plasma membrane. The arachidonic acid is attached to the middle carbon, but can be hydrolytically released from lipids, for example, by the action of phospholipase A_2 (see **Figure 8.9**).

As the prostaglandins are derived from a molecule embedded in a lipid environment, it is not surprising that they are inherently hydrophobic in nature. Although prostaglandins are a relatively large group of chemicals they can be roughly divided into nine different classes. Although originally named after the organ where they were thought to be made, prostaglandins are in fact produced by most cells, with their action usually being local. Their synthesis can be inhibited by several anti-inflammatory drugs, including aspirin, and their effects range from control of platelet aggregation to the induction of uterine contraction.

Lipophilic molecules that are detected by intracellular receptors

Not all hormones are perceived by cell surface receptors. Many are recognized by receptors inside the cell, and therefore such hormones need to be lipophilic, or hydrophobic, to enable them to penetrate the plasma membrane before perception. This class of hormones encompasses the steroid hormones, which include for example, oestrogen, progesterone, androgens and glucocorticoids, as well as the thyroid hormones and retinoids (see Figure 4.3).

Steroid hormones are all derived from cholesterol and are synthesized and secreted by endocrine cells. Progesterone is synthesized by the ovaries and placenta, and is involved in the development of the uterus in readiness for implantation of the new embryo, as well as for the stabilization of early pregnancy and development of the mammary glands. Oestrogens, such as oestradiol, are also involved in the development of female sexual characteristics such as uterus differentiation and mammary gland function. Similarly, testosterone, which is produced by the testis, is responsible for the development and functioning of the male sex organs, as well as development of less obviously useful male characteristics such as hair growth.

Mammals are not the only animals that use such hormones. For example, in insects and crustaceans a similar role in the development of sexual characteristics is fulfilled by a steroid-like compound called α-ecdysone.

Other steroid hormones include cortisol and vitamin D. Cortisol, a glucocorticoid, is produced by the cells of the cortex of the adrenal gland and controls the metabolic rates of many cells. It is formed from progesterone by three hydroxylation steps, at carbons 11, 17 and 21. Vitamin D is synthesized in skin exposed to sunlight. 7-Dehydrocholesterol (provitamin D_3) in the skin is lyzed by ultraviolet light to previtamin D_3, an inactive form, which is activated by hydroxylation in the liver and kidneys with its conversion to calcitrol (1,25 dihydroxycholecalcitrol). Calcitrol is responsible for the control of Ca^{2+} uptake in the gut and lowering Ca^{2+} excretion by the kidneys. Lack of vitamin D in a child's development can lead to the development of the condition known as rickets. Here, the cartilage and bone fail to calcify properly leading to the malformation of the bones and this often leaves the long bones of the patient bent. This is most noticeable in the legs of patients and was seen when children were forced to work long hours indoors or underground away from adequate sunlight, a situation that is fortunately now not common. Dietary vitamin D, such as derived from fish oils, can

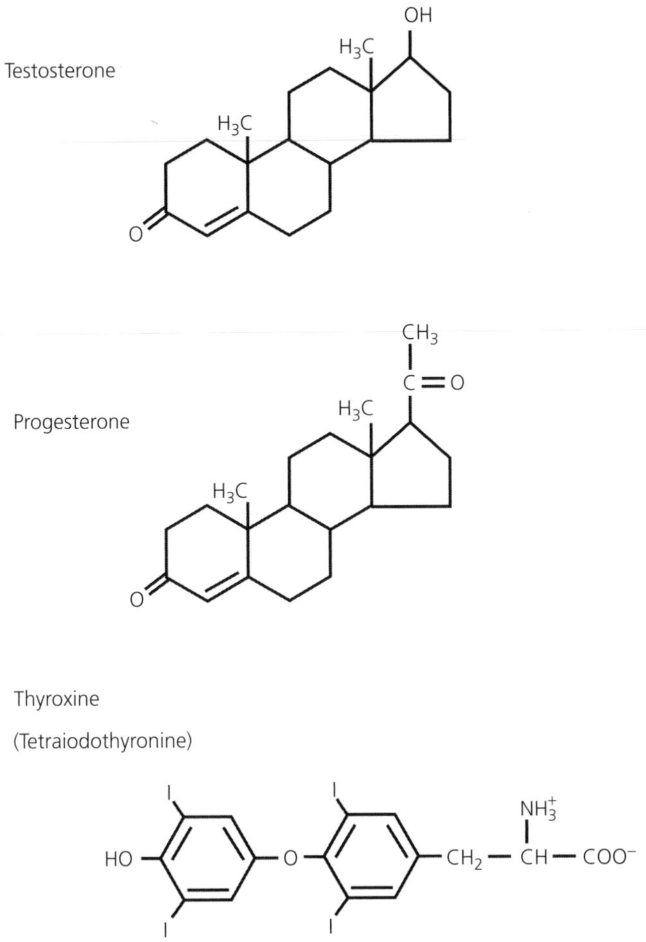

Testosterone

Progesterone

Thyroxine

(Tetraiodothyronine)

Figure 4.3 The molecular structures of testosterone, progesterone and thyroxine: examples of lipophilic molecules detected by intracellular receptors.

overcome the lack of its *de novo* synthesis in the skin and prevent the associated symptoms.

Thyroid hormones that have effects on the metabolism of many cells, including the increase in heat production and production of polypeptides involved in metabolic pathways, are derived from the amino acid tyrosine. A good example is thyroxine, otherwise known as tetraiodothyronine, which is produced by the thyroid gland. The structure of thyroxine is shown in **Figure 4.3**.

Another vitamin that is involved in the synthesis of lipophilic hormones is vitamin A (also called retinol), from which the retinoids are synthesized. For example, retinoic acid is formed by the oxidation of the alcohol group of retinol to a carboxyl group. The effects of retinoic acid include an alteration of gene expression profiles in the receptive cell.

Retinol is also the precursor of retinal, the light sensitive group of rhodopsin, and therefore is instrumental in the photoperception of mammals (see **Chapter 12**). It was thought that a deficiency in vitamin A can lead to an impairment in light detection of the eye, especially under low lighting conditions, a condition commonly called "night blindness", a fact that was used in propaganda during the Second World War to persuade people to eat vegetables, particularly carrots.

Transport between cells of these lipophilic homones is not as simple as that seen for the water soluble ones, as once released these lipophilic molecules are inherently insoluble in water. Therefore, in the blood stream, they need to be stabilized, which is achieved by their association with specific carrier proteins. Dissociation from these carriers then needs to occur before they cross the plasma membrane and enter the cells.

In general, these hormones can persist in the circulation for hours or even days and are involved in long term control. However, once inside the cells they are detected by receptors in the cytosol or in the nucleus (see **Chapter 5**), often culminating in the alteration of specific gene expression, which again might lead to long term effects in the cells and tissues.

Steroid hormones and treatments

Many steroid compounds are used for treatment of skin problems, where they are applied topically. Due to their hydrophobicity they can be taken up by the skin where they are needed. This also limits, to some extent, their take up into the blood stream, and therefore reduces their systemic effects.

4.3 Plant hormones

The existence of plant hormones was first postulated by Darwin, who showed by experimenting on the phototropism of coleoptiles that substances, yet to be identified, must be transported around plants. By 1928, Went had discovered auxin, and now a wide range of substances have been placed under the rather broad umbrella of plant hormones. It is in fact a very loose term and other terms such as "plant growth substances" and phytohormones have been suggested, but never used extensively. In animals, the extracellular signalling molecules have been divided, albeit rather crudely, into classes such as hormones and cytokines: the term "plant hormone" embraces all the substances that act as extracellular signals in plants (see **Figure 4.4** for the structures of some of the common ones). That means they might exert their influence at a site some distance from their site of manufacture, or act within the tissue where they are made, or even on the same cell. For example, ethylene has a very short distance of influence, whereas cytokinins can be transported from the roots to the leaves.

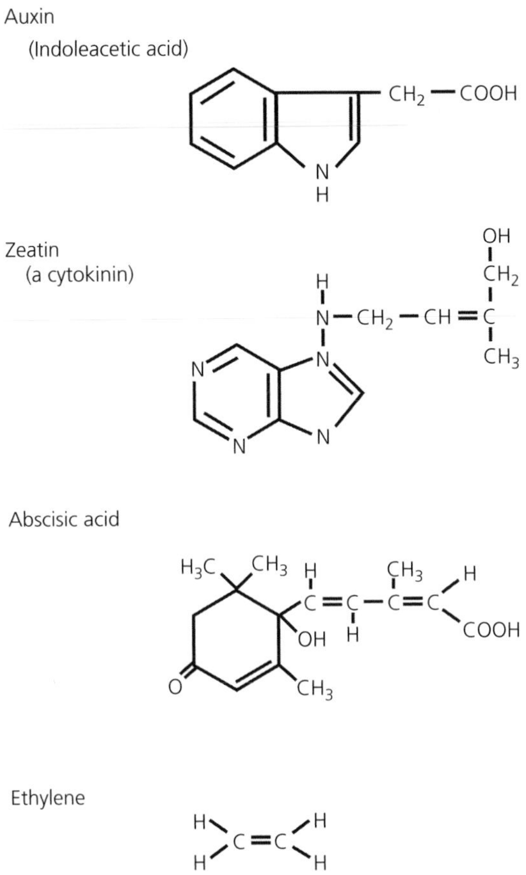

Figure 4.4 The molecular structures of auxin, zeatin, abscisic acid and ethylene: examples of plant hormones.

Auxin

Although the main auxin found in plants is indole-3-acetic acid (IAA; Figure 4.4), several compounds derived from this molecule show auxin-like activity, including indoleacetaldehyde or indoleacetyl aspartate. IAA is manufactured in the leaves, particularly young leaves and in developing seeds from tryptophan or indole. Its transport appears to be primarily from cell to cell, but it can also be found in the phloem. Its effects are diverse, ranging from the stimulation of cell enlargement and stem growth, stimulation of cell division and the differentiation of the phloem and xylem vessels, to the mediation of tropistic responses to light and gravity. Even though auxin was identified a long time ago, and many of its effects have been well characterized, its perception, that is the receptor proteins to which it binds, have yet to be fully characterized, although several have been identified and are the subject of intensive research.

Cytokinins

Cytokinins (CK) are derived from adenine in the root tips and in developing seeds, the most common being zeatin (see **Figure 4.4**). Their transport is via the xylem system and their activities include the induction of cell division, although this requires the presence of auxin, leaf expansion caused by cell enlargement and a delay of leaf senescence.

Gibberellins

The most common biologically active gibberellin in plants is gibberellin A_1 (GA_1), although the term gibberellin encompasses a family of active compounds. Their structures are complex, based on what is known as the *ent-gibberellane* structure. This involves four ring structures with a fifth oxygen containing ring bridging across. They are synthesized from mevalonic acid in young shoots and in developing seeds, whereas their transport is via both the xylem and the phloem vascular systems. The biological activity of gibberellins include the stimulation of flowering, the induction of seed germination and the stimulation of the synthesis of enzymes such as α-amylase. Their perception by cells also leads to cell division and cell elongation, which manifests itself as the elongation of the stems.

Abscisic acid

Like the gibberellins, abscisic acid (ABA, **Figure 4.4**) is synthesized from mevalonic acid, but this time mainly in the roots and mature leaves, although most tissues are probably capable of producing it. Seeds can also synthesize ABA or indeed import it from the parent plant. Its transport is carried out by both the phloem, down to the roots and by the xylem up to the leaves. It is synthesized particularly in times of water stress, and one of the functions of ABA is to mediate stomatal closure and so control transpirational water loss from leaves. Here, its perception invokes a wide range of intracellular responses that include the production of hydrogen peroxide and nitric oxide, and the stimulation of calcium signalling. It also induces the synthesis of storage proteins in seeds and inhibits shoot growth. However, as with auxin, the receptor that perceives ABA has remained elusive, although several likely candidate proteins have come to light recently.

Ethylene

What at first appears to be an odd compound to act as a signalling molecule is the gas ethylene (**Figure 4.4**). However, other gaseous compounds that act as cell signals, such as nitric oxide (NO) have also been discovered, firstly in animals and then in plants, as will be discussed in **Chapter 10**.

Ethylene is synthesized from methionine in most plant tissues, particularly in response to stress. Its transport is by diffusion, although intermediates in its synthesis can be transported around the plant and lead to its appearance at a site distant from the original stimulus. One of the most commonly known roles for ethylene is in the induction of fruit ripening, but it can also lead to shoot and root differentiation and growth, and to opening of flowers, amongst many others. One of the most well-known problems associated with this is the release of ethylene from bananas. It has long been known that bananas should not be stored with other fruit, as ethylene from the bananas will cause premature ripening of those fruit, and storage time is very much reduced. Hence, once purchased, bananas are often hung on "trees" away from apples, pears, etc.

Other plant hormones

Many other compounds come under the broad term of "plant hormone", including jasmonates, (discussed in Chapter 8), polyamines, salicylic acids and brassinosteroids. This last group contains over 60 steroidal compounds, whereas salicylic acid has been implicated in thermogenesis and the production of pathogen-induced proteins.

Other extracellular plant signals, perhaps not classed as hormones, include the reactive species hydrogen peroxide and nitric oxide. Although first identified as instrumental signals in mammals, these compounds have now been found to be important as signals in plants, controlling water loss from leaves by modulating stomatal apertures, and inducing the death of cells by a mechanism similar to that of apoptosis, referred to as programmed cell death (PCD), which is seen as part of the hypersensitive response (HR).

> **Hypersensitive response (HR)** a mechanism used by plants to fight pathogen attack. It often involves the death of cells in the immediate location of pathogen perception, and so limits pathogen spread to the rest of the plant.

4.4 **Cytokines**

A group of peptide molecules that have profound effects on cells, but which are classed separately from the hormones, are the cytokines and chemokines. These extracellular signals have only come to light relatively recently, compared to the work carried out on hormones, and the rate at which new members of these families are being discovered is truly alarming, and no doubt new cytokines and chemokines, and variants thereof, are yet to be found. However, with the completion of the human genome sequencing, it should be possible to identify all of them, even if we do not know what they all do.

The cytokines are a group of peptide molecules that are produced by many cell types, but have their effect on other cells within a short distance, or often even on the cells that produce them. Hence, cytokine effects tend to be local, where they are referred to as being involved in paracrine or autocrine function. Under this definition come those molecules that are principally responsible for the coordination of the immune response of higher animals, and include the interleukin series, tumour necrosis factors and interferons.

With the recent explosion in the number of cytokines identified, classification into groups has been suggested. At present, there are more than 80 peptides that have been classed as cytokines, and these have been put into 16 families, (I–XVI), although classification is not easy for all of them. Some appear to have characteristics common to more than one class, whereas others are not easy to classify at all. However, one classification system is shown in Table 4.1.

Many of the cytokines have names that are historic, in that they were classified as part of a growing family, for example the interleukin series, whereas others are named in a logical way that denotes a function or characteristic, for example, FGF-a, which stands for fibroblast growth factor acidic. Often such naming leads to acronyms, which at least allows for the easy memorization of their names. Excellent examples here include Trail (TNF-related apoptosis inducing ligand), Trance (TNF-related activation induced cytokine) and April (a proliferation-inducing ligand).

The sizes of the cytokines are quite variable, with some being relatively small, weighing in at less than 10 kDa. Others are much larger, with individual subunits as large as 60–70 kDa. However, the majority seem to fall between 15 and 40 kDa. In most cases, the receptors that perceive the cytokines have now been identified, and the intracellular cascades that are invoked are being unravelled. The likely receptors and transduction components involved are discussed in the following chapters.

There is a massive interest world-wide in these peptides, not least because the elucidation of the mechanisms by which they function should lead to the identification of future drug targets, and as they are involved in a plethora of diseases, they are extremely important.

Despite the wide range of cytokines, it is worth focussing on three families, namely the interleukins, interferons and tumour necrosis factors.

Interleukins

The interleukin series has, at the present time, 35 members identified, designated IL-1 to IL-35. The first 18 (IL-1 to IL-18) to be discovered are listed in Table 4.2, but several have been identified very recently, and it is probable that new ones, or new forms of known ones, will be uncovered in future research. Although IL1 was named as one, it in fact exists in two distinct forms, IL-1α and IL-1β, which are coded for by separate genes. However, both are produced as larger precursor molecules and a cleavage event produces the active extracellular form. IL-1α is produced as a 271 amino acid, which is cleaved to 159 amino acids, whereas IL-1β is a 153 amino acid peptide derived from a 269 amino acid precursor. Therefore, similar to the peptide hormones, such as insulin, a major cleavage event has taken place,

Table 4.1 The cytokine groups.

Name of group	Examples	Notes
Interferons	IFN-α IFN-β IFN-γ	Used in response to viral infections. IFN-γ produced mainly by T cells and NF cells. IFN-α produced by leukocytes and epithelia cells.
Interleukins	IL-1 to IL-35 IL-1 has more than one isoform: α and β	Large family of cytokines. Variety of effects which include control of cell growth and differentiation. May control production of other cytokines (IL-10 inhibits). Mainly produced by T cells and phagocytes.
Tumour necrosis factors	TNFα TNFβ TGFβ	Mediates the inflammatory response: activation and adhesion of cells involved. TNF-α produced by macrophages and lymphocytes.
Chemokines	MIP-1α NAP-2 RANTES	Involved in coordination of leukocyte location, for example, transition from blood to tissues. At least four families but mainly grouped in CC or CXC families. Also includes at least one interleukin (IL-8).
Colony stimulating factors	M-CSF G-CSF GM-CSF	Involved in division and development of bone marrow and hence white blood cells. Produced mainly by monocytes/macrophages, endothelial cells and fibroblasts.

removing a substantial proportion of the polypeptide. Such cleavage events are common in the synthesis of the interleukin family. In the case of IL-1, the uncleaved cell associated form still seems to retain biological activity. In fact, several of the cytokine family appear to have membrane bound or membrane associated forms, and are not released freely from the cell.

Several of the genes for the cytokines, for example IL-3, IL-4, IL-5, IL-13 along with GM-CSF, are grouped together on the same region of chromosome 5 in humans, or chromosome 11 in mice, suggesting that they originally arose through gene duplication events. This is quite commonly seen where families of proteins exist. Evolution has allowed the copying of a successful protein, and then its subsequent subtle alteration to fulfil a new function or role, unable to be undertaken by the original protein. Such a process is then repeated, building up a family of subtly different, but related, proteins, each with slightly, but significantly, different roles.

Table 4.2 summarizes some of the characteristics of the interleukins, and indicates the receptors that have been identified.

Table 4.2 The interleukin family and some of their characteristics.

Cytokine	Amino acids precursor	Active form	Molecular weight kDa	Receptor(s) used	Chromosomal location
IL-1α	271	159	17.5	IL-1 RI IL-1 RII	2q13
IL-1β	269	153	17.3	IL-1 RI IL-1 RII	2q13-q21
IL-2	153	133	15–20	IL-2 Rα/β/γc IL-2 R/β/γc	4q26-q27
IL-3	152	133	14–30	IL-3 Rα/βc	5q23-q31
IL-4	153	129	15–19	Combinations IL-13 receptor	5q23-q32
IL-5	134	115	45	IL-5 Rα/βc	5q23.3-q32
IL-6	212	184	26	IL6 R/gp130	7p15-p21
IL-7	177	152	20–28	IL7 R/γc	
IL-8	99	72–77	8–8.9	CXCR-1 CXCR-2	
IL-9	144	126	30–40	IL-9 R/γc	5q31-5q35
IL-10	178	160	39 (dimer)	IL-10 R1/R2	1
IL-11	199	178	23	IL11 Rα/gp130	19q13.3-13.4
IL-12		197 and 306	70 (heterodimer)		
IL-13		132	9, 17	Four combinations, some with IL-4 receptor	5q31
IL-14			53		
IL-15	162	114	14–15	IL-15Rα/IL2 Rβ/γc	
IL-16			14	CD4	
IL-17		136	15, 22	IL-17 R	
IL-18		157	18	IL-18Rα IL-18Rβ	

Most of the interleukins are active in the monomeric state, but interleukin 5 exists as a homodimer of two 115 amino acid polypeptides, whereas interleukin 12 exists as a heterodimer of variably sized polypeptides. Interleukin 8, otherwise known as neutrophil activating protein 1 (NAP-1), is also a homodimer, but here the subunits are of variable length, 72 amino acids up to 77 amino acids, due to truncation of the N-terminal end of the polypeptides. It is interesting that the IL-8 molecules that contain the shorter versions of the subunits appear in some cases to be more potent than the ones with the longer polypeptides.

Cytokines are produced by a variety of blood cells and cells involved in the immune response. IL-2 and IL-3 are, for example, produced mainly by helper T lymphocytes, whereas IL-6 is produced by a variety of cells including T cells, macrophages and fibroblasts.

The biological responses of the IL family are also varied. IL-13, formerly known as P600, seems to induce the growth and differentiation of B cells and inhibits cytokine production of macrophages and their precursor cells, monocytes. IL-5 leads to the activation of eosinophil function, including chemotaxis and eosinophil differentiation. Some of the individual cytokines have a diverse range of biological activities, a good example being IL-6. It stimulates the differentiation of myeloid cell lines, acts as a growth factor on B cells, modulates the responses of stem cells to other cytokines and has effects on non-haematopoietic cells, including affecting the development of nerve cells.

These cytokines all work in a concerted way to coordinate the whole response. They can exert coordinating responses on the cells themselves or regulate the synthesis of each other. For example, IL-5 enhances the IL-4 effect on B cells, enhancing the IL-4 induced synthesis of IgE and the expression of CD23, whereas IL-11 has synergistic effects with both IL-3 and IL-4. On the other hand, IL-6 can induce the production of IL-2 in T cells, but IL-10 inhibits the synthesis of several other cytokines, including IL-1, IL-6, IL-8, IL-10 and IL-12. Some cytokines produce similar responses. In the case of IL-13 and IL-4, no additive effect is seen if both of these cytokines are added together, and it is possible that they share a common receptor, or share the same signal transduction pathway. Likewise, IL-15 and IL-2 may share common elements in their receptors.

An interesting interleukin is IL-1ra, which acts as an antagonist to IL-1 by competing for the receptor binding sites, so reducing the effect of IL-1. The production of IL-1ra may itself be under the control of IL-13.

Interferons

The interferons (IFNs) fall into two main groups. Type I includes IFNα, IFNβ and IFN-Ω, whereas type II includes γ-interferon (IFN-γ). IFN-γ rose to prominence in the press as an anti-cancer agent and was branded as "a wonder drug". Although the interferons do have profound effects in some conditions, they are still not widely distributed by the medical profession, perhaps in some cases because of their terrific cost.

Like the interleukins, interferon-γ is a peptide. Here, the polypeptide is 143 amino acids long and the interferon exists as a dimer in the active form, or even multimers. However, the peptide is made from the expression of a single gene on the long arm of chromosome 12 in humans, the sequence including a signal sequence of 23 amino acids. The α/β forms on the other hand are monomers, coded for by genes on chromosome 9.

Produced by T cells amongst others, activity of γ-interferon includes the induction of expression of the class II histocompatibility antigens on epithelial, endothelial and connective tissue. It also acts as a macrophage activating factor, causing specific gene expression in these cells. This leads to the enhanced ability of these cells to be cytotoxic to tumours and to kill parasites.

Tumour necrosis factors

Tumour necrosis factor-α (TNF-α), originally known as cachectin, is coded for by a single gene, which in humans is found on chromosome 6, but in the soluble active form it is a homotrimer of 157 amino acid subunits (17.5 kDa form). There is also a 26 kDa form that remains membrane anchored.

TNF is produced by monocytes and macrophages. Its activity is closely coupled to that of interferon-γ, and it has been shown to cause necrosis of tumours, hence its name, and to be cytotoxic to transformed cells *in vitro*.

TNF-β (also known as lymphotoxin α, LT-α) is also a homotrimer and is encoded for by a gene that is close to the TNF-α gene. They both have similar biological activity and even bind to the same receptors, TNF RI and TNF RII. However, TNF-β can also form membrane surface trimers with LT-β, and bind to the membrane protein herpes virus entry mediator.

Other cytokines, chemokines and receptors

Many of the small protein signalling molecules come under the grouping of the chemokines. This is a large family, or superfamily, of peptides, which are grouped together because of their structural relationships rather than any functional similarity. At present there are at least 36 peptides that can be grouped under the banner of chemokine.

There are two main sub-groups of the chemokine family, characterized by a particular structural motif containing four cysteine residues. The first sub-group are known as the C-C chemokines and these contain the sequence:

N-terminus-X(10-11)-**Cys-Cys**-X(22-23)-Cys-X(15) Cys-X(18-24)-C-terminus

where the numbers in brackets denote the number of amino acid residues that are likely to be present at this point in the primary structure. Members of this group include Rantes (regulated on activation, normal T-cell expressed and secreted), monocyte chemoattractant proteins (MCP1-4), macrophage inflammatory proteins (MIPs) and eotaxins. As the names suggest, like the cytokines, these peptides are commonly involved in immune responses.

The second class of chemokines are termed the C-X-C chemokines, and they contain the sequence:

N-terminus-X(6-12)-**Cys-X(1)-Cys**-X(23-24)-Cys-X(15-16)-C-X(15-53)-C-terminus

where the first two cysteine residues are separated by one amino acid. This group contains peptides such as interleukin 8 (IL-8), monokine induced by IFN-γ (MIG), platelet factor 4 (PF4) and neutrophil activating peptide 2 (NAP-2).

Some chemokines fall outside of this CC or CXC classification. Amongst these are lymphotactin (LTN) and fractalkine, the latter classed as a CX_3C chemokine.

Chemokines have been alternatively referred to by the names intercrines, the small cytokine family, or scy, and the small inducible secreted cytokines, or SIS. Many of these chemokines have been discovered not by their functional activity but by their sequence homology when their cDNAs have been cloned.

Many of the receptors for the cytokines have been found to contain some commonality in their structure and primary amino acid sequence. Four cysteine residues amongst a hydrophobic region, along with a hydrophobic region encompassing runs of positively and then negatively charged residues, have been found in several of the receptors cloned. Downstream of the receptor, the intracellular signalling on binding of the cytokine usually involves tyrosine phosphorylation, and this will be discussed in **Chapters 5 and 6**.

4.5   **Growth factors**

The term growth factor here is used to define those compounds that have been shown to have specific functions in the regulation of the growth and differentiation of cells. Other works may well categorize members of this group under the other headings used above, in particular for discussions here, groups VIII, XI, and XII. This emphasizes the vagueness of the terms used to classify extracellular signals. At present, over 50 known proteins that possess growth factor-like activity have been reported, with at least 14 different receptor families involved in their detection.

Platelet-derived growth factors

Platelet-derived growth factors (PDGF) are dimeric proteins, which may contain two related polypeptides, A or B. The active growth factor has a molecular weight of between 25 and 29 kDa, and is made up of either two of the same subunits, for example AA or BB, or it may be a heterodimeric protein, AB. The individual subunits have been shown to have a molecular weight of 12-18 kDa, and are coded for by two separate genes. They are held together by disulphide bonds, and in fact all the cysteine residues in the polypeptides are involved in either inter- or intra-molecular bonding. Interestingly, the receptors for PDGF are also dimeric, composed of two identical subunits, $\alpha\alpha$, or $\beta\beta$, or heterodimeric $\alpha\beta$, and the different forms of the receptor have different binding properties to the different forms of PDGF (see **Table 5.2**).

PDGF-BB, with two B subunits can bind and activate all the receptor forms, whereas PDGF-AA, with two A subunits can only activate the αα receptor. The heterodimeric PDGF-AB has an intermediate activity.

PDGF has been shown to induce both cell migration and cell proliferation, and has been implicated in several disease states including fibrosis and arteriosclerosis.

Epidermal growth factor

A larger group of growth factors comes under the heading of epidermal growth factors (EGF). These include EGF itself, transforming growth factor-α (TGF-α), betacellulin, and heparin binding EGF. They are very small peptides, rat EGF being only 5.2 kDa, although their precursors are very large. The EGF precursor is in fact a transmembrane protein of 1168 amino acids, of which only 53 are cleaved off to create the active EGF molecule. Likewise, TGF-α is only 6 kDa (50 amino acids), although there are larger members of the family – sensory and motor neuron derived factor (SMDF) being 296 amino acids.

These growth factors are characterized by the presence of at least one domain containing six cysteine residues, which are involved in the formation of three disulphide bonds. The growth factors also appear to have several aromatic residues, which are exposed to the aqueous medium. This is unusual in proteins, which, in general, try to hide these hydrophobic side chains in the interior of the polypeptide folding structure. It has been proposed that there are interactions between these aromatic groups and the protein surface may contain an aromatic domain, which is vital for its function. Proteins with domains related in sequence to EGF have been found in *Drosophila* and in sea urchin embryos. The proposed signal transduction cascade leading from EGF receptors is discussed in Chapter 6 and illustrated by Figure 6.8.

Fibroblast growth factor

Fibroblast growth factor (FGF) represents a family of molecules that are involved in the regulation of proliferation, differentiation and cell mobility. In mammals the family comprises many members, including FGF-a, FGF-b and FGF 3–19.

The FGFs are structurally related proteins of 20–30 kDa, with the genes encoding them probably arising from gene duplication of an ancestral gene. However, as with the expression of many genes, further isoforms can arise from alternative splicing or the use of alternative initiation codons for translation. Alternative post-translational modification leads to further diversification of the molecules within the family, where some are glycosylated, phosphorylated or even methylated or cleaved.

Some of the FGF family have been identified as proto-oncogenes, where the oncogene variants lead to the transformation of cells, where proliferation is uncontrolled resulting in the formation of tumours.

Species differences

In general mammals are grouped together and assumed to be similar to each other, but here it should be noted that there appears to be no FGF-15 in humans. It is erroneous to assume that all mammals are the same, and caution in interpretation of data is required.

4.6 **Neurotransmitters**

In the axon termini of the presynaptic cells of the nervous system there are storage vesicles called synaptic vesicles, which contain neurotransmitters. Voltage gated Ca^{2+} channels are opened on the arrival of an action potential, leading to a sharp increase in the concentration of intracellular Ca^{2+}. This, in turn, leads to exocytosis from the synaptic vesicles, releasing the neurotransmitters into the space between the nerve cells, with receptors on the postsynaptic cells detecting the presence of these compounds, leading to the propagation of the signal or the response.

These transmitters fall into two main groups. The first group is made up of a group of small molecules and contains neurotransmitters including acetylcholine, GABA or γ-aminobutyric acid, and dopamine. This group also includes some molecules already discussed above, such as epinephrine and histamine (see **Figure 4.5**).

Also included here are amino acids that can act as signals, such as glycine and glutamate, and derivatives of amino acids. For example, dopamine is derived from tyrosine, serotonin is a derivative of tryptophan, and GABA is derived from glutamate. Also included are nucleotide-derived compounds such as ATP and adenosine.

The second group are the neuropeptides. These include amongst their number substance P, β-endorphin and vasopressin, the latter also being classed amongst the hormones.

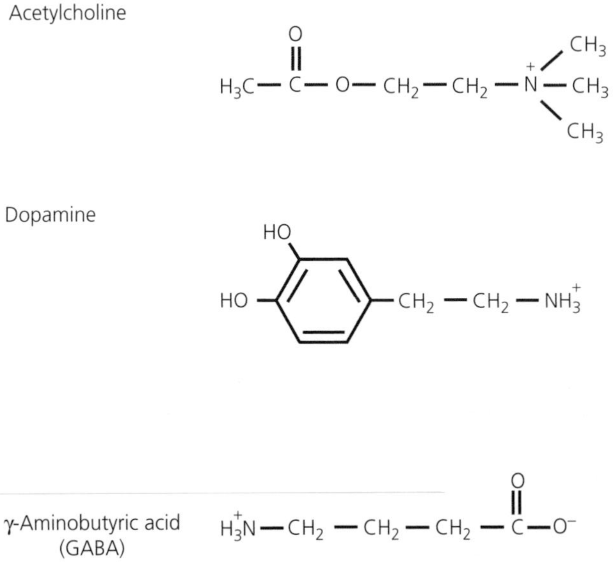

Figure 4.5 The molecular structures of acetylcholine, dopamine and GABA: examples of neurotransmitters.

■ Substance P was the
first neuropeptide to be
discovered.

The effects of the neurotransmitters are local, acting on the post-synaptic cell (across the synaptic cleft), or at least within a short distance. The receptors on the post-synaptic cell themselves fall mainly into two classes: G protein linked or ligand-gated ion channels (see Chapter 5).

4.7 ATP as an extracellular signal

Although ATP is well known as the molecule that acts as an ubiquitous supply of energy in cells, extracellular ATP can arise. ATP can be made in small amounts by the anaerobic metabolism of glycolysis, although this is not usually enough to sustain an active or busy cell, and therefore the majority of ATP is produced by aerobic respiration of mitochondria or from photosynthetic pathways in chloroplasts. Even though it is so crucial for the survival of the individual cells, extracellular release from cells can occur, for example, in cells involved in the secretion of neurotransmitters, or through storage granules from cells such as those of the adrenal medulla cells, or lymphocytes. Cell death and breakage can also lead to the release of intracellular ATP into the extracellular medium, and therefore the local concentrations of ATP may be nanomolar or even micromolar. However, three different ectonucleotidases are responsible for the sequential hydrolysis of ATP to adenosine, so removing the ATP from solution. Even so, many cells have been shown to possess purinoceptors on their surface, capable of detecting ATP, including platelets, neutrophils, fibroblasts, smooth muscle cells and cells of the pancreas. These receptors fall into two groups, P_1 and P_2 depending on their specificity for the adenosine compound, P_2 having the highest affinity for ATP. P_2 receptors themselves can be subdivided into four groups depending on their action, some acting through G proteins with a commensurate effect on phosphatidylinositols via phospholipase C (see Chapter 8), whereas others act through the operation of cation channels. The presence of extracellular ATP and its detection by the appropriate receptors is thought to have ramifications for the process of apoptosis for example, and therefore should not be dismissed as a strange anomaly.

The slime mould *Dictyostelium discoideum* can exploit another nucleotide that is more normally associated with having an intracellular function but as an extracellular signal, that is, the adenosine compound cAMP (see Chapter 7 for discussion of cAMP as an intracellular signal). Here, cAMP is used for controlling differentiation and cellular aggregation. If a food source becomes scarce, cAMP is released into the extracellular medium and acts as a chemoattractant, signalling to the normally free swimming cells to aggregate into a slug. This is also accompanied by changes in gene expression and culminates in the formation of a fruiting body.

4.8 **Pheromones**

Karlson and Lüscher first defined pheromones in 1959. They are substances excreted by an individual that have their effect on another individual of the same species. The word pheromone comes from the Greek, where pherein means "to transfer" and hormon means "to excite".

Many bacterial species use pheromones in communication, the effects usually being seen if the cells grow to a particular density. Responses induced by pheromones include the production of light, or luminescence, the production of virulence factors, the development of fruiting bodies and plasmid transfer.

The chemical structure of bacterial pheromones seems to be quite variant and includes amino acids, short peptides, proteins and branched-chain fatty acids. One of the largest groups of pheromones is the so-called AHLs, or N-acetyl-L-homoserine lactones. Many AHLs not only induce a response in another individual, but also induce the genes for their own production. The transcriptional activators used in the AHL response are grouped together in what is called the LuxR superfamily of response regulators, of which at least 15 members of the family are known. Similarly, a superfamily of enzymes that produce AHLs have been defined, the 10 members of this family being known as the LuxI superfamily.

It is not just prokaryotes that use pheromones in their organism to organism communication, but such systems are extensively used also by higher organisms. The so-called water mould, *Allomyces*, uses a compound called sirein as a sexual attractant. This compound is related to a cyclic organic compound called 2-carene, which is found in pine resins. Another slime mould *Achlya*, uses two steroid pheromones, one produced by the male and one produced by the female. Detection of male pheromone by the female is essential for the development of the female's sexual machinery, and likewise the male needs to detect the female's pheromone.

The sea anemone *Anthopleura elegantissima* uses a positively charged organic compound called anthopleurine. This is an interesting pheromone as it is distributed by a second species. The sea anemones are eaten by a sea slug, and therefore the sea slug also ingests the pheromone from the sea anemone. The pheromone is then released by the sea slug as it moves around and this acts as a warning to other sea anemones that a predator is coming. The sea anemones near the advancing sea slug will then retract as a defence.

Much of the early work on pheromones was carried out by Adolph Butenandt, a German organic chemist. He worked with the silk-worm moth, *Bombyx mori*. This insect uses a compound that has been named bombykol, a long C_{16} unsaturated fatty acid. Much work has been done subsequently to study the pheromones of other insects.

Higher organisms also use pheromones in their silent communications, including fish, amphibians and even mammals, and although it is impossible to discuss them all here in detail, the principles are the same. One individual communicating to another by means of chemical messengers, in a way analogous to the communication between cells in a single individual.

■ Summary

- Cells commonly communicate by the release and detection of signalling molecules, often over relatively large distances.

- Released compounds have often been referred to as hormones, taken from the Greek meaning "to arouse" or "to excite".

- Other released molecules are now categorized separately and are known as cytokines, chemokines or growth factors.

- Such classification does not appear to follow hard and fast rules, allowing the same compounds to be classified differently, a problem that might lead to confusion.

- It is important to remember that, like all signals, extracellular signals need to relay a specific message to specific targets at the correct time. That might require their transportation between cells within an organism, or even between organisms. Either way, their recognition and their ability to cause a defined response is the key to their success as messengers.

- Hormones are a diverse group of molecules, which may be small and water soluble peptides or lipophilic molecules.

- Hormones are detected either by a cell surface receptor or an intracellular receptor, as seen with steroid hormones.

- Plant hormones are a very diverse group of chemicals,

including amongst them compounds that are derived from amino acids, lipids or even being gases, such as ethylene or nitric oxide.

- Cytokines are a group of peptides that include the interleukins, currently having 35 members in the group, the interferons and tumour necrosis factors.

- Cytokines can be classified into 16 functional or structural groups, with such molecules being important in the orchestration of the immune response and the development of the cells used in animal host defence.

- Chemokines are a large group of extracellular signals, often involved in control of the immune system.

- Chemokines can be mainly split into two families, the CC chemokines and the CXC chemokines.

- Growth factors are molecules that are involved in the regulation of the growth and differentiation of cells.

- The growth factor family is known to contain at least 50 proteins, but as with the cytokines and chemokines, undoubtedly more are yet to be discovered.

- Other extracellular signals include the energy storage compound ATP and a compound usually associated with intracellular signalling, cAMP, as seen in control of the aggregation of the slime mould *Dictyostelium*.

→ Further reading

Introduction

Baulieu, E.E. and Kelly, P.A. (1990) *Hormones: from molecules to disease*. Chapman and Hall, London.

Chaplin, D.D. (2006) 1. Overview of the human immune response. *Journal Allergy Clinical Immunology*, 117, S430–435.

Le Roith, D., Shiloach, J., Roth, J. and Lesniak, M.A. (1980) Evolutionary origins of vertebrate hormones: Substances similar to mammalian insulins are native to unicellular eukaryotes. *Proceedings of the National Academy of Science USA*, 77, 6184–6188.

Hormones

Henry H. and Norman A.W. (2003) (eds) *Encyclopedia of Hormones*. Academic Press, Amsterdam. ISBN 0123411033.

Cytokines

Arai, K.-I., Lee, F., Miyajima, A., Miyatake, S., Arai, N. and Yokota, T. (1990) Cytokines: coordination of immune and inflammatory responses. *Annual Review of Biochemistry*, 59, 783–836.

Minami, Y., Kono, T., Miyazaki, T. and Taniguchi, T. (1993) The IL-2 receptor complex: its structure, function and target genes. *Annual Review of Immunology*, 11, 245–267.

Moore, K., O'Garra, A., de Waal Malefyt, R., Vierra, P. and Mosmann, T. (1993) Interleukin-10. *Annual Review of Immunology*, 11, 165–190.

Scott, P. (1993) IL-12: initiation cytokine for cell mediated immunity. *Science*, 260, 496–497.

Thomson, A. ed., (2003) *The Cytokine Handbook*. 4th ed. Academic Press, Amsterdam. ISBN 0126896631 [an excellent review of the area, but 1800 pages].

Growth factors

Mason, I.J. (1994) The ins and outs of fibroblast growth factor. *Cell*, 78, 574–552.

Oefner, C., D'Arcy, A., Winkler, F.K., Eggimann, B. and Hosang M. (1992) Crystal structure of human platelet-derived growth factor BB. *EMBO Journal*, 11, 3921–3926.

ATP as an extracellular signal

El-Moatassim, C., Dornand, J. and Mani, J.-C. (1992) Extracellular ATP and cell signalling. *Biochimica Biophysica Acta*, 1134, 31–45.

Pheromones

Agosta, W.C. (1992) *Chemical Communications: the Language of Pheromones*. Scientific American Library, New York. USA.

Kell, D.B., Kaprelyants, A.S. and Grafen, A. (1995) Pheromones, social behavior and the functions of secondary metabolism in bacteria. *Trends in Ecology and Evolution*, 10, 126–129.

Wirth, R., Muscholl, A. and Wanner, G. (1996) The role of pheromones in bacterial interactions. *Trends in Microbiology*, 96, 96–103.

5

Detection of extracellular signals: the role of receptors

Chapter 4 discussed the array of signals that are used by cells to communicate to each other. However, regardless of how many signals are present on the outside of a cell, that cell will only respond if it has the capacity to recognize a signal, and then to respond to it. Therefore, this chapter discusses the mechanisms used by cells to perceive signals, and the immediate action taken when they do. Most signals are perceived at the cell surface by a variety of protein receptors, but some signals are able to penetrate the cell, and therefore need to be recognized on the inside.

As with dysfunction of the signals themselves, dysfunction of the receptors can, and does, lead to disease. For example, if the insulin receptor cannot recognize the presence of insulin, then a cell cannot take the appropriate action when insulin is released into the blood, and diabetes will result. Therefore, an understanding of the types of receptor, their mechanisms and actions, and how they lead to the signalling cascades inside the cell is vital to the understanding of cell signalling mechanisms.

The study of binding characteristics of receptors is discussed, as an understanding of the ability of a cell to bind to a signalling ligand will reveal the capability of a cell to respond, or not. Furthermore, the capacity of a cell to recognize a signal may not be constant, with the number of available receptors changing over time. The mechanisms of how a cell may modulate its complement of receptors is also discussed. Cells can become less responsive to drugs, as well as endogenous signals, and therefore it is important to understand the way in which cells can change their array of receptors.

5.1 **Introduction**

Many signals to which a cell needs to respond reach it from the extracellular environment, whether that is via diffusion in the growth medium or actively through a vascular system, as discussed in Chapter 4 (Section 4.1). Such extracellular signals may be at extremely low concentrations, but even when they rise to a level to which the cell needs to react, the concentration could still commonly be very low, perhaps in the order of 10^{-8} M. Therefore, if the cell is to respond to the presence of such a signal, the cell must have the capacity to detect the signal molecule, even at these low concentrations, and have the capacity to act upon it. Secondly, as discussed in Chapters 1 and 15 (Sections 1.1 and 15.2), cells are usually awash with a plethora of signalling molecules, and it is only to a selection of these that they need to respond.

The role of detection of signals arriving from outside of the cell is usually fulfilled by the presence of specific receptors, either on the cell surface or inside the cell (either in the cytoplasm or in the nucleus). However, four crucial criteria have to be met by the functioning of a receptor:

- A receptor has to have specificity, detecting only the signalling molecule (or range of molecules) that the cell wishes to perceive.
- The binding affinity of the receptor must be such that it can detect the signalling molecule at the concentrations at which it is likely to be found in the vicinity of the cell.
- The receptor must be able to transmit the message that the signalling molecule conveys to the cell, usually by the modulation of further components in a signalling cascade.
- Usually the receptor needs to be able to be turned off again once the "message" is received and acted on.

Therefore, receptors usually have a high binding affinity that is in the concentration range of their ligand, and binding of the ligand to the receptor will then stimulate the required intracellular response, usually via a complex signal transduction pathway. Detection of the signalling molecule has to be precise and a cell has to have the ability to have a repeated response to the same signal or even simultaneous responses to several signals.

The receptor must have the capacity to transmit the message on to the next component in the signalling cascade. Often this involves binding of a ligand on the outside of the membrane and an interaction with protein on the inside of the cell, the message being transmitted through the membrane. As discussed below, the membrane receptors typically have α-helices (perhaps one, often seven, sometimes more) passing through the membrane. Therefore, on ligand binding a series of molecular events will ensue. Ligand binding itself will induce a conformational change in the outer domain of the receptor; this change will be transmitted through the membrane spanning helices, and this will induce a conformational change in the intracellular domain of the receptor. This will either activate or inhibit the receptor's intrinsic activity, if it has one, or perhaps disrupt the receptor's association with other proteins, such as a G protein. By this significant conformational change through the receptor, the ligand binding can effect events on the other side of the membrane. Even if the receptor has no membrane location, as with steroid receptors, conformational change on ligand binding is still a key event.

The binding ability of the receptor to its ligand, which will determine if a cell does actually respond, will ultimately depend on the three-dimensional shape of the receptor protein. The amino acid sequence of the receptor will determine this 3-D shape, and it has to be appropriate to "recognize" the hormone or molecule it is present to sense. Furthermore, the interaction may involve the attraction or even repulsion of charged groups, so the amino acids in the binding site of the receptor not only have to be in the correct orientation in space, but also need to have the correct charge, and therefore acid or amino groups for example have to be dissociated or associated with their respective protons, which will be ultimately determined by the isoelectric point, pI.

Of course other interaction forces are at work here too, not just electrostatic interactions, and all the possible factors that influence whether there is a relatively stable interaction of the receptor to the ligand have to be considered. This may involve van der Waals forces and hydrogen bonding, as well as whether a ligand will simply physically fit into the binding space offered.

> **Isolectric point** The isoelectric point (pI) of a molecule is the pH at which no net electrical charge is carried. Altering the pH may lead to the gain or loss of protons, so altering the charge. For example, a –COOH group may lose a H$^+$ to become –COO$^-$, and so bestowing vastly different properties to that group in respect to attraction or repulsion of interacting partners, such as a ligand.

5.2 Types of receptors

To further fulfil the requirements outlined above, a wide number of receptors have evolved to fill the vital role of the detection of extracellular signals. However, despite the vast array of extracellular molecules that need to be detected by a single cell, including hormones, cytokines and chemokines (see Chapter 4), most receptors fall mainly into five classes (see Figure 5.1):

- G protein-coupled (GPCRs).
- Ion channel linked.
- Containing intrinsic enzymatic activity.
- Tyrosine kinase linked.
- Intracellular.

However, like all classifications, this is somewhat over-simplified and examples that do not readily fall into these categories can be found in the literature.

A. Ion channel linked receptor

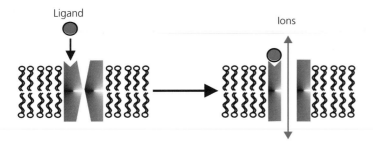

B. G protein-coupled receptor

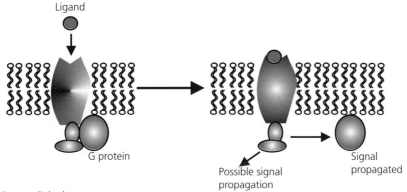

C. Enzyme linked receptor

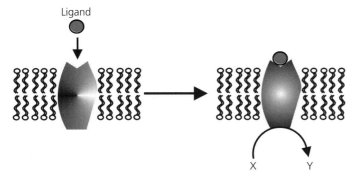

Figure 5.1 Three different types of cell receptor. A. An ion channel linked receptor. B. A G protein-coupled receptor. C. An enzyme linked receptor.

G protein-coupled receptors

G protein-coupled receptors (GPCRs) are part of a very large group of receptors, which all share a common protein structure, and would probably more accurately be termed the seven-transmembrane receptors (7TM). Many of them have a common mode of action, that is, they interact with G proteins, hence the name G protein-coupled receptors, but some of them control signal transduction pathways in different ways.

In mammals there are 800–1000 genes that encode for 7TMs, which have been grouped into families:

- Family A: this is the largest group of these receptors, and includes adrenergic receptors and rhodopsin (see **Chapter 12**).

- Family B: this includes receptors for glucagon, calcitonin and parathyroid hormone. These receptors appear to all lead to increases in cAMP through activation of adenylyl cyclase.

- Family C: these receptors have a large N-terminal extracellular domain, and include GABA$_B$ receptors and glutamate receptors.

GPCRs are an incredibly important group of receptors, and as their name suggests the response to their binding to a ligand is mediated by G proteins. When a GPCR is activated by binding to its ligand, the result is the activation of a heterotrimeric G protein that conveys the message to the next component in the signal pathway (see **Chapter 7**). There are a vast number of receptors that work in this way, and they include those that have a specificity for hormones such as epinephrine (adrenaline), serotonin and glucagon along with those listed in **Table 5.1**. These receptors are of great interest to the pharmaceutical industry as they can be targets of therapeutic interest. In fact, it has been estimated that a third of all drugs have direct binding to this type of receptor, whereas many more are targeting events down-stream from these receptors. For example, drugs that have their action on these receptors include some anti-histamines, anti-cholinergics, inhibitors of the β-adrenergic receptor and some opiates.

Length of membrane spanning α–helices

If an α-helix is to span the membrane it has to be long enough. Most membranes have a similar thickness, as they are a lipid bilayer, so any protein structure that crosses them has to span that distance. Often in protein structure prediction studies a researcher will look for α-helices, and if they are too short the prediction can rule them out as being membrane spanning.

In general, the topology of these receptors is such that they contain seven regions of approximately 22–24 amino acids, which form hydrophobic α-helices spanning the plasma membrane, hence these proteins being alternatively referred to as 7TMs (see **Figure 5.2**). Therefore, their structure is analogous to that described for bacteriorhodopsin and rhodopsin (see **Chapter 12**, and **Figure 12.3**). With G protein-coupled receptors, the N-terminal end of the polypeptide is on the exterior face of the plasma membrane, whereas the C-terminal end is on the inside. This means that there are four cytoplasmic loop regions, the third of which is probably the site for G protein binding, along with the cytoplasmic C-terminal tail.

The class of G protein involved in binding to these receptors and relaying the signal along the transduction pathway is that known as the trimeric, or

Table 5.1 A list of a few of the ligands that act through G protein-coupled receptors.

Acetylcholine
Bradykinin
Calcitonin
Dopamine
Epinephrine (adrenaline)
Glucagon
Histamine
Leukotrienes

heterotrimer   sed of three
subunits of d   important
class of sign   iscussed in
Chapter 7.

In many c   re they are
fully function   , but not all
GPCRs. In p   tors. These
receptors are   bunits, and
they need to   ome of the
GCPRs resp   mers.

As well as   with other
receptor sub   ociate with
GPCRs, whi   ing-modify-
ing protein (ly have one
transmembra   ple of their

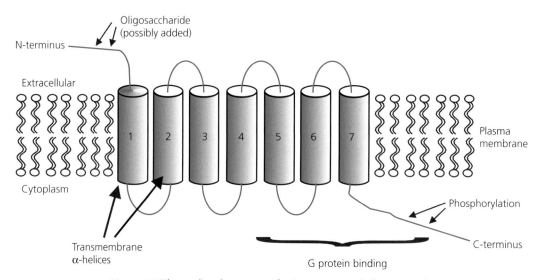

Figure 5.2 The predicted structure of a G protein-coupled receptor, showing the seven α-helices spanning the membrane and the likely G protein binding region.

function is their interaction with the calcitonin receptor-like receptor (CRLR) protein. If CRLR is associated with RAMP1, it is a receptor for calcitonin-gene-related-peptide. If CRLR associates with RAMP2 or RAMP3, it acts as an adrenomedullin receptor. On the other hand, a CRLR on its own, i.e. as an un-associated peptide, is non-functional, showing that interaction of a RAMP is essential for its function.

As well as being activated by binding of the relevant ligand, the activity of some GPCRs is also altered by phosphorylation. Phosphorylation can be catalyzed by cAMP-dependent protein kinase (PKA), or by a class of kinases known as G protein-coupled receptor kinases (GRKs). GRKs are known to phosphorylate these receptors on multiple sites, using threonine and serine residues as targets. Phosphorylation deactivates the receptor, as well as allowing for the interaction of the receptor with an inhibitory protein known as β-arrestin. Such mechanisms are involved in the termination of the signal and receptor desensitization. As will be discussed in Chapter 12, one of the classical examples of such a system is seen with rhodopsin, used in light perception in the eye. One of the most studied examples of these kinases is β-adrenergic receptor kinase (βARK), and these kinases and their actions will be further discussed in Chapter 6.

Orphan proteins

With the advent of whole genome sequencing, once a consensus sequence has been determined for a class of proteins, other putative members of a protein family may easily be identified in the genome of interest. However, biochemical and functional genomic analysis is required to determine if the "new" protein actually does anything, or in some cases even exists.

Not all 7-pass receptors are actively mediated through a G protein pathway. For example, the receptor Frizzled (Fz) contains seven transmembrane α-helices, but forms a complex with other proteins such as LDL receptor-related lipoprotein (LRP) and Dishevelled (Dsh). This is part of the Wnt pathway, which has a major role in developmental biology as will be discussed further in **Chapter 13**. Another example is the protein Smoothened (Smo), again a pathway instrumental in development and discussed in **Chapter 13**. 7TMs have also been found to signal through mitogen-activated protein kinase (MAPK) cascades, and through Janus kinases, both in mechanisms that involve direct interactions of the receptors with components of these pathways, and not through heterotrimeric G proteins.

Ion channel linked receptors

These receptors are often involved in the detection of neurotransmitter molecules, and are sometimes referred to as transmitter-gated ion channels. Binding of the ligand to the receptor changes the ion permeability of the plasma membrane, as the receptor undergoes a conformational change that opens or closes an ion channel, allowing the efflux or influx of specific ions. However, this is only a transient event, the receptor returning to its original state very rapidly.

These receptors make up a family of related proteins, but with one distinguishing feature being that they contain several polypeptide chains that pass through the membrane. For example, the acetylcholine receptors are composed of two identical polypeptides that contain acetylcholine binding sites along with three different ones, giving an α, α, β, γ, δ subunit structure (see **Figure 5.3**). These five polypeptides are, therefore, encoded by four separate genes, but interestingly they show a large degree of homology suggesting that they probably arose from a gene duplication event during evolution. In the membrane the proteins are arranged in a ring, in a similar fashion to that seen with the gap junctions (see **Figure 1.5**), and therefore there is a water filled

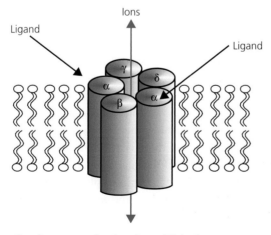

Figure 5.3 The predicted structure of an ion channel linked receptor.

channel that runs through the middle from one side of the membrane to the other, allowing the passage of ions. The ions that pass through the hole made by the acetylcholine receptor are usually positively charged, such as Na^+, K^+ or Ca^{2+}, as the presence of negatively charged amino acids at the ends of the hole bestow some selectivity to the channel.

Receptors in this class with different ligand binding specificities, and subsets within these groups of receptors, all contain polypeptides with high levels of sequence similarity, and are encoded for by genes with high levels of homology or from genes where the expression involves alternative splicing events. Therefore, a vast array of receptors can be made using combinations of these different gene products. Often the presence of a particular receptor subset is tissue specific. It is this type of receptor, i.e. the ion channel linked receptor, that is the target of many drugs, such as barbiturates used in the treatment of insomnia, depression and anxiety. Their interest is extended by the implication in a wide range of diseases, including schizophrenia, Parkinson's disease, Alzheimer's disease, epilepsy and autism.

> **Alternative splicing** a process where different exons within a gene are used, in a variety of combinations, when the gene is expressed, so that one gene may give rise to many different transcripts and therefore many different proteins. A good example of this mechanism of gene expression is seen with antibody production.

Receptors containing intrinsic enzymatic activity

These receptors make up a quite heterogeneous class of receptors, which are characterized by the presence of a catalytic activity integral within the receptor polypeptide, and it is this catalytic activity that is controlled by the ligand binding event. The ligand binding domain is found on the extracellular side of the membrane, with usually a single span of the membrane leading to the catalytic domain on the cytoplasmic side. The catalytic activity, for example, may be a guanylyl cyclase (as discussed further in Chapter 7), a phosphatase or a kinase. The receptors may contain a serine/threonine kinase activity and be referred to as receptor serine/threonine kinases, or they may contain a tyrosine kinase activity and be referred to as receptor tyrosine kinases or RTKs (these are also discussed further in Chapter 6).

RTKs represent a class of these receptors, which have been extensively studied partly because of their potential role in cancer and tumour formation. If they are over-active for any reason, either through impaired signalling or perhaps because the gene encoding them is mutated, then loss of control of cell proliferation may result. With RTKs, ligand binding leads to activation of the kinase activity of the receptor, which causes phosphorylation of the receptor itself on tyrosine residues. This leads to the creation of new binding sites, particularly for proteins that contain protein binding domains, for example, SH2 domains (see Chapter 1). Such binding proteins may be adaptor proteins, such as GRB2 of mammals or Drk in *Drosophila*, which contain both SH2 and SH3 domains. On phosphorylation of the receptor (by itself), the new binding sites are recognized by the binding protein, and binding of such adaptor proteins to the receptor stimulates the formation of protein complexes. Such complexes may also include guanine nucleotide releasing factors, for example, Son of Sevenless (Sos). Here, the result would be the activation of a G protein. This in turn may lead to a transduction cascade, which could include further kinases, such as MAP kinases. Such signalling is further discussed in Chapter 6 (see Figure 6.8) and Chapter 11 (see Figure 11.4).

However, phosphorylation by the receptor is not confined only to autophosphorylation and other proteins may also be phosphorylated leading to further propagation of the signal. For example, the IRS-1 protein is phosphorylated on multiple sites by the insulin RTK as discussed in Chapter 11.

So far, over 50 RTKs have been identified and these can be grouped into at least 14 different families. Most classes of RTK are monomeric in nature, but some share the insulin receptor's tetrameric topology, that is, they have an $\alpha_2\beta_2$ structure held together by disulphide bonds. Some RTKs contain cysteine rich extracellular domains, whereas others contain extracellular antibody-like domains. They all, however, seem to share the characteristics of having their N-terminal ends on the extracellular side of the membrane and the polypeptides only cross the membrane once, through the use of a single hydrophobic α-helix.

Although in the inactive state these receptors are often monomers, on activation they are often found as dimers, the ligand binding leading to the dimerization event. However, some receptors are constitutively dimers/tetramers, as seen with the insulin receptor. Interestingly, the insulin receptor is coded for by one gene that leads to the formation of one mRNA, but the protein product undergoes post-translational modification, including a cleavage event. This leads to the formation of two polypeptides, α and β, where the two polypeptide chains are held together by the formation of cystine or disulphide bridges. This structure then dimerizes to form a tetrameric receptor structure (see Chapter 11 for more discussion of this receptor). Having said that, dimerization is a common theme in the activation of this type of receptor. The dimer may be a homodimer, that is, containing two identical receptor subunits, or a heterodimer, where the complex is composed of two different subunits from the same receptor family. Subunit structure may also involve other, or accessory, proteins. An example here is the involvement of the protein gp130. The formation of heterodimers allows the creation of a wider diversity of receptor specificities, as seen with platelet-derived growth factor receptors where the receptor dimers can be $\alpha\alpha$, $\beta\beta$ or $\alpha\beta$, each having a different specificity to isoforms of growth factor (see Table 5.2).

Receptors linked to separate tyrosine kinases

Several receptors do not themselves contain a tyrosine kinase domain, but on activation by ligand binding they cause the stimulation of a tyrosine kinase.

Table 5.2 The binding specificities of the platelet-derived growth factor receptor, showing how dimerization allows for subtleties of ligand binding.

Type of receptor	Isoforms of ligand that are recognized
αα	AA
	AB
	BB
αβ	AB
	BB
ββ	BB

Such kinases are normally resident in the cytoplasm of the cell, but will recognize and bind to an activated receptor, which on ligand binding has adopted a new conformation. This class of receptors is commonly referred to as the cytokine receptor superfamily, as they are involved commonly in the recognition of cytokines and growth factors.

As discussed above, binding of the ligand to the receptor often induces dimerization to occur. Here, for example, interferon-γ (IFN-γ) binding leads to homodimerization of its receptors. Other ligands, on the other hand, cause heterodimerization on binding to the receptor protein, or dimerization may involve accessory proteins, for example, gp130. However, the receptor for tumour necrosis factor β (TNF-β) forms trimers, whereas other ligands can lead to the formation of hetero-oligomers of three different polypeptides.

Once activated, the receptor needs to recruit and activate the relevant protein kinase, and it is often the soluble protein tyrosine kinases called Janus kinases (JAK) that are involved. These kinases, which contain two catalytic domains, enable the propagation of the signal along the signal transduction cascade, and are discussed further in Chapter 6.

Intracellular receptors of extracellular signals

Not all extracellular signalling molecules are detected on the surface of the cell by plasma membrane-borne receptors. Many very important signals are released by cells but the receptors for their perception are inside the target cells, not on their surface. These include receptors for steroid hormones, thyroid hormones, retinoids, fatty acids, prostaglandins and leukotrienes, the receptors for which are all intracellular. Steroid hormones are derived from cholesterol and include cortisol, vitamin D and steroid sex hormones. The amino acid tyrosine is the base for the thyroid hormones, whereas the retinoids are derived from vitamin A: such signals are discussed further in Chapter 4 (Section 4.2). These signalling pathways are also an important target for many therapeutic regimes, with steroid-based drugs being used commonly.

Intracellular ligand binding for these extracellular signalling molecules means that they have to move through the plasma membrane, and commonly these signals are small and hydrophobic, hence allowing them to get access to the intracellular receptors. However, as these molecules are readily soluble in the hydrophobic environment of the membrane, allowing their free passage into the cell, they are therefore inherently insoluble in the aqueous fluids outside the cells, such as the blood stream. Therefore, these extracellular signalling molecules need their solubility in water to be increased, which is facilitated by their association with specific carrier proteins. Dissociation from the carrier occurs before the signalling molecules can enter the cell.

The receptors found inside cells are commonly referred to as the intracellular receptor superfamily or steroid hormone receptor superfamily. The activation of such receptors often leads to effects in the nucleus of the cell, commonly the alteration of transcription rates of specific genes, or sets of genes. Therefore, the signal can either enter the cell and bind to a cytoplasmic receptor that then moves to the nucleus to have an effect, or the signalling molecule itself can move directly to the nucleus and bind to the receptors there. Both scenarios exist in cells, and receptors for these extracellular signals can be found either in the cytoplasm or in the nucleus of the cell.

There are numerous nuclear receptors in cells. There are known to be 48 different nuclear receptors in humans, 21 in the fly *Drosophila melanogaster*, and hundreds in the nematode *Caenorhabditis elegans*, highlighting the importance of this type of signalling to cells. It is certainly not an oddity, but an important part of the signalling that takes place in an organism. The steroid hormone receptor superfamily represents, in fact, the largest known family of transcription factors described for eukaryotes. However, these include putative receptors identified through the use of sequence similarity analysis for which the ligand specificity has not yet been determined, so-called orphan receptors, as well as numerous isoforms of known receptors.

Orphan receptors

With the advent of modern molecular biological techniques and whole-genome sequencing, once a consensus sequence has been determined for a class of proteins, such as receptors, other putative members of a protein family may easily be identified in the genome of interest. With receptors, if no ligand is known to bind to such proteins that have been identified, they are termed orphan receptors. However, biochemical and functional genomic analysis is required to determine if the novel protein actually does anything, what it might bind, or in some cases if it even exists.

The structures of the intracellular receptors are very conserved and it is thought that nearly all these receptors have evolved from a common ancestral receptor, the oestrogen receptor. In general, the amino acid sequences show that these receptors can be divided into several domains. At the N-terminal end of the polypeptide is a variable region known as the A/B domain containing an activation function 1 (AF-1) region, which is a transcription activator and is involved in gene activation. The length of the A/B domain is very variable between receptors, ranging from less than 50 to over 500 amino acids in humans. The next domain along the polypeptide is the C domain or DNA-binding domain (DBD), often containing two zinc fingers. This domain is relatively short, being around 60 amino acids, and is responsible for DNA recognition and also partly for dimerization of the receptor. The next domain, the D domain, consists of a variable hinge region and may contain sequences responsible for the localization of the receptor to the nucleus. This domain is also relatively short in most cases, and can be as little as 18 amino acids.

Ligand binding is the responsibility of a large E domain (often referred to as the ligand binding domain (LBD)). This region is approximately 200–250 amino acids in length and is also responsible for association with heat-shock proteins (Hsps), as well as being involved in receptor dimerization. The E domain also contains an activation function 2 (AF-2) region, which is needed for the transcriptional activation brought about by these receptors. At the C-terminal end is the last region, called the F domain, but not all of the intracellular receptors seem to contain this region and little is known of the function of this area of the protein.

The holo-structure of intracellular receptors also often involves other proteins. The receptors that are found in the cytoplasm exist there as an inactive complex, which involves Hsps, usually Hsp90, Hsp70, and Hsp56. On binding of ligand to the receptors, the latter undergoes a conformational change and dissociation from the inhibitory heat-shock protein complex (see **Figure 5.4**). Receptor proteins may then enter the nucleus through nuclear pore complexes, which span the nuclear membrane. Once in the nucleus, binding of the receptor can occur to specific sequences of bases on DNA, so altering the rates of transcription of specific genes, and therefore these receptors can be classified as transcription factors, even though they were originally found in the cytoplasm. The receptors bind to the DNA as either homodimers or heterodimers.

The second group of these receptors include those already found in the nucleus. The ligands for such receptors include those that are endogenously produced by the organism, such as androgens, progesterone and oestradiols, along with compounds that might be in the diet, such as isoflavanoids and phyto-oestrogens. Of concern is the fact that some other compounds encountered in the environment, such as pesticides, plasticizers and polychlorinated biphenyls, may also control the functioning of this class of receptors. Such receptors that are already located in the nucleus in an inactive state may also be associated with Hsps, for example, Hsp90, Hsp70 and Hsp40, but other proteins may also be involved. Such protein–protein binding interactions probably control the access of the steroid to the ligand binding site.

Intracellular receptors control the expression of genes once they are activated, but for there to be a controlled cellular response to the presence of a ligand there must be a specific gene or set of genes targeted by these receptors. So, as well as characterizing the receptors themselves, the region of DNA recognized by the receptors is also important to determine. The region of DNA

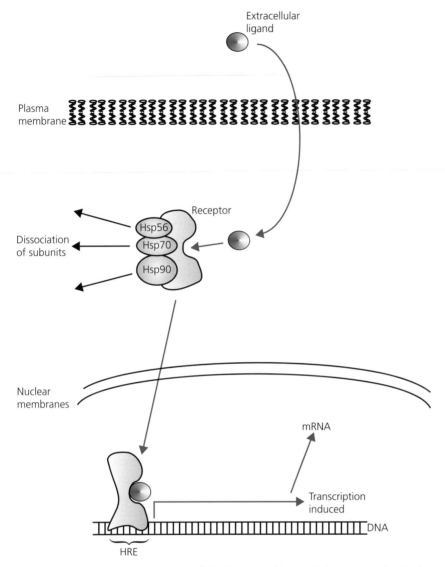

Figure 5.4 A schematic representation of the detection of lipophilic hormone molecules by an intracellular receptor.

forming the target for intracellular receptors is called the hormone-response element (HRE). For example, for glucocorticoid receptors the HRE has been found to be two short imperfect inverted repeats with three nucleotides separating them. Other HREs have been found to be similar to this pattern, although, of course, the exact nucleotide sequence involved bestows the specificity on the receptor–DNA interaction and, therefore, the specificity of the cellular response to the original stimulus (i.e. presence of the specific ligand).

Ligand binding is not the only mechanism by which the activity of the intracellular receptors is modulated. Phosphorylation also alters the activity of these receptors. Phosphorylation usually takes place on serine or threonine amino acids in the N-terminal domain or in the DNA binding domain, and may alter the ability of the nuclear receptor to bind its interacting proteins, or modulate DNA binding itself, or may be involved in the turnover of the receptor.

It should be noted that activation of an intracellular receptor does not always increase the rate of expression of a gene, as receptor activation can lead to decreased transcription rates too. Furthermore, the genes expressed might encode proteins that themselves alter the rates of transcription, a point further discussed in Chapter 15. So, the overall response to the activation of an intracellular receptor could be complex and very profound to the workings of the cell.

A study of receptors in the cytoplasm and nucleus will, therefore, highlight that it is not only the receptors on the plasma membrane that are important for ligand binding and control of cellular activity, but also those receptors present inside the cell. Many of these intracellular receptors, such as those in the nucleus, are already present in the cell organelle where the response is to take place, so bypassing the need for the complex pathways used by plasma membrane receptors. These receptors often control transcription of many genes, and therefore are in direct control of the future cellular complement of proteins, and hence future cellular activity.

5.3 Ligand binding to their receptors

Ligands that bind to receptors can be classified under general headings. A ligand termed an agonist is one that binds to a receptor and results in the activation of that receptor. Alternatively, a ligand referred to as an antagonist will bind the receptor but not result in activation of the receptor, and furthermore, the presence of an antagonist may interfere and stop the action of an agonist. However, a receptor may be active in the absence of a ligand, in which case it would be said to be constitutively active, a situation seen with some oncogenes.

Receptors are, and need to be, highly specific for their ligand and usually have a high affinity for that ligand. However, in molecular terms, the binding site of the receptor can in many ways be viewed as being like an active site of an enzyme. It is a specific local environment, determined by the presence of specific amino acids, which are held in a three-dimensional orientation by other amino acids within the protein. It is the make-up of the binding site that determines both the specificity and affinity for the ligand. The individual forces involved in holding the ligand onto the receptor are generally weak, being ionic attractions, van der Waals forces, hydrogen bonding or hydrophobic interactions, as discussed above.

Ligand binding is usually a reversible reaction, often allowing the receptor to be used and re-used over a long period of time. Therefore, the binding reaction can be written as follows: where L = ligand; R = receptor.

$$L + R \Leftrightarrow LR$$

Therefore, as with enzyme kinetics, where it is useful to determine the concentration of substrate at which the reaction proceeds at half the maximal rate, i.e. K_m, one of the calculations that is useful when characterizing ligand binding is the determination of concentration of ligand at which half of the receptors are bound, with half the receptors in the unbound state. This value is called K_d and can be defined by the following equation:

$$K_d = \frac{[R][L]}{[RL]}$$

where [R] is the concentration of receptor, [L] is the concentration of ligand and [RL] is the concentration of receptor bound to ligand as a complex. The lower the K_d value, the higher the affinity of the receptor for its ligand. Usually the K_d values approximate to the physiological concentrations of the ligand, allowing the receptor to have the highest sensitivity to changes in the ligand concentration in the usual concentration range found. This, of course, would be ideal for the cell to respond quickly and efficiently as ligand concentrations fluctuate.

To perform the calculation above, the amount of ligand actually bound to the receptor needs to be experimentally determined. Ligand binding can usually be studied using a ligand that has been labelled or tagged, and therefore the binding to its receptor can be followed. To visualize the ligand binding on, for example, a cell surface receptor, a fluorescent label may be used in conjunction with a fluorescence microscope, a laser-scanning confocal microscope or a fluorescence activated cell sorter. However, as fluorescence is hard to quantify accurately, a radiolabel is usually employed to quantify ligand binding. The most common radiolabels are either iodine-131 or iodine-125. However, these isotopes have extremely short half-lives, approximately 8 days and 60 days respectively. Furthermore, the presence of a large iodine molecule may well interfere with the ligand/receptor interaction. Alternatively, ^3H can be used as a label in many instances; one advance being its long half-life of approximately 12 years.

If we consider a cell surface receptor, a usual experiment would quantify the amount of ligand bound as the concentration of the ligand was increased. This would result in the total binding curve as depicted in **Figure 5.5**.

Binding measurements under the conditions adopted usually means that binding has come to equilibrium. However, it should be noted that the speed of binding may be affected by the pH of the solution and/or the temperature of the reaction. Furthermore, some of the larger ligands might well have binding that is very slow, and therefore care needs to be exercised if a true measure of binding is to be obtained.

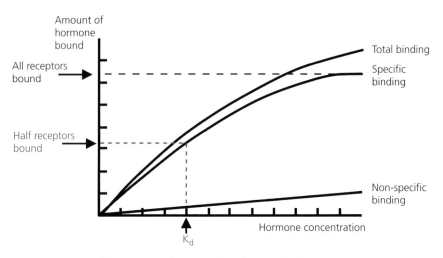

Figure 5.5 A plot of the amount of hormone bound versus the hormone concentration, showing the total binding and the proportions made up by the specific binding and the non-specific binding.

Once the ligand binding has been quantified, the total amount of ligand bound does not necessarily mean that all the ligand has bound to the receptor, and the amount bound will almost certainly include an element of non-specifically bound ligand. The contribution of non-specific binding to the total is usually estimated by the inclusion of control experiments in which a huge excess of unlabelled ligand, approximately a hundredfold excess, is added. This means that the high affinity receptor sites will be saturated with unlabelled ligand, and, therefore, any labelled ligand remaining bound will be due to binding to non-specific sites. The non-specific binding is, in general, linear in a concentration-dependent manner. Once these values are subtracted from the total binding curve, the binding specific to the receptor can be seen (see **Figure 5.5**). This binding starts off being very much greater than the non-specific binding, due to the high affinity of the receptors, but, like typical enzyme kinetics, the receptors become saturated and the binding curve tails off to a maximum, beyond which no more ligand binding can occur despite the addition of more ligand. Therefore, from this graph, the total specific binding sites can be estimated, as can the K_d value.

Estimating the binding characteristics from a curve can be difficult and not very meaningful, and, therefore, in an analogous way to the analysis of enzyme activity, the ligand binding data need to be mathematically manipulated to give a linear relationship. Such analysis should give a better insight into the characteristics of the binding, for example, whether the binding shows any cooperativity, that is whether the binding of the second ligand is more or less favourable because of the bound first ligand. The two common methods employed are derived from those developed by Hill and Scatchard. The original development of the Hill plot was to analyze the binding of oxygen to haemoglobin, but the same rationale can be used here.

The Scatchard plot is derived by dividing the concentration of bound ligand by the concentration of free ligand and then subsequently plotting this against the concentration of bound ligand, that is:

$$\frac{[\text{Bound ligand}]}{[\text{Free ligand}]} \text{ vs} [\text{Bound ligand}]$$

As illustrated in **Figure 5.6**, from this plot the K_d value can be obtained, as the slope of the line will be $-1/K_d$ and the maximum amount of ligand bound (B_{max}) can be determined by extrapolating the line to cross the X-axis.

Although, theoretically, Scatchard analysis should give a straight line, this is quite often not the case. Disruption of the ligand interaction with the receptor, for example, by the interference of the label used on the ligand may result in a curvature of the line. However, as well as for artifactual reasons, the Scatchard analysis may not be linear. The binding of the ligand to the receptor may show what is termed as negative cooperativity, that is where binding of the first ligand to the receptor causes the receptor to have a reduced affinity for the second ligand. Alternatively, there may be more than one type of receptor present, each with different binding characteristics, or the interaction of an intracellular subunit with the receptor, such as a G protein, reduces the affinity for the ligand. If the Scatchard analysis is done when any of these factors is involved the resultant line will be curved and lie under the expected straight line (**Figure 5.7B**). Alternatively the line obtained may be curved but lie above the expected linear plot (**Figure 5.7A**). This may be an indicator of positive cooperativity, where binding of the first ligand increases the receptor's affinity for a second ligand, very much the same as binding of oxygen to haemoglobin. Such effects may be due to conformational changes within the protein on ligand binding and may involve more than one polypeptide, for example, interactions of receptor molecules that would be possible with receptor dimerization or complex formation.

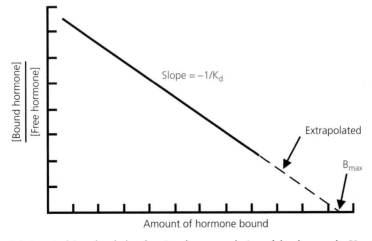

Figure 5.6 A typical Scatchard plot showing the extrapolation of the slope to the X-axis.

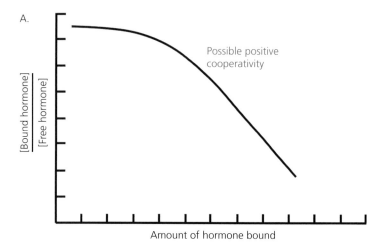

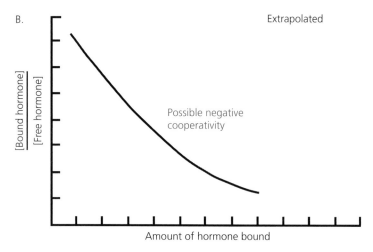

Figure 5.7 Typical Scatchard plots where the ligand binding is not straightforward. A curve is obtained for both positive cooperative binding (A) and negative cooperative binding (B).

5.4 Receptor sensitivity and receptor density

The concentration, or density, of receptors on the surface of a cell is not necessarily constant, and in fact rarely is, and it is very apparent that a cell may become more sensitive or less sensitive to a given concentration of extracellular ligand.

An increase in ligand sensitivity, or sensitization, may occur by an increase in the amount of a receptor on the cell surface. Here, a cell is maximizing its chance of detecting the ligand and so responding to it. A real increase in the receptor available can be achieved by the synthesis of new receptor molecules, their recruitment from intracellular stores, such as from vesicles, and a decrease in the rate of removal of the receptor from the cell surface, with one or more of these methods being responsible.

In many cases, if a cell has been exposed to a specific ligand, and has shown a given response, but shortly afterwards has a second exposure to that same ligand, the response seen is very much reduced. The cells are said to have become refractory to the second dose of ligand (**Figure 5.8**). If the cell retains a normal response against other ligands, that is, only one receptor seems to be involved in the refractory state, the phenomenon is called homologous desensitization. An example of this is seen with the β_2-adrenergic receptor and its response to adrenaline. The receptor becomes desensitized extremely rapidly, and it has been shown that both Mg^{2+} and ATP are required, suggesting the involvement of a phosphorylation step. It is now clear that there are in fact two separate phosphorylation events. Firstly, the β_2-adrenergic receptor causes a rise in cAMP and subsequent activation of cAMP-dependent protein kinase (PKA), leading to phosphorylation of the receptor on a serine residue, and hence disruption of the activation of a G protein by the receptor. Secondly, once activated, the receptor becomes a target for a specific kinase, β-adrenergic receptor kinase, which phosphorylates the receptor on several threonine and serine residues towards the C-terminal end of the polypeptide. This phosphorylated polypeptide then binds to a protein called β-arrestin, which stops the receptor activating its associated G protein, and so prevents propagation of a response. It is interesting to note that rhodopsin, which shows structural similarity to the β-adrenergic receptors, also undergoes a similar desensitization involving an arrestin type of mechanism (discussed further in **Chapter 12**).

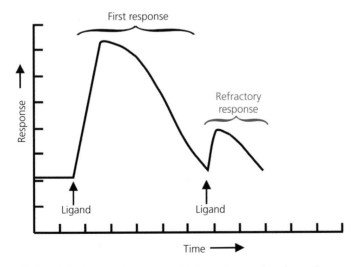

Figure 5.8 A plot of the hormonal response with time showing that the application of a second dose of the same amount of ligand, e.g. a hormone, does not necessarily invoke the same magnitude of response as the first dose.

Arrestins

Although the classical role of arrestin is still important, these proteins have been found to be involved in coupling GPCRs to other pathways. β-Arrestin has been found to control MAP kinase cascades, and even translocate to the nucleus where it can interact with transcription factors and alter gene expression.

As $β_2$-adrenergic receptors can be phosphorylated in response to rises in intracellular cAMP levels, which are controlled by many other receptors as well, $β_2$-adrenergic receptors may well be desensitized in response to one of these other receptors binding to its respective ligand. Such down-regulation is referred to as heterologous desensitization, as it is not caused by the presence of the ligand normally associated with that receptor.

As well as direct control of receptor function by phosphorylation, desensitization of a cell to a ligand may also occur by removal of the receptor from the cell surface. A common sequence of events on ligand binding may be as follows. Once a cell surface receptor has been activated by a particular ligand and has transmitted its message to the next element of the signal transduction cascade, for example, a G protein, the receptor is internalized into the cell through endocytosis. The membrane bound receptor and ligand complex become an integral part of the vesicle formed during endocytosis, and will become part of an endosome. On the cell surface, the receptor faces outwards, that is its ligand binding site is on the outside of the cell, but in the formation of the vesicle the receptor ends up facing the inside of the endosome and is exposed to the environment of the inside of the endosome. This is usually an acidic environment, unlike that to which the receptor would have been exposed on the outside of the cell, which generally is neutral. The change in pH on arriving at the endosome often causes a change in the conformation of the receptor, and so alters its affinity for the ligand, usually reducing it. Therefore, often the receptor/ligand complex dissociates. The receptor binding site is now empty, enabling the receptor to potentially be re-used. It only needs to be transported by the vesicular system of the cell, this time back to the plasma membrane where it can again be used to detect the extracellular presence of the ligand. The ligand that has been left behind in the endosome is usually delivered to lysosomes, where it is degraded. Hence, cells can effectively remove ligands from extracellular fluids and cause ligand-induced signals to be turned off, unless of course the ligand continues to be released from its source and continues to activate the receptors as they are recycled.

However, not all receptor/ligand complexes dissociate to allow the recycling of the receptor back to the plasma membrane, and, in many cases, both the receptor and ligand are transported to the lysosomes and destroyed. Maintaining the receptor concentration on the cell surface therefore requires *de novo* protein synthesis.

The process of internalization of the receptors is commonly preceded by their relocalization in the plane of the membrane into clusters (**Figure 5.9**),

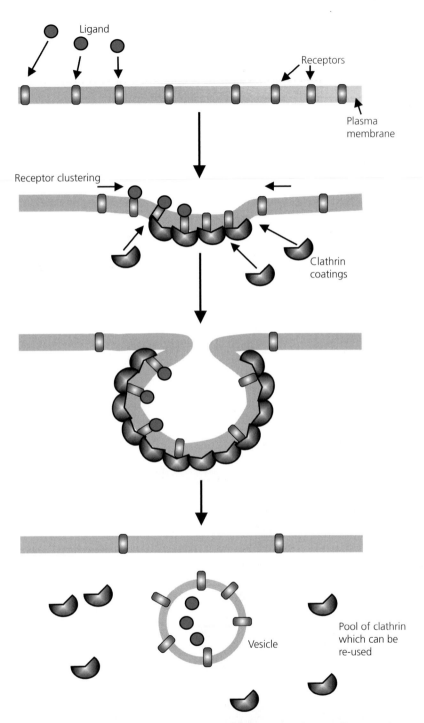

Figure 5.9 The process of endocytosis. Receptors firstly move to the site of invagination and then a vacuole is formed, aided by the presence of clathrin molecules. The capture of ligand by the receptor is shown down the left side of the figure, whereas on the right empty receptors are shown.

a process called capping. This is followed by invagination of the membrane and the formation of what is known as a coated pit. This is so-called because the cytoplasmic side of the vesicle that is forming is covered, or coated, with a protein, in this case clathrin. Clathrin is composed of two polypeptides, a light chain and a heavy chain, three of each of these chains come together to form a structure known as a triskelion. This is a three-legged structure that can further polymerize to form a basket-like matrix, which will form a scaffold around the vesicle (**Figure 5.10**).

The clathrin probably performs two main roles. Firstly, it propagates the formation of the invagination of the membrane and stabilizes the vesicle formation, as the clathrin complex naturally takes up a concave shape.

Transcytosis

Sometimes a receptor may bind a ligand at one point on the plasma membrane and be internalized, but then subsequently be relocated in a different region of the plasma membrane, so transporting its ligand across the cell. This is a process referred to as transcytosis, and is seen in cells lining the gut in mammals. The process may be important in vascular disease, and also important in breast feeding as it is a mechanism by which the new-born infant can gain things from its diet, and therefore from its mother.

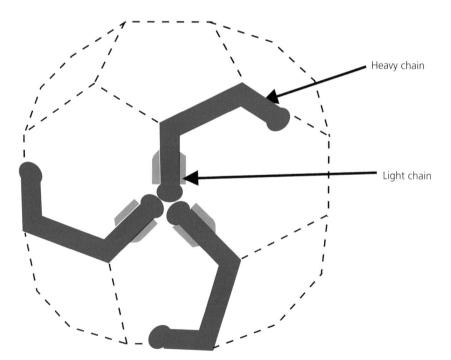

Figure 5.10 The likely quaternary structure of the clathrin protein, showing how the heavy chains and light chains align giving the appearance of a Manx flag, but the overall shape encloses a sphere.

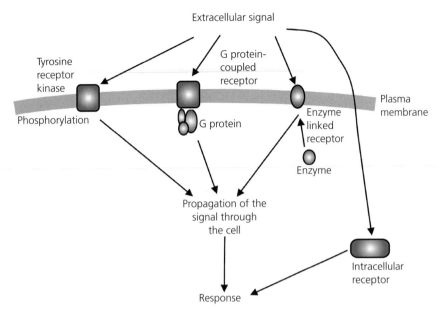

Figure 5.11 A simplified overview of the role of receptors. Four different types of receptor are shown in blue, but all lead to signalling pathways and responses.

Secondly, it may be involved in the encapture of the receptor molecules. Many receptors contain on their cytoplasmic side a short stretch of four amino acid residues, which acts as a signal for endocytosis. This short polypeptide region is recognized by a group of proteins known as adaptins. These proteins recognize both the receptor and the clathrin protein and will facilitate the association of the receptor with clathrin, and hence the uptake of the receptor into the cell. Different adaptins will recognize different receptors and so coordinate the internalization process, giving specificity to the receptor clathrin association.

Once formed the vesicles will shed their clathrin coat, as illustrated in Figure 5.9. This process probably involves ATP and Hsps, such as Hsp70. The control of uncoating is also thought to involve the concentration of Ca^{2+} in the cell, where local rises in concentration may be encountered as the vesicle is transported deeper into the cell.

■ Summary

- The detection of extracellular signals and the transmission of these signals into the cell is the responsibility of proteins known as receptors.

- Receptors are commonly found on the plasma membrane of the cell (see Figure 5.11), where they are ideally placed to be in contact with their extracellular ligand.

- Steroid receptors, however, are found inside the cell, with the ligand itself transversing the membrane.

- Membrane receptors fall into several groups:

  - G protein-coupled receptors, which lead to activation of the trimeric class of G proteins;

  - ion channel linked receptors leading to changes in ion movements across the membrane;

- receptors that contain intrinsic enzyme activity, such as the receptor tyrosine kinases;

- receptors that recruit separate enzymes, for example, tyrosine kinases such as the JAK proteins.

- Ligand binding to the receptor can be viewed as a reversible event, quantified by the K_d value, that is, the concentration of ligand at which half the receptors are bound. However, the binding of the ligand to a cell usually includes two elements: specific binding to the receptor and non-specific binding.

- Binding curves can be linearized by the use of Scatchard analysis, where the slope of the line is defined as $-1/K_d$.

Such analysis can indicate if any cooperativity is involved in receptor–ligand binding.

- A cell's sensitivity to the concentration of extracellular ligand can be varied by altering the density of receptors available at its surface.

- New receptor molecules can be synthesized and routed to the membrane or the response can be down-regulated by capping and endocytosis, resulting in the internalization of the receptor molecules. Such internalized receptors can be returned to the membrane for the further perception of ligand, or alternatively destroyed.

→ Further reading

Types of receptors

Alexander, S.P.H., Mathie, A. and Peters, J.A. (2008). Guide to Receptors and Channels (GRAC), 3rd edn. *British Journal of Pharmacology*, 153 (Suppl. 2): S1–S209.

Allen, S.J., Crown, S.E. and Handel, T.M. (2007) Chemokine:Receptor structure, interactions, and antagonism. *Annual Review of Immunology*, 25, 787–820.

Chen, H., Jinghong, M., Li, W., Eliseenkova, A.V., Xu, C., Neubert, T.A., Millar, W.T. and Mohammadi, M. (2007) A molecular brake in the kinase hinge region regulates the activity of receptor tyrosine kinases. *Molecular Cell*, 27, 717–730.

Dani, J.A. and Bertrand, D. (2006) Nicotinic acetylcholine receptors and nicotinic cholinergic mechanisms of the central nervous system. *Annual Review of Pharmacology and Toxicology*, 47, 699–729.

Evans, R.M. (1988) The steroid and thyroid hormone receptor superfamily. *Science*, 240, 889–895.

Franco. R., Casadó, V., Cortés, A., Mallo, J., Ciruela, F., Ferré, S., Lluis, C. and Canela, E.I. (2008) G-protein-coupled receptor heteromers: function and ligand pharmacology. *British Journal of Pharmacology*, 153, S90–S98.

Haga, T. and Takeda, S. (2005) *G Protein-Coupled Receptors: Structure, Function, and Ligand Screening. Methods in Signal Transduction Series*. Vol. 6 CRC Press, London, ISBN: 9780849327711.

Heldin, C.-H. (1995) Dimerization of cell surface receptors in signal transduction. *Cell*, 80, 213–223.

Jacobs. M.N. and Lewis, D.F.V. (2002) Steroid hormone receptors and dietary ligands: a selected review. *Proceedings of the Nutrition Society*, 61, 105–122.

Ledda, F. and Paratcha, G. (2007) Negative regulation of receptor tyrosine kinase (RTK) signaling: a developing field. *Biomarker Insights*, 2, 45–58.

Lemmon, M.A. and Schlessinger, J. (1994) Regulation of signal transduction and signal diversity by receptor oligomerization. *Trends Biochemical Science*, 19, 459–463.

Lieberman, B.A. (2001) *Steroid Receptor Methods Protocols and Assays: Methods in Molecular Biology Series*, Vol 176, Humana Press, Totowa, New Jersey, ISBN 978-0-89603-754-0.

Pierce, K.L., Premont, R.T. and Lefkowitz, R.J. (2002) Seven-transmembrane receptors. *Nature Molecular Cell Biology*, 3, 639–640 [see also references listed within].

Strader, C.R., Fong, T.M., Tota, M.R. and Underwood, D. (1994) Structure and function of G protein-coupled receptors. *Annual Review of Biochemistry*, 63, 101–132.

Tsai, M.-J. and O'Malley, B.W. (1994) Molecular mechanisms of action of steroid/thyroid receptor superfamily members. *Annual Review of Biochemistry*, 63, 451–486.

Turner, A.J. (ed.) (1996) *Amino Acid Neurotransmission*. Portland Press, London UK.

Willars, G.B. and Challiss, R.A.J. (eds) (2004) *Receptor Signal Transduction Protocols: Methods in Molecular Biology Series*, Vol. 259, Humana Press, Totowa, New Jersey, ISBN 978-1-58829-329-9.

Woodward, A.W. and Bartel, B. (2005) A Receptor for Auxin. *The Plant Cell*, 17, 2425–2429.

Ligand binding

Davenport, A.P. (ed.) (2005) *Receptor Binding Techniques, Methods in Molecular Biology Series*, Vol. 306, Humana Press, Totowa, New Jersey, ISBN: 978-1-58829-420-3.

Hill, A.V. (1913) The combination of haemoglobin with oxygen and carbon monoxide. *Biochemical Journal*, 7, 471–480.

Scatchard, G. (1949) The attraction of protein for small molecules and ions. *Annual New York Academy of Sciences*, 51, 660–672.

Receptor sensitivity and receptor density

Anderson, R. (1992) Dissecting clathrin-coated pits. *Trends Cell Biology*, 2, 177–179.

Brodski, F.M., Hill, B.L., Acton, S.L., Nathke, I., Wong, D.H., Ponnambalan, S. and Parham, P. (1991) Clathrin light chains: array of protein motifs that regulate coated-vesicle dynamics. *Trends Biochemical Science*, 16, 208–213.

Keen, J.H. (1990) Clathrin and associated assembly and disassembly proteins. *Annual Review of Biochemistry*, 59, 415–438.

Lefkowitz, R.J. (1993) G-protein-coupled receptor kinases. *Cell*, 74, 409–412.

Ma, L. and Pei, G. (2007) Beta-arrestin signaling and regulation of transcription. *Journal Cell Science*, 120, 231–218.

Mills, I.G. (2007) The interplay between clathrin-coated vesicles and cell signalling. *Seminars in Cell & Developmental Biology*, 18, 459–470.

Nathke, I.S., Heuser, J., Lupas, A., Stock, J., Turck, C.W. and Brodsky, F.M. (1992) Folding and trimerisation of clathrin subunits at the triskelion hub. *Cell*, 68, 899–910.

Palczewski, K. and Benovic, J.L. (1991) G-protein-coupled receptor kinases. *Trends Biochemical Science*, 16, 387–391.

Pearce, B.M. (1989) Characterisation of coated-vesicle adaptins: their assembly with clathrin and with recycling receptors. *Methods Cell Biology*, 31, 229–246.

Predescu, S.A., Predescu, D.N. and Malik, A.B. (2007) Molecular determinants of endothelial transcytosis and their role in endothelial permeability. *American Journal Physiology Lung Cell Molecular Physiology*, 293, L823–L842.

6

Protein phosphorylation, kinases and phosphatases

Phosphorylation is a central and extremely important theme of cell signalling. Signalling often involves the alteration of the activity of a protein, seen as either an increase or a decrease. A protein's ability to interact with another protein may also be altered, either allowing or stopping a signalling cascade. Such alterations of a protein's function are often, although not always, brought about by the addition or removal of a phosphate group: phosphorylation and dephosphorylation. **Chapter 6** discusses the mechanisms used to carry out this protein alteration.

Phosphorylation occurs at many points in cell signalling cascades. It may be right at the beginning, at the level of the receptor, or it might be at the end, with the phosphorylation of a transcription factor for example. Alternatively, phosphorylation may occur at a point in between, and there are also cascades of phosphorylation events. The end of a signalling response also may involve phosphorylation or dephosphorylation.

As with most signalling mechanisms, dysfunction of phosphorylation can lead to disease, but furthermore, a vast number of drugs are directed against the modulation of phosphorylation of proteins in cells. Therefore, it is important to understand the mechanisms involved in phosphorylation, and many of the main proteins involved are discussed in this chapter.

This chapter will also briefly consider other covalent modifications, although oxidation and reduction along with nitrosylation will be discussed in more detail in **Chapter 10**. Some modifications looked at here allow proteins to become associated with membranes, such as the addition of a hydrophobic group, whereas others result in the degradation of the protein, i.e. the ubiquitin-proteasome system.

6.1 **Introduction**

The arrival of an extracellular signal at a cell's surface and its detection by its receptors, must lead to further events inside the cell if the cell is going to respond to the presence of that ligand. The activation of a receptor allows it to interact with and/or activate several different types of protein, which lead to a wide range of intracellular signalling cascades as will be discussed in subsequent chapters. However, a crucial event in all cell signalling pathways is the modification of the activity of enzymes or the alteration of the function of control factors.

An enzyme's specific activity may be altered by an alteration in its conformation, that is, the folding of its polypeptide chain. Sometimes this involves the alteration of the primary structure of the protein, where parts of the polypeptide might be cleaved off for example, but much more commonly this is usually seen as an alteration in its tertiary structure, or for multipolypeptide enzymes it may also involve the quaternary structure of the protein. The altered spatial arrangement of the active site amino acids will then reduce or increase the substrate binding and/or the catalytic action of the protein.

There are several ways of facilitating the change in the conformation of a protein. A change in pH may alter the interaction capabilities of amino acids and so disrupt the way that the polypeptide is held in three-dimensions. Such changes were discussed in the previous chapter, especially seen when receptors are internalized and a conformational change leads to dissociation of the ligand from the receptors. A change in the reduction/oxidation environment of a protein has recently been suggested to have profound changes on the way a protein might function, as further discussed in **Chapter 10**. However, the most common way of modifying protein structure is by the addition, or removal, of one or more phosphate groups to the primary amino acid sequence of the polypeptide, processes known as phosphorylation or dephosphorylation respectively.

For phosphorylation to be an effective control mechanism allowing the activity of an enzyme to be both increased and decreased, the overall reaction has to be reversible (**Figure 6.1**). If, for example, the need for glycogen breakdown was

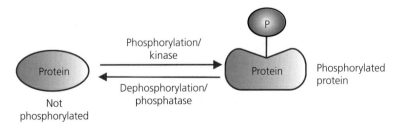

Figure 6.1 The phosphorylation and dephosphorylation of the protein, the former being catalyzed by a kinase, and the latter by a phosphatase. Note: phosphorylation usually causes a conformational change in the protein.

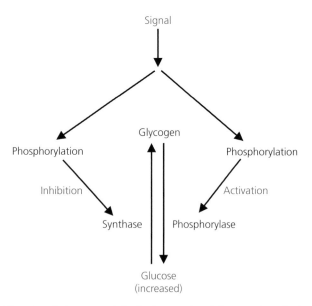

suddenly increased due to a necessity to supply the muscles with glucose, and hence energy, the enzyme responsible, phosphorylase, will become phosphorylated with a concomitant increase in its activity. However, on cessation of the muscle's activity, the requirement for glucose would drop and therefore the rate of glycogen breakdown would need to be reduced. A quick and energetically favourable way to do this would be dephosphorylation of phosphorylase, thus lowering its activity, rather than for example destruction of the phosphorylase polypeptide to remove its activity. (The mechanism of control of glycogen metabolism is further explored in **Chapter 7**, see **Figure 7.1**.)

The coordinated control of metabolic pathways also requires that futile cycles are avoided. Again using glycogen storage as an example, to facilitate glycogen usage phosphorylase is phosphorylated causing an increase in its activity, as already mentioned. However, there would appear to be little point in increasing glycogen breakdown if its synthesis continued unabated, or in fact increased due to the rise in the concentration of intracellular glucose resulting from the enhanced glycogen breakdown. Indeed, this is coordinated. The enzyme responsible for synthesis of glycogen, glycogen synthase, is phosphorylated at the same time as phosphorylase, but in the case of the synthase phosphorylation causes a lowering of activity of the enzyme, hence slowing the production of glycogen. Therefore, by phosphorylation of the two key enzymes in the pathway at the same time, both the breakdown is increased and the synthesis decreased, resulting in the necessary effect (**Figure 6.2**).

Some enzymes that are involved in metabolic pathways, and which are controlled by phosphorylation and dephosphorylation are listed in **Table 6.1**.

Futile cycle When the production of a compound and the breakdown of the same compound are carried out at the same time, resulting in little change in the steady state concentration of the compound. Such cycles would be wasteful to cells, and mechanisms are usually in place to avoid this happening – discussed again in **Chapter 7**.

Figure 6.2 Phosphorylation controls both glycogen breakdown and synthesis (see also Figure 7.1).

Table 6.1 Some of the enzymes controlled by phosphorylation/dephosphorylation and the metabolic pathway in which they are involved.

Enzyme	Metabolic pathway
Phosphorylase kinase	Glycogen usage
Glycogen phosphorylase	Glycogen usage
Glycogen synthase	Glycogen storage
Acetyl-CoA carboxylase	Lipid metabolism
Lipase	Lipid breakdown
Isocitrate dehydrogenase	Citric acid cycle/in prokaryotes

The addition of phosphate groups to proteins, and their removal, are enzyme catalyzed events. The enzymes that catalyze protein phosphorylation are known as protein kinases and there is a reciprocal group of enzymes that carry out dephosphorylation, called phosphatases. However, phosphorylation usually only takes place on one of three amino acids in the primary sequence of the polypeptide, either on a serine, threonine or tyrosine (Figure 6.3), although as we shall see there are exceptions. Even so, the amino acid targets of the kinases are used in their classification, being grouped according to which amino acid they are specific for. The two main groups are:

- Serine/threonine kinases that add the phosphate to serine and/or threonine.
- Tyrosine kinases that only use tyrosine as acceptors of the phosphate.

The phosphate group itself is supplied by ATP, the third phosphoryl group (γ) of the chain being transferred to the hydroxyl group of the acceptor amino acid, with the subsequent release of ADP. Phosphorylation usually takes place inside the cell and here ATP is generally abundant, being supplied mainly by the mitochondria in eukaryotes. Dephosphorylation is the simple removal of the phosphoryl group from the amino acid with the regeneration of the hydroxyl side chain and the release of orthophosphate. Each enzyme catalyzes their relevant reaction in an irreversible manner, although the overall reaction of phosphorylation and the subsequent return of the protein to its original state is effectively accompanied by the hydrolysis of ATP and hence results in a favourable free-energy change.

Phosphorylation of the protein is not restricted to a single site on the polypeptide chain and indeed a protein may be phosphorylated by more than one kinase, allowing in many cases the convergence of several signalling pathways. Glycogen synthase can be phosphorylated by protein kinase A (cAMP-dependent protein kinase or PKA), phosphorylase kinase and Ca^{2+}-dependent kinase. Each phosphorylation event might have a different effect on the protein, or may have no effect at all. For example, phosphorylation of serine 10

■ Other groupings include the splitting of the kinases into five families: cyclic nucleotide/phospholipid-dependent (AGC); calcium-dependent (CaMK); cyclin-dependent (CMGC); tyrosine kinases (PTK); and others.

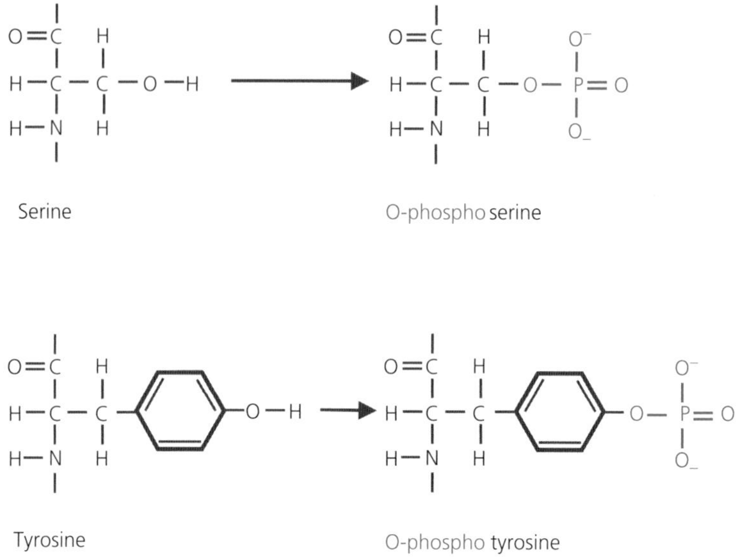

Figure 6.3 The phosphorylation of serine and tyrosine yields the altered residues phosphoserine and phosphotyrosine.

Bioinformatic analysis The use of computers in biological studies. Computers have allowed rapid and elaborate analysis of biological systems, and in particular the analysis of DNA, RNA and protein sequences. Many annotated databases of sequences are now available on the internet, for example Genbank.

on cAMP-dependent protein kinase appears to serve little purpose as phosphorylation seems to bestow little alteration of activity on this polypeptide, although future research will no doubt reveal the reasons.

Using bioinformatic analysis, once the amino acid sequence of a protein is known it is relatively easy to predict if a protein can be phosphorylated by particular kinases, but that does not indicate the effect on the protein of the phosphorylation, or whether it actually takes place in the cell (see discussion below).

The phosphorylation of a polypeptide can alter the enzymatic activity of that molecule and this might be achieved for several reasons. The added phosphoryl group is a relatively large group that needs to be accommodated, but it also adds negative charge to an enzyme that can disrupt electrostatic interactions or may be instrumental in the formation of new interactions. Similarly, the phosphoryl group can form hydrogen bonds, which may favour a new conformation. The free-energy change involved in phosphorylation may also help to push the equilibrium from one conformational state to another.

Phosphorylation is ideal as a means of regulation in response to a cellular signal as it can occur in under a second, as can dephosphorylation, therefore making the system ideal for the fast interaction often needed in the alteration of metabolic rate. Conversely, the process may have kinetics over a matter of hours, which might be needed in other physiological conditions. Furthermore, one of the basic needs of a signalling system in the cell is the ability to have amplification of a signal. As discussed in Chapter 1, a few molecules arriving on the outer surface of the cell might need to alter the activity of many enzyme molecules on the inside. The activation of a single kinase molecule will result in the phosphorylation of many enzymes and so the process of phosphorylation plays a major role in the amplification of intracellular signals.

If kinases are to control the activity of enzymes within cells, then they themselves have to be under some sort of control. The modulation of a cell's activity may be through various different routes, including alteration of calcium concentrations, cAMP, cGMP and inositol phosphate metabolism. These different signalling pathways appear to end commonly with the activity of kinases with their own degree of specificity towards the enzymes which they control. However, not all kinases are controlled by second messengers, a classic case of messenger-independent protein kinases being the casein kinases. These enzymes are widely distributed throughout the plant and animal kingdoms, where they are used for the phosphorylation of acidic proteins.

In fact, some kinases might themselves be controlled by a phosphorylation event. For example, phosphorylase kinase is phosphorylated by cAMP-dependent protein kinase, whereas mitogen-activated protein kinase cascades show a series of phosphorylation events controlling a series of kinases.

Some kinases appear to be specific for one protein, for example, phosphorylase kinase, whereas others have a more generic action, where they are capable of phosphorylating many proteins, for example, protein kinase C. The specificity of the kinase is determined by the specific amino acid sequence either side of the target amino acid residue, which is destined to receive the phosphoryl group. In many cases, consensus sequences for the sites of phosphorylation for different kinases have been identified using bioinformatics, but, the appearance of a consensus sequence in the protein's primary sequence does not automatically mean that the site will be phosphorylated. Many will be buried deep inside the structure of the protein and not be accessible to kinase action, and will never be used. Using such information, once the gene encoding a protein has been cloned and sequenced, a researcher can more easily predict whether that protein can be phosphorylated by a variety of kinases. With such predictions, biochemical experiments can be carried out to determine which kinases actually modify the protein.

Consensus sequence

If the sequences of several related proteins are known, bioinformatic analysis allows for the sequences to be aligned and an "average" or consensus sequence to be determined, incorporating the common features found in all the sequences known. Positions in the sequence where the identity of a single amino acid cannot be determined in a consensus, usually because it can be a variety of amino acids across a group of sequences, are simply marked as an "X".

Although the kinases have been separated into these two broad classes, the actual catalytic sites within most of them seem to be quite well conserved, and contain common characteristics amongst these enzymes. Most protein kinases contain a catalytic core domain of approximately 250 amino acids in size, which is relatively conserved between them. Within this domain there appear to be two particular regions which have had consensus sequences or signature patterns assigned.

The first region is located at the N-terminal extremity of the catalytic region. It is characterized by a lysine residue believed to be involved in ATP binding, which is close to a stretch of glycine residues. In the central part of the catalytic region, the second conserved region identified contains an aspartic acid residue believed to be important in the catalytic activity of the kinase. However, the amino acids around this residue appear to differ for the two classes of kinase, and, therefore, two separate signature patterns can be deduced.

Not all kinases contain these regions. Interestingly, the signature pattern specific for the tyrosine kinase catalytic site has homology to some bacterial phosphotransferases. These are thought to be evolutionarily related, and, in fact, do contain some structural homology to protein kinases.

The identification of protein signatures and consensus sequences will of course aid in the identification of kinase-like active sites in newly discovered proteins, but, like all consensus searches, just because it is found does not necessarily mean that it is active and bestows any functionality on the protein. It has, in fact, been estimated that the human genome may contain as many as 2000 genes coding for different kinase polypeptides, leaving us many still to be discovered.

6.2 Serine/threonine kinases

The kinases that preferentially phosphorylate the amino acids serine or threonine within polypeptides, and therefore come under the classification of serine/threonine kinases, encompass a large group of phosphorylating enzymes, including cAMP-dependent protein kinase, cGMP-dependent protein kinase, protein kinase C, Ca^{2+}-calmodulin-dependent protein kinases, phosphorylase kinase, pyruvate dehydrogenase kinase and many others. Not all these can be treated in detail here, but important representative examples will be examined that should give an overall picture of the activity of these ubiquitous enzymes.

Oncogenes commonly encode for proteins that contain a cell signalling function, as discussed in Chapter 1, and it is no surprise that several oncogenes appear to function because they encode proteins that contain serine/threonine kinase activity. These include the products of the genes *mil*, *raf* and *mos*. Such activity highlights the importance of phosphorylation by these kinases in the control of cellular functions, and also highlights the importance of such functionality being tightly controlled.

Raf An extremely important serine/threonine kinase found at the top of mitogen-activated protein kinase cascades. See later chapters for further discussion.

cAMP-dependent protein kinase

cAMP-dependent protein kinase, otherwise referred to as protein kinase A, cAPK, or PKA is widespread in eukaryotes being found in animals, fungi and as a slightly different form in plants.

Many processes within the cell are controlled through the activation of cAMP-dependent protein kinase. These include the regulation of the rates of metabolism and control of gene expression. Examples of the former include the control of phosphorylase kinase, and hence activation of phosphorylase. This results in the increased breakdown of glycogen. As discussed above and in Chapter 5, the activation of PKA not only causes the activation of phosphorylase kinase and so phosphorylase, but PKA also catalyzes the concomitant phosphorylation and deactivation of glycogen synthase, so preventing a futile cycle (also see Figure 7.1).

Gene expression can also be controlled through phosphorylation by PKA, causing the activation of transcription factors such as CRE-binding protein (CREB). CREB in its active state binds to CRE (cAMP-response element) regions of the DNA. Genes containing such control elements include those that encode enzymes of gluconeogenesis in the liver.

As its name suggests, cAMP-dependent protein kinase is controlled by the levels of cAMP present in the cell — see further discussion on the production of cAMP in Chapter 7. In the inactive state, that is, when the levels of cAMP are low, cAMP-dependent protein kinase is found as a tetramer containing two catalytic subunits (C) and two regulatory subunits (R). If the concentration of cAMP rises, it causes activation of PKA. cAMP binds to the regulatory subunits causing a conformational change in the structure of the protein and an alteration in the affinity of the regulatory subunits for the catalytic subunits, causing the complex to dissociate. The regulatory subunits remain as a dimer, with the release of the two active monomeric catalytic subunits (Figure 6.4).

It is thought that virtually all of a cell's responses to cAMP are mediated by the activity of the catalytic subunits of cAMP-dependent protein kinase. As a monomer, this catalytic polypeptide has a molecular weight of around 41 kDa, but at least three isoenzyme forms have been identified. The isoforms Cα and Cβ in mammals differ by less than 10% when their amino acid

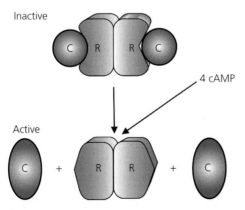

Figure 6.4 The activation of cAMP-dependent protein kinase, showing the dissociation caused by the binding of cAMP.

sequences are compared, and they seem to be highly conserved between species. However, unlike Cα, which appears to be expressed constitutively in most cells, the expression of Cβ is tissue-specific. The catalytic core of these PKA subunits also shares homology with other known kinases and includes defined regions for peptide binding, ATP binding and a catalytic site (see discussion above). Such molecular studies of enzymes have revealed details of the mechanisms involved. Analysis of the catalytic regions of these PKA polypeptides, both by the use of fluorescence analogues of ATP and by studying the modification of lysine residues by acetic anhydride in the presence and absence of MgATP, has shown that lysine residues at amino acid positions 47, 72 and 76 are important for the functionality of these proteins. Sequence comparisons with other kinases showed that lysine 72 is totally invariant and site-directed mutagenesis has shown its vital importance in the catalytic cycle of kinases. To the N-terminal side of this lysine lie three highly conserved glycine residues, found at positions 50, 52 and 55, which are probably involved in the binding of the phosphate in the nucleotide. Interestingly, point mutations in an analogous glycine-rich region may be responsible for the constitutive activity of the *v-erbB* oncogene product, which resembles the EGF receptor but is in a permanently turned-on state. However, it should be noted that the *v-erbB* oncogene protein also lacks the extracellular EGF binding domain.

Once activated, PKA needs to be able to phosphorylate the next component in the cell signalling cascade. Therefore, if it has many potential targets, it should be possible to identify some commonality amongst such targets. cAMP-dependent protein kinase typically phosphorylates peptides that contain two consecutive basic residues, usually arginine, which lie at positions 2 and 3 towards the N-terminal end of the protein from the site of phosphorylation. This would give the consensus sequences as:

-Arg-Arg-X-Ser- X- or
-Arg-Arg-X-Thr- X-.

The residue designated as X between the Arg doublet and the phosphorylation site is usually a small amino acid, whereas the other amino acid designated as X is usually hydrophobic in character. The arginines may also be replaced by lysines, and therefore the consensus sequence above does not make a hard and fast rule.

As well as the lysines identified above, e.g. lysine 72, carboxyl groups within the sequence of the catalytic subunit may well be responsible for this specificity, so allowing the recognition of peptide substrates. Carboxyl groups, particularly at positions 184, an aspartate, and 91, a glutamate, may also be responsible for ligation to the Mg^{2+} in the MgATP substrate. Like lysine 72, these two groups appear to be invariant throughout the kinase domains studied so far. Further carboxyl groups at position 170, and six at the C-terminal end of the protein have also been implicated in the recognition of the peptide substrate.

As with all enzymes, kinetic analysis of the catalysis can be determined and here for cAMP-dependent protein kinase such studies have found that PKA usually has a K_m in the region of 10–20 µM substrate, with a V_{max} of 8–20 µmol/min/mg. The first step in the binding of the target peptide sequence to the enzyme probably involves ionic interactions between the peptide's arginine residues and the enzyme, which is followed by recognition of the amino acid, e.g. serine, which will ultimately accept the phosphate group. Binding of MgATP, which supplies the phosphate needed for phosphorylation, to the catalytic subunit enhances the binding of the peptide, with large conformational changes across the enzyme being induced by substrate binding.

The catalytic subunit of cAMP-dependent protein kinase also contains two phosphorylation sites itself, one at threonine 197 and another at serine 338, and once phosphorylated the phosphate groups are not readily removed by phosphatases. A serine residue at position 10 can also be autophosphorylated, that is the phosphate group is added by PKA itself, although it is not known if this has any physiological role.

Other modifications to PKA can also occur, besides phosphorylation. For example, the catalytic subunit can be covalently modified by the myristoylation of the N-terminal end. It was thought that this might serve as a signal for translocation of the subunit to the plasma membrane, an event in which myristoylation is often associated, but after such modification the catalytic subunit remains soluble, and myristoylation appears not to be essential for catalytic activity either.

Therefore, much is known about the functioning and mechanisms involved with the activity of the catalytic subunits. But what of its controlling peptide, the regulatory subunit? In the presence of low concentrations of cAMP, the regulatory subunits of PKA have the function of binding to the catalytic subunits, so rendering them inactive. As with many signalling proteins, there is more than one isoform of this peptide. Two main groups of the regulatory subunits have been found, called type I and type II, which differ in their amino acid sequences but also in their function. Type II subunits can be autophosphorylated, whereas type I are not autophosphorylated, but do contain a binding site for MgATP, which binds with relatively high affinity. Isoforms of the different types of subunit have been cloned, and expression patterns have been found to differ too. Some isoforms seem to be fairly ubiquitously expressed, whereas others are expressed in a tissue-specific way. Furthermore, the expression of some isoforms is inducible, whereas others appear to be expressed constitutively.

In general, the regulatory subunit exists as a dimer, with all the subunits sharing the same structural features (**Figure 6.5A**). These include two consecutive gene-duplicated sequences at the C-terminal end of the molecule, which are the cAMP binding sites. Type I subunits are held together covalently by two disulphide bonds, with the two polypeptide chains running antiparallel. In type II, the subunits are held together by interactions between the N-terminal amino acids in the chains. Another general feature of the

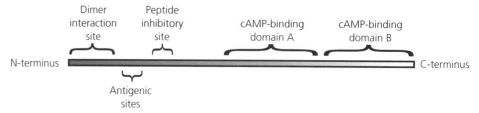

Figure 6.5A The domain structure of the regulatory subunit of cAMP-dependent protein kinase.

subunits is what is referred to as the "hinge region" towards the N-terminal end, which is sensitive to proteolytic cleavage. This region encompasses the amino acids that are most likely to be involved in the interaction with other proteins, as well as being a highly antigenic part of the molecule.

The most important part of the regulatory subunit is a region that controls the activity of the catalytic subunit. The hinge region contains an amino acid sequence, known as the peptide inhibitory site, which resembles that of the substrate for the catalytic subunit. In type II regulatory subunits there is an autophosphorylation site here, whereas type I subunits have a sequence that contains the two arginines needed for recognition by the catalytic subunit active site, but lack the serine or threonine that would normally accept the phosphate group. This so-called pseudophosphorylation site contains an inert alanine or glycine residue instead. Several lines of evidence, including limited proteolytic cleavage, affinity studies of the regulatory subunit to the catalytic subunit and site-directed mutagenesis support the idea that the regulatory subunit maintains the catalytic subunit in an inactive state, by pseudophosphorylation or the inhibitor site of the regulatory subunit occupying the peptide binding site of the catalytic subunit, so preventing the binding of the correct protein substrate.

It can be seen, therefore, that here in PKA, a sequence which mimics the true substrate of the enzyme can inhibit by competing for the active site of the enzyme. In the inactive state, the conformation of the subunits is such that binding to the pseudo-substrate, that is the regulatory subunit, is preferred, but on activation this is dissociated, allowing the binding of the true substrate. In PKA the inhibitory site and the catalytic site are on separate polypeptide chains, but this general principle of a regulator region controlling the activity of the catalytic region is used by several other protein kinases. However, usually in other kinases, both sites are part of a single polypeptide chain. This includes the related kinase, cGMP-dependent kinase, as well as other kinases such as myosin light chain kinase (MLCK) and protein kinase C. With MLCK it is the binding of Ca^{2+}-calmodulin that causes a major conformation change, preventing the inhibitor site from blocking the active site of the catalytic region. In many kinases, a mechanism more akin to that of the type II cAPK regulatory subunits is used, in that the regulatory site is actually autophosphorylated. Enzymes using this type of mechanism include cGMP-dependent kinase and Ca^{2+}-calmodulin-dependent kinase II.

A representation of an example of the kinase is shown in **Figure 6.5B**. This is a type II enzyme, where the structure was solved by X-ray diffraction so that it could be compared to type I enzymes, and therefore mechanisms could be determined. Looking at structures in such detail can give a real insight into the conformational changes that may be taking place as the enzyme functions.

In cAMP-dependent protein kinase, the conformational change needed to release the inhibitor site from the catalytic active site is induced by the binding of cAMP. Each regulatory site has two cAMP binding sites, both of high affinity. These two sites, A and B, show high sequence homology to each other, but their binding to various cAMP analogues varies. They also show high sequence homology to catabolite gene activator protein (CAP) of *E. coli*.

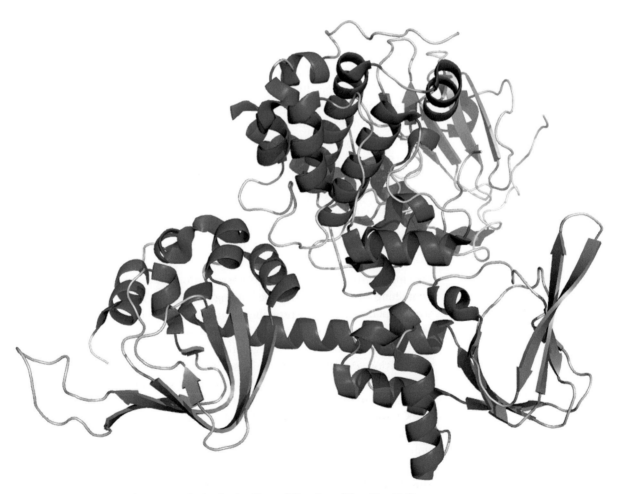

Figure 6.5B The proposed structure obtained using X-ray diffraction of Type IIa cAMP-dependent protein kinase (mouse). Structure was obtained from the RCSB Protein Data Bank (www.rcsb.org/pdb/) PDB ID: 2qvs. (Wu, J., Brown, S.H.J., von Daake, S. and Taylor, S.S. (2007) PKA type II alpha holoenzyme reveals a combinatorial strategy for isoform diversity. *Science* 318, 274–279.)

This protein is involved in the cAMP-dependent regulation of the lactose operon, turning on gene expression when the cells are depleted of glucose and allowing them to survive on a new carbon source. In CAP it was found that two amino acid residues were of crucial importance to the functioning of the protein: an arginine that interacts with the negative charge of the phosphate of cAMP and a glutamine residue that hydrogen bonds to the ribose ring. These two residues are found in PKA regulatory subunits as well as in the cGMP binding sites of cGMP-dependent protein kinase.

Studies where parts of the protein have been deleted and the functionality analyzed have shown that the N-terminal region of the regulatory subunit, along with the cAMP binding site B, can be removed and still the polypeptide is able to bind to the catalytic subunit in a cAMP-dependent manner. Removal of the cAMP binding site A has also been shown to produce a molecule that is cAMP-dependent and can still bind to the catalytic subunit. Furthermore, removal of either cAMP binding site seems to have little effect on the affinity for cAMP of the other binding site in isolated regulatory subunits. However, in reality, both cAMP binding sites probably participate in activation of type I PKA. Both sites need to be occupied for dissociation of the native holoenzyme and binding of the two cAMP binding sites shows cooperativity. cAMP probably binds to the site B first, causing conformation changes that increase the accessibility of the A site to cAMP. Binding at the A site causes a further conformational change, altering in particular the hinge region, and so resulting in the activation of the enzyme.

To be inactivated, once the need for the phosphorylation catalysis of PKA has passed, the enzyme needs to be re-associated into its original tetrameric form. For the type II regulatory subunit containing enzymes, re-association of the cAPK complex probably involves the use of phosphatases, allowing the inhibitor site to once again enter the catalytic site and cause deactivation of the enzymes' activity. However, the re-association for regulatory subunit type I containing enzymes involves the binding of MgATP. This binding shows positive cooperativity, which probably ensures that the holoenzyme is a tetramer and that trimers with only one catalytic subunit are not formed.

The N-terminal end of the regulatory subunits is also the target for phosphorylation, in fact by several kinases, including casein kinase II and glycogen synthase kinase. It may be that phosphorylation here is involved in the association of the regulatory subunits with other proteins.

PKA is usually found as a soluble enzyme in most cells, but it has been found that some forms of the kinase are associated with the membrane fractions of cells. Such associations are probably mediated by the regulatory subunits binding to an integral membrane protein. The catalytic subunits do not appear to be involved here, as they can be released and found in the soluble fractions of cells following activation. The exact location of the enzyme and its activity is, of course, important for controlling exactly where in the cell cAMP will be produced and found, as is discussed below in a section on compartmentalization (Section 7.7).

cGMP-dependent protein kinase.

cAMP is not the only nucleotide to be found to be responsible for the activation of a kinase. cGMP is also an important cell signalling molecule in cells (see Chapter 7 for further discussions). The levels of cGMP in cells, like that of cAMP, can be used to control phosphorylation, often via cGMP-dependent protein kinase, otherwise referred to as cGK or PKG—an enzyme with many similarities to PKA. cGMP-dependent protein kinase has been found to be abundant in many mammalian tissues, including smooth muscles, heart, lung and brain tissues. Although mainly cytosolic in location, following cell sub-fractionation analysis the enzyme has also been found in particulate fractions.

cGK, such as that purified from lung and heart, has been found to be a dimer of identical 76 kDa subunits, with the holoenzyme having a molecular weight of approximately 155 kDa. The two subunits are held together in an antiparallel arrangement by disulphide bonds. Just as was found in PKA, this enzyme also has an inhibitor sequence. In this case the orientation of the dimer allows the inhibitor site of the regulatory domain of one subunit to act as the inhibitor of the catalytic domain of the other subunit.

Sequencing of cGK has revealed that it can be thought of as being six segments, making up four functional domains. The segment at the N-terminal end of the polypeptide contains the sites used to maintain the dimer structure, a hinge region as seen in cAMP-dependent protein kinase, and the inhibitor site which will undergo autophosphorylation. Also, like cAMP-dependent protein kinase, the next two segments have high sequence homology to the CAP protein of *E. coli* and make up the two cGMP binding domains of the polypeptide. The rest of the molecule is the catalytic domain, with the fourth and fifth segments showing high homology to other protein kinases.

Also, like cAMP-dependent protein kinase, cGMP-dependent protein kinase shares very similar preferences for the peptide sequences that are phosphorylated. The enzyme typically phosphorylates a serine, or threonine, in the peptide that lies at positions 2 and 3 towards the C-terminal end of the protein from two consecutive basic residues, usually arginine. Therefore, like cAMP-dependent protein kinase, the consensus sequences would be:

-Arg-Arg-X-Ser- X- or
-Arg-Arg-X-Thr- X-.

However, in cGK, phosphorylation is enhanced by the presence of a proline residue N-terminal to the Ser/Thr phosphorylation site and a basic residue on the C-terminal side. Hence, such differences give cGK a different substrate specificity to cAMP-dependent protein kinase.

Amongst other similarities with PKA is the fact that both nucleotide binding sites appear to need to be occupied for activation. However, unlike cAMP-dependent protein kinase, cGK does not dissociate on activation.

Although activated primarily by cGMP, cGK can potentially also be controlled by other mechanisms. It can bind, for example, to cAMP and cIMP, albeit at much higher concentrations than seen with cGMP, whereas cGK has also been shown to be phosphorylated. However, the physiological relevance of such mechanisms is not clear, but would allow for the convergence of signalling in cells.

Protein kinase C

One of the most important protein kinases is one called protein kinase C, or PKC. PKC was originally thought to be just a single protein, but more recent research has revealed that it is, in fact, a family of closely related protein kinases. These different polypeptides are encoded for by different genes, or in some cases are derived from the alternate splicing of a mRNA transcript from single genes. However, not all cells express all the variants, but cells often express more than one form.

The protein kinase C proteins can be broadly split into two families. The first to be cloned was the group containing the α, βI, βII and γ subspecies, but later cloning revealed a group containing δ, ε and ζ subspecies. The genes for the α, β and γ forms have been located to different chromosomes, with the variation in the βI and βII forms coming from alternate splicing from one gene. Other isoforms described include η, θ and λ, making a total of at least ten isoforms that have been described.

In general, the protein kinase C enzymes are monomeric in nature, having between 592 and 737 amino acids, which gives molecular weights between 67 kDa and 83 kDa. The two groups of PKC each have a common structure, which shows close similarity to the other group, although distinct differences are also seen between the two families. The first group, containing the α, βI, βII and γ forms have four conserved regions (C_1–C_4), along with five variable regions (V_1–V_5). The second group, containing δ, ε and ζ lacks the C_2 conserved region, although overall their molecular weights appear to be similar across the families.

The polypeptides can be roughly divided into two domains, a regulatory domain and a protein kinase domain (Figure 6.6). The regulatory domain contains the C_1–C_2, and V_1–V_2 regions of the first group, but only the C_1 region of the δ, ε group. The C_1 region of both groups contains a highly cysteine-rich region, which resembles the consensus sequences of a cysteine-zinc DNA-binding finger, although there appears to be no DNA binding. The second domain contains the rest of the molecule and is referred to as the protein kinase domain. Sequence homology between this region and other protein kinases has been reported, and it is the C_3 region that contains an ATP binding sequence. This motif is repeated in the C_4 region, although is probably not used.

X-ray absorption studies suggest that protein kinase C may contain four zinc ions within its structure, which are coordinated mainly by sulphur atoms supplied by the cysteine residues. However, it is likely that these zinc ions probably serve a role in the structural stability of the polypeptide rather than having a direct role in the catalytic cycle.

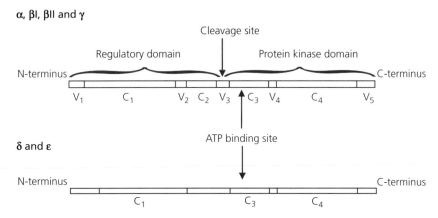

Figure 6.6 The domain structure of the two families of protein kinase C.

As with all cell signalling components, protein kinase C has to have its activity controlled. It can be activated in a wide range of manners, including proteolytic cleavage. In many cases its activity can also be controlled by intracellular Ca^{2+} concentrations, which is how it derived its name, that is, as a calcium-dependent protein kinase. However, some isoforms are Ca^{2+} ion-independent, such as δ, η, θ and ε.

Of particular importance is the fact that protein kinase C can also be controlled by phospholipids, and, in particular, by diacylglycerol (DAG). DAG is derived from inositol phosphate metabolism simultaneously with $InsP_3$ by the action of PLC. $InsP_3$ formation leads to an increase in intracellular calcium, and therefore the action of PLC has two potential ways of controlling PKC activity. Inositol metabolism is discussed in further detail in **Chapter 6**. The response to DAG is not homogeneous across the isoforms of PKC, the exact effect differs between isoenzyme forms, due to the fact that some seem to lack a DAG binding site, for example, ζ and λ. Furthermore, it has been suggested that some protein kinase C subspecies become activated at different times during a cellular response, orchestrated by a series of phospholipid metabolites such as diacylglycerol, arachidonic acid or other unsaturated fatty acids.

Activation of some PKC forms can also be through the PtdIns 3-kinase pathway. This kinase, as discussed in **Chapter 7**, forms phosphorylated inositides, which are now thought to be extremely important in many pathways, such as insulin signalling to GLUT4 glucose transporters (see **Chapter 11**). Phosphoinositides formed are able to activate a kinase known as 3-phosphoinositide-dependent kinase (PDK1), and this kinase has been found to be able to phosphorylate PKC, particularly isoforms ζ and λ.

Chemical stimulants, such as the tumour-promoting phorbol esters, for example phorbol 12-myristate 13-acetate (PMA) otherwise known as 12-O-tetradecanoyl-phorbol-12-acetate (TPA), are widely used in the laboratory to cause activation of protein kinase C, as they work by their action as diacylglycerol analogues. However, their exact effect differs between isoenzyme forms, because, as mentioned above, some lack the DAG binding site. Caution is needed when using phorbol esters in experimental work, not only due to their toxicity to the users but also because they are only very slowly metabolized from the cells, unlike DAG. New stimulators that are more specific than phorbol esters, such as Sapintoxin A, can also be used in the laboratory to activate protein kinase C.

On looking at the molecular mechanism of the enzyme, it can be seen that here, as in cAMP-dependent protein kinase, it appears that protein kinase C has an inhibitor site, or pseudophosphorylation site, near the amino acid terminus of the polypeptide. Again, this site contains an inert alanine instead of the threonine or serine that would normally be found in the substrate. Binding of PKC activators such as Ca^{2+}, diacylglycerol and phosphatidylserine will cause a conformational change within the protein, releasing this pseudophosphorylation site and resulting in activation of the kinase.

Activation by proteolytic cleavage can occur through the action of enzymes such as calpain. Cleavage occurs within the variable V_3 region, releasing a catalytically active fragment. It may be that the active form of PKC is the target for calpain, which itself is active in micromolar Ca^{2+}, although, unlike the γ form, not all subspecies of PKC are susceptible to rapid cleavage, for example, subspecies α is relatively resistant. It is possible that cleavage of protein kinase C is, in fact, not part of its activation, but perhaps the first step to its degradation and removal from the cell.

Having discussed the activation of PKC, and the presence within it of a pseudophosphorylation site as part of its enzymatic mechanism, it is necessary to investigate the likely substrate targets of this enzyme. Under physiological conditions, once activated, protein kinase C preferentially phosphorylates a polypeptide on a serine or threonine residue found in close proximity to a C-terminal basic residue. Therefore, the consensus sequence would thus be:

-[Ser/Thr]-X-[Arg/Lys]-.

Additional basic residues, either on the C-terminal or N-terminal side of the target amino acid, may enhance the V_{max} and lower the K_m of the phosphorylation reaction.

The exact role or roles of protein kinase C within the cell, however, remain surprisingly obscure. Phosphorylation experiments *in vitro* have shown that a large variety of polypeptides become phosphorylated by PKC, and it has been implicated in activation of Ca^{2+} ATPases and the Na^+/Ca^{2+} exchanger, controlling Ca^{2+} levels within the cell as well as the phosphorylation of receptors such as the EGF receptor and the interleukin-2 receptor. Clearly, future work will establish the exact role of each isoform and how they fit into the complex web of cellular signals.

■ Phorbol 12-myristate 13-acetate (PMA) can cause tumour formation, probably through its actions on PKC, which highlights the role of PKC in the control of cell function.

Ca^{2+}-calmodulin-dependent protein kinases

One of the major ways of controlling the functioning of enzymes and metabolic pathways in a cell is by alteration in the intracellular concentration of Ca^{2+}, [Ca^{2+}]$_i$, as discussed further in **Chapter 9**. Many kinases have been found to be dependent on or regulated by Ca^{2+}, including protein kinase C (discussed above), myosin light chain kinase and phosphorylase kinase. Besides these more specific enzymes, there is a group of kinases that are controlled by Ca^{2+} called the multifunctional calcium/calmodulin-dependent protein kinases. Several classes of these have been identified, but the most characterized is known as calcium/calmodulin-dependent protein kinase II. This enzyme is also referred to as CaM kinase II, Type II CaM kinase or sometimes simply as just kinase II.

Although kinase II is widespread throughout several tissues, it is most abundant, up to 20–50 times more concentrated, in brain and neuronal tissue compared to non-neuronal tissues. The enzyme prepared from brain tissue consists of an α subunit of approximately 50 kDa and a protein doublet (β/β') of approximately 60 kDa. Both types of subunit are able to bind to calmodulin and are autophosphorylated in a Ca^{2+}/calmodulin-dependent manner. Both subunits, therefore, contain both regulatory and catalytic domains. The α and β/β' subunits are very closely related, the N-terminal halves of the two subunits being 91% identical, with over three-quarters of the other half also being found to be identical when the genes were cloned from rat brain. The holoenzyme has been reported to have a molecular weight of 500–700 kDa in most tissues that have been investigated, but a smaller complex of 300 kDa has been found in liver. Therefore, subunits must come together. The complex probably assumes a dodecamer structure in which two hexameric rings are stacked on top of each other. The ratio of α subunits to β/β' subunits seems to vary enormously between tissues. In some tissues, subunits of approximately 20 kDa have been reported, for example in spleen tissues, suggesting other components too. In fact, isoenzymes cloned include δ and γ subunits that are of similar molecular weight to the β/β' subunits, and which appear to have a wide tissue distribution. The sequences of these isoforms show that they are again closely related to the α subunits.

The subunits can be split into three distinct functional domains. The N-terminal section is the catalytic domain, with the centre section encompassing the regulatory domain and a domain essential for the formation of the holoenzyme complex at the C-terminal end of the polypeptides known as the association domain (see **Figure 6.7**). Proteolytic cleavage can release an active catalytic domain, and here homology to other calmodulin regulated kinases is seen, such as phosphorylase kinase. The fact that proteases can cleave here suggests the presence of an exposed hinge region between the catalytic and regulatory domains.

As with other kinases, the regulatory domain contains an autoinhibitory site, as discussed above, but here the regulatory domain also has a calmodulin binding site. Binding of Ca^{2+}/calmodulin to the enzyme, a process that shows

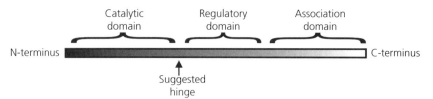

Figure 6.7 The domain structure of calcium/calmodulin-dependent protein kinase II.

positive cooperativity, will cause an overall conformational change in the kinase's three-dimensional structure, which releases the autoinhibitory site from the active site and allows kinases to have full activity. In the inactive state, the autoinhibitory site not only prevents peptide substrate binding but also prevents the binding of the donating ATP. The autoinhibitory site involves autophosphorylation, as seen with other kinases, and it is thought that this autophosphorylation potentiates the effects of the controlling messenger. A rapid spike of intracellular Ca^{2+} can activate the enzyme, not only allowing it to phosphorylate its normal substrate, but also causing autophosphorylation, which may slow down the calmodulin dissociation from the enzyme. This will have a consequence of lengthening the time that the kinase has activity, even after the Ca^{2+} concentrations have returned to their basal levels.

The subcellular location of the enzyme also seems to vary. Sometimes it is membrane associated, sometimes it is soluble, or sometimes it is bound to the cytoskeletal structures of the cell.

As with all kinases, there are likely to be specific targets for their action. The substrates for kinase II are quite wide ranging, and include tryptophan hydroxylase, glycogen synthase from skeletal muscle, synapsin I, proteins associated with the microtubules, ion channels and transcription factors. The specific part of the proteins targeted, that is the peptide specificity, in some cases seems to be similar to that identified for cAMP-dependent protein kinase or myosin light chain kinase, but in other cases the peptides that are phosphorylated are distinct from those recognized by other kinases. In general, it has been found that an arginine residue, three amino acids towards the N-terminal end of the polypeptide from the serine or threonine which is to accept the phosphate group, seems to be essential. It has also been suggested that acidic residues within two residues of the phosphorylated amino acid, or a hydrophobic residue on the C-terminal side of the phosphorylation site, might also be important.

Other multifunctional calcium/calmodulin-dependent protein kinases, differ from kinase II substantially. Kinase I, for example, is a monomer of molecular weight of only approximately 40 kDa. Kinase I and kinase III, an elongation factor-2 kinase, also seem to have a much narrower substrate specificity than kinase II. In fact, kinase I has a substrate specificity more closely related to cAMP-dependent protein kinase than to kinase II.

A fourth Ca^{2+}/calmodulin-dependent kinase, kinase IV, is expressed primarily in brain tissue, and in the thymus, but has also been seen in the spleen and testis. However, it appears to be absent from several tissues studied. It is expressed as two splice variants, and has been purified from rat brain to be two bands on SDS-PAGE, of 65 kDa and 67 kDa. Within the cell, it is mainly found in the nucleus, but it does reside in the cytoplasm too. Again, as discussed above, its activation involves autophosphorylation, here of its Ser/Thr rich N-terminal end, and it has been shown to be involved in the control of gene expression, mainly through its phosphorylation and activation of transcription factors, for example, CREB.

Although calcium kinases are clearly regulated by the calcium signalling, some of them are now known to be part of a cascade. A CaM kinase kinase, itself activated by elevated Ca^{2+} ion concentration, can activate both Kinase I and Kinase IV through phosphorylation. CaM kinase kinase exists in two forms, α and β, which are found in the cytoplasm and nucleus respectively. Interestingly, CaM kinase kinase can also phosphorylate and therefore control the kinase PKB, allowing divergence of the signalling.

G protein-coupled receptor kinases

An important class of kinases has been discovered that are involved in the down-regulation of the receptors, which act through the heterotrimeric class of G proteins. These cytosolic kinases are known as G protein-coupled receptor kinases (GRKs), of which the best studied are probably β-adrenergic receptor kinase (βARK) and rhodopsin kinase (the latter is discussed further in **Chapter 12**).

At least six GRKs are known, of which two are classed as βARks. Towards the C-terminal ends of these proteins are pleckstrin homology (PH) domains (see **Chapter 1**), which are capable of binding the βγ subunit complex from the heterotrimeric G protein. It was originally thought that the βγ subunits of these G proteins contributed little to cell signalling pathways, but clearly here they are involved in kinase activation. An interaction between the G protein subunits and the kinase results in translocation of the kinase to the membrane where it can interact with, and phosphorylate, the receptor. With the kinases GRK2 and GRK3, lipids are also seen to stimulate this activity, including phosphatidylserine, although the activity was inhibited by $PtdInsP_2$ at high concentrations. It has been suggested that the lipids and βγ G protein subunits may compete for the same binding site on GRK type kinases.

Further to the regulation by lipids and G proteins, the β-adrenergic receptor kinase has been shown to be phosphorylated by protein kinase C. Again, the translocation of the GRK to the membrane was enhanced, along with an increase in kinase activity. Phosphorylation of the kinase was towards the C-terminal end of the polypeptide.

Of the GRKs not classed as βARKs, GRK4 is expressed most highly in the testis but is actually a family of four GRKs, which arise from alternative splicing of the mRNA involving exons II and XV. Meanwhile, GRK5 appears to be a unique member of the family of GRKs. It is capable of phosphorylation of rhodopsin, m2 muscarinic cholinergic receptor and β(2)-adrenergic receptor, and it is inhibited by heparin.

What is being recognized and phosphorylated by these kinases though? In the β(2)-adrenergic receptor, the phosphorylated residues are all located within a 40 amino acid stretch at the extreme C-terminal end of the receptor polypeptide. GRK5 will phosphorylate threonines at positions 384 and 393, and serines at 396, 401, 407 and 411, whereas GRK2 will phosphorylate serines 396, 401, 411 and threonine 384. A similar phosphorylation pattern is seen with rhodopsin. It is important to note that it is only the activated receptors which are phosphorylated, so the conformation of the protein must be very important in the kinase/receptor recognition.

The role of the kinase in control of the receptor is thought to work in the following way. Once the receptor has bound ligand, the receptor undergoes a conformational change and the G protein is activated. The Gα transmits its signal to, for example, adenylyl cyclase, whereas the βγ complex activates the GRK. GRK will phosphorylate only those receptors that are bound to ligand and deactivate the receptor, stopping its further activation of G proteins. Furthermore, phosphorylation of the receptor allows its interaction with a protein known as β-arrestin, which further inactivates the receptor. Therefore, the receptor is desensitized and the cell shows adaption to a prolonged exposure to ligand (also see Chapter 5).

Protein kinase B

A serine/threonine kinase that, although it was discovered about 15 years ago, has not had much prominence in cell signalling literature is protein kinase B (PKB). PKB is otherwise known as c-Akt, and is the cellular homologue of a virally encoded oncogene v-Akt. However, like many proteins, its importance is being increasingly recognized and it is now known to be used in a variety of signalling pathways, including those of glycogen metabolism and responses to insulin and growth factors.

PKB is a 57 kDa protein that exists in three isoforms in mammals, α, β and γ (also known as Akt1, Akt2 and Akt3). This family of proteins shows some homology to PKA and PKC, hence the adoption of its name. When looked at in more detail it can be seen that PKB contains a pleckstrin homology domain (PH domain: see Chapter 1) at the N-terminal end, a kinase domain in the middle and a regulatory domain at the C-terminal end.

Activation is thought to be through the PH domain, where it interacts with 3-phosphoinositide lipids created in the membrane by the action of PtdIns 3-kinase (see Chapter 6). Activation also involves translocation of PKB from the cytoplasm to the plasma membrane, and once there it is phosphorylated. One site of phosphorylation of PKB is in the kinase domain, whereas another is in the regulatory domain. The phosphorylation of the kinase domain is thought to involve an enzyme called 3-phosphoinositide-dependent kinase (PDK1), and PKB autophosphorylation is also thought to occur.

AMP-activated protein kinase

A protein kinase that has recently been shown to be important in a variety of pathways is 5'-AMP-activated protein kinase (AMPK). In the active state it phosphorylates, and inactivates, several metabolic enzymes which are involved in the synthesis of fatty acids and cholesterol, and it has also been shown to be important for insulin signalling (discussed in **Chapter 11**) where inhibition of AMPK in the presence of glucose activates insulin secretion. It is thought that AMPK may be a sensor for cellular energy status in some cells. The mammalian kinase is a heterodimer, which consists of a catalytic α subunit and noncatalytic β and γ subunits, which exist in different isoforms and aid in the control of the enzyme. The presence of either control subunit increases kinase activity, but both of these non-catalytic subunits have been found to be needed for full activity of the kinase, the subunits having a positive and synergistic effect on the control of the enzyme. Experimentally, AMPK can be activated by the addition of 5-amino-4-imidazolecarboxamide riboside (AICAR). Physiologically, AMPK is controlled as its name suggests by 5'AMP, but also by phosphorylation. The target for phosphorylation within AMPK is Thr-72, phosphate being added by AMPK kinase (AMPKK).

Interestingly, the mammalian AMPKs are related to the yeast SNF1 kinase (sucrose non-fermentor kinase), although the control of the two enzymes by their respective interacting proteins, (mammalian β and γ, and yeast Sip1p/Sip2p/Gal83p proteins and Snf4p) seem to be subtly different.

Haem regulated protein kinase

Haemoglobin

Although haemoglobin is known as an extremely important protein, it should be noted that it contains a haem prosthetic group. It is this haem group that is crucial to its functionality, and therefore haem and protein need to come together to produce a working haemoglobin molecule.

It is not only the enzymes in a metabolic pathway that are under the control of specific kinases, but also protein synthesis can be regulated in this way. The classical example of this is the work done in reticulocytes, where it has been shown that globin synthesis is regulated by haem. It would be a waste for a cell to produce a large quantity of globin polypeptides if there is no protohaem available for completion of the holoenzyme. This mechanism is regulated by phosphorylation of one of the protein synthesis initiation factors, eIF-2, used to bring Met-tRNA to the ribosome to start the new polypeptide chain. If phosphorylated, this initiation factor forms a stable complex with a guanine nucleotide exchange factor (GEF: see **Chapter 7**), preventing the exchange of GDP for GTP on the initiation factor and so preventing another round of protein synthesis initiation. However, if haem is present, it inhibits the phosphorylation event and allows protein synthesis and formation of holoproteins.

Phosphorylation of eIF-2 is catalyzed by a kinase known as haem-regulated protein kinase or haem-controlled repressor (HCR). The purified kinase has a molecular weight of approximately 95 kDa under denaturing conditions, but shows an apparent molecular weight of about 150 kDa under non-denaturing conditions. The target sequence recognized by the kinase appears to be -Leu-Leu-**Ser**-Glu-Leu-Ser-, where the first serine is the site of phosphorylation. The only known substrate for the enzyme is the initiation factor eIF-2, although the enzyme also undergoes autophosphorylation. Both phosphorylation events are inhibited by the presence of haem. The activity of the kinase is also affected by the state of its sulphydryl groups. If its sulphydryl groups are oxidized (see discussion in **Chapter 10**) or modified the enzyme is found in the active state even in the presence of haem. Other factors might also be involved in the activation of this kinase.

A similar control of translation is seen by an interferon-induced kinase, suggesting that phosphorylation might be a commonly used pathway for the control of protein synthesis.

Plant specific serine/threonine kinases

Although the text above has been devoted to the discussion of kinases as characterized in animal tissues, it should be noted that plants and fungi also use protein phosphorylation in their control of cellular functions, and, in fact, plants have been found to contain several kinases which are very analogous to those described above and are under similar control mechanisms. However, several plant specific kinases have also been identified and partially or completely purified. One of the best characterized is a kinase that phosphorylates the light-harvesting chlorophyll a/b complex (LHC). LHC is phosphorylated on at least one threonine residue, activating it when exposed to low levels of red light. The kinase, found in thylakoid membranes has an approximate molecular weight of 64 kDa, and its activation probably involves the redox state (see **Chapter 10**) of plastoquinone.

With growing evidence that plants also contain kinases analogous to those found in animal tissues, one of the exciting areas of research now is the search for animal-like proteins and control mechanisms in plants. For example, several effects of phorbol esters have been seen in plant tissues suggesting the presence of a protein kinase C-like enzyme. It is not only with kinases of course, but many areas of similarity in cell signalling can be seen between animals, plant and organisms of other kingdoms, and the research in one area will no doubt inform the future of research elsewhere.

Tyrosine kinases

The second major class of protein kinases includes those that add a phosphoryl group to tyrosine, as opposed to serine or threonine. Phosphoryl addition to tyrosine is less common than the modification on serine or threonine, and an analysis of the phosphoamino acid content of a cell in 1980 revealed that only approximately 0.05% was phosphotyrosine. Even though much more is known about phosphorylation events now, 30 years later, this still suggests that tyrosine phosphorylation is not that prevalent, but such numbers should not belittle their importance. The tyrosine kinases that carry out this phosphorylation are themselves found in two broad groups, those that are soluble and those that are part of a receptor.

The general phosphorylation site that has been characterized as the target for these kinases has a lysine or an arginine residue, seven amino acids to the N-terminal side of the tyrosine. An acidic amino acid such as aspartate or glutamate is quite often found three or four residues also to the N-terminal side of the tyrosine, giving the consensus sequences:

-[Lys/Arg]-X-X-[Asp/Glu]-X-X-X-Tyr or

-[Lys/Arg]-X-X-X-[Asp/Glu]-X-X-Tyr.

However, as with most of these types of signatures, there are exceptions that do not seem to fit into this neat pattern.

Receptor tyrosine kinases

The receptors that fall into this broad class of kinases can themselves be split into 14 groups characterized by their general structural patterns, although they all share the same basic topology. They all possess an extracellular ligand binding domain, a single transmembrane domain and a cytoplasmic domain, which contains the kinase activity. These receptors are also discussed in **Chapter 5**, and later, for example in **Chapter 11**.

In all cases, binding of the ligand to the extracellular binding site of the receptor causes a conformational change to the protein, which activates the kinase activity on the cytoplasmic side of the receptor. Phosphorylation by this kinase activity then leads to intracellular signalling by the activation of, what may be common, signalling pathways. Enzymes that are turned on by this type of kinase include, amongst many others, phosphatidylinositol 3-kinase involved in part of the inositol pathway, GTPase-activating protein, involved in G-protein signalling, and mitogen-activated protein kinase cascades (MAP kinases).

Not only does the ligand binding turn on the tyrosine kinase activity towards the proteins in the signalling pathway, but also these receptor kinases autophosphorylate. In many cases the receptors also dimerize, for example, the epidermal growth factor receptor family of kinases. Where the receptor exists in different isoforms, the dimers can be made up of two of the same polypeptide or may be a mixture of the isoforms. A good example of this is seen with the platelet-derived growth factor receptor, which can have dimers that are two of the same subunit, $\alpha\alpha$, $\beta\beta$, or one of each subunit as a heterodimer, $\alpha\beta$, with each combination having a different specificity towards the precise ligand that can be bound (see Chapter 5). However, the insulin receptor is a tetramer having a $\alpha_2\beta_2$ conformation. Although this is a tetramer, it is effectively a dimeric-type structure, with each part of the "dimer" comprising two others (see Chapter 11).

Autophosphorylation involves the addition of phosphate groups to tyrosine residues on the cytoplasmic side of the receptor. This autophosphorylation may be intramolecular, where the polypeptide chain phosphorylates itself, or, in many cases, the autophosphorylation is intermolecular, where one polypeptide in a dimer phosphorylates the other and vice versa. This latter case is seen, for example, with the fibroblast growth factor receptor family.

It is often the autophosphorylation of the receptor itself that is the signal that enables the message to be relayed further down the pathway. The addition of the phosphate group creates a binding site for other cytoplasmic proteins. As discussed in Chapter 1, proteins that contain SH2 domains are able to recognize and bind to phosphotyrosine residues. Therefore, creation of such covalent modifications within proteins creates binding sites for these SH2 domains. Observations that have arisen from the use of synthetic peptides that can be used to block the interactions of polypeptides by binding to recognition sites and stopping the binding of the true protein partner, have helped to unravel such mechanisms. These studies have also shown that amino acids, other than the phosphotyrosine, are involved in the polypeptide interactions. For example, for the binding of phosphatidylinositol 3-kinase a methionine residue is required three residues to the C-terminal side of the phosphotyrosine. Residues to the N-terminal side of the tyrosine seem to be less important than those on the C-terminal side. The importance of the creation of these binding sites by autophosphorylation may well be to create high affinity binding sites for the proteins that are to be phosphorylated. This allows efficient phosphorylation of proteins that might be in a very low abundance within the cell, and, once bound to the receptor protein kinase, the bound protein will itself be phosphorylated on tyrosine so altering its activity. Alternately, it may be the simple act of binding of the protein to the receptor that alters the conformation of that protein, so allowing it to relay the message on without the need for it to be phosphorylated. A good example here is the use of the linker or adaptor protein. These commonly contain SH2, and perhaps SH3 domains. Examples of such adaptor proteins are GRB2 in mammals or Drk in the fly *Drosophila*. These proteins are often associated with guanine nucleotide releasing proteins (GNRPs), and the interaction of the adaptor with the activated receptor through the SH2 domains and newly formed phosphotyrosine residues leads to the association of the GNRP with the membrane proteins, and, ultimately, the activation of its respective G protein. Such a scheme is illustrated in Figure 6.8 and is discussed further in Chapter 11.

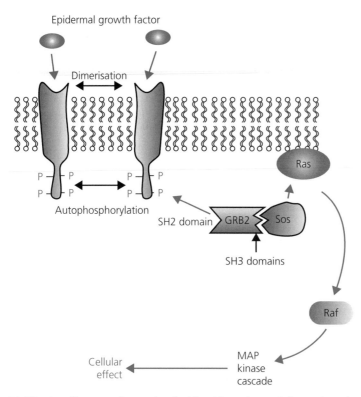

Figure 6.8 The signalling cascade associated with epidermal growth factor detection as an example of signalling from RTKs. Binding of the growth factor to the receptor causes dimerization and autophosphorylation of the receptor. Phosphotyrosines formed create binding sites for SH2 domains of the GRB2, adaptor protein. This protein is found associated with Sos, a guanine nucleotide releasing protein through SH3 domains. The new receptor/GRB2/Sos complex causes activation of the G protein Ras, which leads to the activation of a kinase cascade and ultimately the cellular effect.

In several of the classes of tyrosine receptor kinases, common areas of homology have been characterized that will be useful in the identification of new kinases and assigning kinase activity to proteins, which may at present have no function assigned to them.

In many of these receptor kinases, it is not only tyrosine that can be phosphorylated but phosphorylation also occurs on serine or threonine. In the insulin receptor family, serine/threonine phosphorylation occurs on the β-subunit of the receptor as well as on a number of intracellular substrates such as the protein kinase Raf-1. However, the exact significance of this is yet to be determined.

Cytosolic tyrosine kinases

Kinases that contain tyrosine phosphorylation capacity are not only membrane bound, like the receptor kinases, but may also be soluble and reside in the cytoplasm of the cell, or be associated with membrane proteins. One such group of kinases are the Janus kinases, otherwise known as JAKs. Janus in Roman mythology was the god of doors and gateways, and signified beginnings that ensured good endings. The statue of Janus has two heads, which are seen to be gazing in opposite directions, hence the name of the first month of the year as January. Janus kinases contain tandem, but non-identical, catalytic domains, and although involved in interactions with other proteins they lack both SH2 and SH3 domains. They are involved in the signal transduction pathways that lead from the plasma membrane to deep within the cell, such as those which are cytokine-induced, for example in leukocytes and lymphocytes. In fact, there are around 40 cytokine receptors that have their signalling mediated by this system. The kinases associate with their receptor, which undergoes a conformational change on binding to its ligand, which activates the system. Once in the active state, the JAK proteins phosphorylate each other, and also subsequently phosphorylate the receptor protein. This allows them to associate with other proteins, and the JAKs can also phosphorylate such proteins. One type of protein phosphorylated by JAKs is intracellular proteins such as STAT proteins (signal transducers and activators of transcription), which leads to the activation of transcription (**Figure 6.9A**). As these cytosolic STAT proteins contain SH2 domains, once they are phosphorylated by JAKs they can dimerize by association of the new phophotyrosine of one STAT protein with the SH2 domain of another (**Figure 6.9B**). Such activated dimers then cause the alteration of transcription observed. A representation of the structure of a STAT protein is shown in **Figure 6.9C**, and highlights the striking symmetry of the protein – such symmetry is often associated with DNA binding.

In humans there are at least seven different STAT proteins (STAT-1, 2, 3, 4, 5a, 5b, and 6), whereas members of the JAK family include JAK1, JAK2, JAK3 and Tyk2 and different members of the family have been implicated in message transfer from different receptors. For example, activation of the erythopoietin receptor leads to activation of JAK1, whereas JAK1 associates with JAK2 in the interferon-γ transduction pathway. JAK3 has been implicated in the signalling from interleukins 2 and 4. JAK3 is a polypeptide of approximately 120 kDa, although it probably exists as at least three variants, but other Janus kinases are slightly bigger, around 140 kDa. However, they all share similar topology, having seven JAK homology domains.

■ Very recently STAT3 has been found in mitochondria where it is thought to have a function in regulating the activity of the electron transport chain.

6.4 Mitogen-activated protein kinases

Many external factors, including cytokines and growth factors, can lead to the phosphorylation and activation of another family of kinases, other than the Janus kinases. These are often referred to as extracellular signal-regulated kinases (ERKs), or mitogen-activated protein (MAP) kinases. Activation of these serine/threonine kinases can lead to their translocation to the nucleus and phosphorylation and activation of transcription factors, leading, for example, to promoted growth, differentiation and altered gene expression profiles.

These kinases are themselves activated by phosphorylation but in an unusual manner, that is, they are phosphorylated on tyrosine and threonine through the action of specific threonine/tyrosine kinases known as MAPK kinases (MAPKK). This enzyme is alternatively known as MAPK/ERK kinase (MEK). The sequence commonly recognized and phosphorylated within the MAPK by MAPK kinase is -Thr-Glu-Tyr-, where both the threonine and tyrosine residues receive phosphoryl groups leading to the activation of the MAP kinase. However, the glutamate residue may be variant in some cases, giving the consensus target for the MAPK kinase as –T-X-Y-.

The MAPK kinase is itself also regulated by phosphorylation, catalyzed this time by MAPKK kinase, otherwise known as MAPKKK or MEK kinase

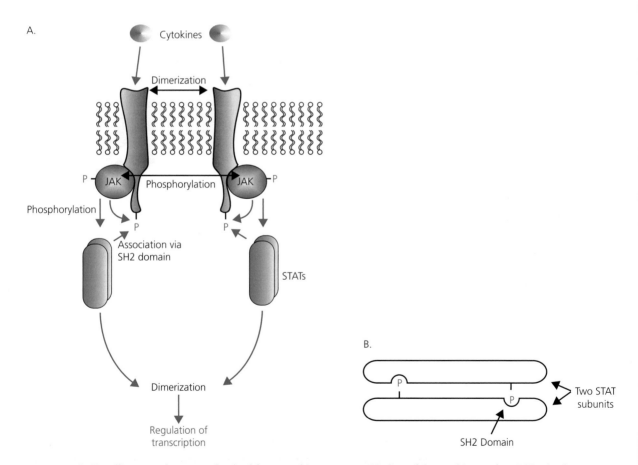

Figure 6.9 A. Signalling cascade proposed to lead from cytokine receptors. Binding of the cytokine, such as INF-γ, leads to dimerization of the receptor, and the activation of soluble tyrosine kinases, JAKs. Phosphorylation of the JAKs and subsequently of the receptors, and other proteins, STATs, takes place. STAT proteins themselves dimerize and carry the signal to the nucleus where transcription is regulated. B. A simple schematic of how the STATs dimer comes together.

C.

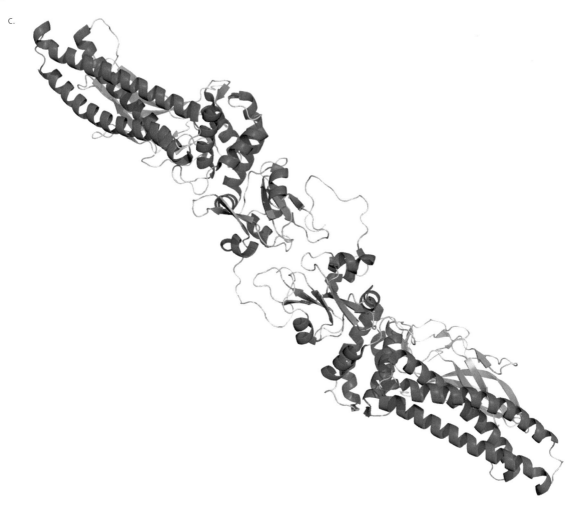

Figure 6.9 (Continued) C. A proposed structure for a STAT dimer (from *Dictyostelium*), obtained by X-ray diffraction. Note the symmetry that is created, which would be ideal for DNA binding. Structure from the RCSB Protein Data Bank (www.rcsb. org/pdb/) PDB ID: 1uur. (Soler-Lopez, M., Petosa, C., Fukuzawa, M., Ravelli, R., Williams, J.G. and Muller, C.W. (2004) Structure of an activated Dictyostelium STAT in its DNA-unbound form. *Molecular Cell*, 13, 791–804.)

(MEKK). This is in fact a group of related serine/threonine kinases. One of the proteins that can catalyze this phosphorylation is Raf. The MAPKKKs (MEK kinases) from mammalian systems show conservation of the catalytic domains, but interestingly they show divergence of a regulatory domain towards the N-terminus. A typical MAP kinase cascade is illustrated in a simplified form in **Figure 6.10**.

The initial signal for this cascade of phosphorylations may originate in the activation of a receptor tyrosine kinase activity, or in lower organisms, in the activation of a two component system as described below. Activation of the MAPK cascade can also involve the activation of G proteins, in some cases through the action of a trimeric G protein, either the Gα subunit, or in other cases through the action of the Gβγ dimer. Or activation may be through a monomeric G protein such as Ras as shown in **Figure 7.8** and **Figure 11.7**. MAP kinases have also been shown to be activated by reactive oxygen species (ROS) as discussed in **Chapter 10**, but whether this is a direct effect or not needs to be established.

To show how MAP kinases fit into a pathway, the activation by epidermal growth factor serves as a good example as illustrated in **Figure 6.8**. The tyrosine kinase receptor leads to formation of binding sites for, and the activation of, an adaptor protein, GRB2, which contains both SH2 and SH3 domains. The SH2 domains are responsible for the adaptor protein binding to the receptor once it has been autophosphorylated on tyrosine residues, whereas the SH3 domains are involved in its interaction with a guanine nucleotide exchange factor and the subsequent activation of Ras. Once Ras has been converted to the GTP bound, and is in the active form, it is able to cause the activation of the kinase Raf and phosphorylation catalyzed by Raf leads to activation of the rest of the cascade.

As well as the specific example above, it has been found that MAPK pathways show remarkable conservation across species and even kingdoms. Although the details are different, the principles of a MAPKKK activating a MAPKK and so on are used in mammals, yeast and plants. Selected examples are shown in **Figure 6.11A**.

It has been postulated that other proteins are also involved in the MAP kinase pathway, some of which are themselves kinases, but others that are proposed to have a function as scaffold proteins rather than possessing kinase activity (**Figure 6.11B**). An example of the latter is the protein Ste5 from *Saccharomyces cerevisiae*, which contains a zinc-finger-like conformational domain and has been shown to interact with other components of the MAP kinase pathway. The existence of such scaffold proteins is really important, even though they often appear to have no obvious inherent function. The same kinase components have been seen to function in more than one pathway, and more than one pathway seems to have the same end effect in some cases. The existence of such a scaffold structure would indeed be an attractive hypothesis to explain the relative lack of interaction between different MAP kinase pathways in the same cell. At least six distinct MAPK pathways have been identified in yeast, and it is difficult to see how the signals, the "messages", remain distinct with such similar and related components. Surely, they should just interact with each other? The existence of a physical scaffold and kinase complexes would help explain these observations. However, if there is a defined scaffold and physical architecture here, then the potential for the amplification of the signal would be severely limited. Perhaps, this is a pay-off that cells might have evolved to allow defined "messages" to reach their targets appropriately?

It should be noted here that some MAPK cascade components can act as scaffold proteins but themselves have a kinase domain, as seen with the

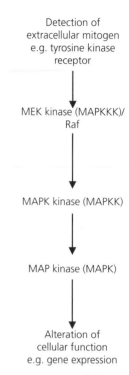

Detection of extracellular mitogen e.g. tyrosine kinase receptor

↓

MEK kinase (MAPKKK)/ Raf

↓

MAPK kinase (MAPKK)

↓

MAP kinase (MAPK)

↓

Alteration of cellular function e.g. gene expression

Figure 6.10 A simplified schematic representation of the MAP kinase cascade.

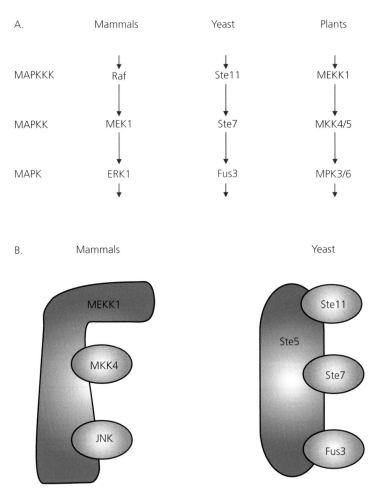

Figure 6.11 MAP kinase cascades. A. Selected MAP kinase cascades from mammals, yeast and plants, showing the similarity and conserved nature of these pathways. B. MAP kinase cascades are held by scaffold proteins, some of which act only in this capacity, as with the yeast Ste5, whereas others have scaffold function and have a kinase domain.

mammalian protein MEKK1 and the yeast protein PbS2. These proteins would have the dual role of keeping the correct components together, as well as relaying the signal down the pathway (Figure 6.11B).

Once the MAP kinase has been phosphorylated, and therefore activated, it must have a role in phosphorylating further proteins causing an alteration in their activity. Transcription factors are known to be targets for MAP kinase, with phosphorylation causing an increase in the rate of transcription of specific genes, for example, in the pheromone response of yeast. However, there is a translocation issue here. MAP kinases are usually found in their inactive state in the cytoplasm of cells, whereas their transcription factor targets are in the nucleus. On activation, the MAP kinases are phosphorylated and then often translocated into the nucleus, where, while active, they can have their effect.

However, MAP kinases are deactivated by dephosphorylation, and once this has taken place they are recognized by the trafficking mechanisms of the cell and re-translocated back to the cytoplasm, where they are ready in the inactive state for a further round of signalling. However, it should be noted that there are also cytosolic targets of some MAP kinases, so not all signals here lead to the nucleus.

As discussed, like all signalling pathways, once turned on, the response has to be turned off again, and here the reversal is by dephosphorylation. The removal of the phosphate group is catalyzed by phosphatases. A group of phosphatases of which the synthesis is inducible by growth factors and stimuli of the MAP kinase pathway, has been shown to be involved in switching off the pathway itself. These phosphatases seem to be specific for MAP kinase, whereas amino acid sequences appear to have some homology to a dual specificity tyrosine/serine phosphatase from *vaccinia* virus, VH1. One such phosphatase, CL100, also shows some homology to another dual-specificity phosphatase, cdc25. Another phosphatase, PAC-1, has been found to be localized to the nucleus, where MAP kinase has its effect.

Other candidates likely to be involved in the cessation of the MAP kinase cascade are less specific phosphatases such as PP2A or CD45.

6.5   Histidine phosphorylation

Although the most common phosphorylations of proteins are based on the amino acids serine, threonine or tyrosine, they are not the only amino acids to participate in this type of reaction and so cause an alteration in the activity of the protein, and hence the propagation of a cellular signal. Histidine and aspartate are found to be phosphorylated in bacteria, and in some higher organisms. These are relatively high energy phosphoamino acids and are involved in what has been termed the two component signalling system. Detection of a stimulus leads to phosphorylation of the detector protein on a histidine residue, using ATP as the donor for the phosphate group, and the phosphoryl group is subsequently transferred to a second protein, the response regulator, which becomes phosphorylated on an aspartate residue. These events lead to an alteration in the functioning of this latter protein. For example, it may have enhanced DNA binding capacity and so alter DNA transcription rates.

In general, the first phosphorylation step is an autophosphorylation event, so the proteins must contain kinase and target domains. Histidine kinases belong to a group of kinases, which share in common the presence of several conserved domains. Five main domains have been identified, but any particular protein may not contain all of them, and may also contain other functional domains.

■ Some histidine kinases appear to be associated with the endoplasmic reticulum, but still are activated by the arrival of extracellular signals, for example, histidine kinases that sense the presence of ethylene. Clearly, there is more to understand about the exact role of such proteins.

However, most contain a conserved region in which there is a histidine that receives the phosphoryl group. Two glycine-rich regions, termed, G1 and G2, contain the kinase activity as well as phosphatase activity and nucleotide binding ability, whereas two other regions, termed N and F, that are conserved have also been identified. The initiating stimulus can either be detected by a receptor region that is part of the kinase polypeptide itself, or by a separate polypeptide that activates the kinase by a protein–protein interaction. An integral receptor domain would probably be extracellular, with the polypeptide having a trans-membrane region, and the other domains would reside on the other, intracel-lular, side of the membrane. However, other histidine kinases are soluble and are found in the cytoplasm of the cells. These histidine kinases are terrific exam-ples of modular proteins, as different domains seem to be "glued" together to give the catalytic activity and specificity required.

The response regulators, which receive the phosphoryl group onto their aspartate residues, share a common domain containing two aspartate resi-dues and a conserved lysine residue. Phosphorylation will lead to an altera-tion of the protein's function. The protein can be subsequently re-instated back to the inactive form by an intrinsic phosphatase activity or by the activ-ity of a separate phosphatase enzyme.

Although, in many cases, the kinase domain and the response domains reside on separate polypeptide chains, there are many examples where both parts are on the same protein. These are referred to as hybrid kinases. Therefore, histi-dine kinases can be grouped into two main classes: hybrid and non-hybrid.

Systems in bacteria that use this two component type of signalling pathway include those which monitor osmolarity, temperature, pH and cell density. In fact, over 50 such systems have been identified.

Using the conserved sequences of the histidine kinases as a basis, researchers have looked for such kinases in higher organisms. Such a strategy has revealed sequences homologous to histidine kinases in the fungi *Neurospora crassa*, in the slime mould *Dictyostelium discoidium* and in the yeast, *Saccharomyces cerevisiae*, whereas several proteins sharing homology to non-hybrid and hybrid histidine kinases have been found in the higher plant *Arabidopsis thaliana*. Here, for example, the histidine kinases have been found to be instrumental in the recognition of, and signalling induced by, ethylene. Ethylene has been found to bind to the membrane spanning domain of at least one of them, whereas the kinase and response domains reside in soluble domains. Interestingly, this kinase has been shown to dimerize, but has a cellular location in the endoplasmic reticulum.

It is very likely that histidine and aspartate residue phosphorylations will be discovered to be more widespread and abundant than first thought in eukaryotic systems, and to be involved in responding to or even sensing many signals. For example, one two component system has been found to be involved in oxidative stress responses in yeast. However, there is little evidence of sequences encoding histidine kinases in mammalian genomes, including the human genome.

6.6 **Phosphatases**

No treatment of phosphorylation would be complete without a discussion on how the phosphoryl group is removed to reverse the cycle, that is dephosphorylation, and return the protein to its previous state, be that active or inactive (refer to Figure 6.1). Dephosphorylation is carried out by a group of enzymes called phosphatases. Like the kinases, these again are split broadly into two groups, those which remove the phosphoryl group from serine or threonine residues, the serine/threonine phosphatases, and the ones that remove the phosphate groups from tyrosine residues, the tyrosine phosphatases. In a similar fashion to the kinases, the number of phosphatases encoded for in the human genome has been estimated, research suggesting that it could be as many as 1000.

As with the kinases, although most of the research in this area has been carried out on animals, plants and fungi also contain phosphatases analogous to those characterized, but some plant-specific phosphatases have also been found, such as the LHC phosphatase associated with the thylakoid membranes.

Serine/threonine phosphatases

Several different types of these enzymes have been identified, referred to as PP1, PP2A, PP2B, PP2C, PP4 and PP5. Interestingly, it has been estimated that a cell's total serine/threonine phosphatase activity is approximate to a cell's total serine/threonine kinase activity, suggesting that a balance of these activities exists within cells. Such a balance would make much sense, as otherwise the activation of a pathway may be preferential compared to the inactivation, as the reversal, dephosphorylation, might not be able to keep pace with overwhelming kinase activity. Indeed, further to this, the cellular concentrations of the major forms of these two enzymes are also comparable.

PP1 has a broad substrate specificity, and in mammals two closely related isoforms have been identified, PP1α and PP1β. These two enzymes arise from alternative splicing of the same gene. The catalytic subunits, of 37 kDa, are under the control of two thermostable proteins that are able to inhibit the activity of PP1: inhibitor 1 and inhibitor 2. PP1 appears to be involved in the regulation of glycogen metabolism, where it is associated with a glycogen binding protein (R_{G1}). This glycogen binding protein is itself under the control of phosphorylation. Phosphorylation of this latter protein by cAMP-dependent protein kinase makes it unable to bind to the catalytic subunit of the phosphatase, showing that the phosphatase PP1 is under the control of the cAMP signalling pathway. Inhibitor 1 is also only active in inhibiting the PP1 when it is in the phosphorylated state, again phosphorylated by cAMP-dependent protein kinase. Therefore, the action of the cAMP pathway is twofold, that is, in deactivating PP1 and thus preventing dephosphorylation of proteins, which are phosphorylated by kinases, which are turned on directly or indirectly by the rise in cAMP. (This idea fits well with the prevention of futile cycles as already mentioned in preceding chapters.)

PP2A is a trimeric protein, and shows some similarity to PP1. PP2A has two identified isoforms of the catalytic subunit, PP2Aα and PP2Aβ, but here they are separate gene products. The other subunits involved are a structural subunit (A) of 65 kDa and a regulatory third subunit (B), which shows a degree of variability. At least 13 genes have been found encoding variants of the B subunit, and splice variants of these also have been reported. Roles for PP2A include the control of DNA replication and apoptosis.

PP2B, also known as calcineurin, is a calcium-dependent enzyme. In the presence of Ca^{2+}/calmodulin, the activity of PP2B is increased. It exists as a heterodimer of a catalytic subunit (A subunit) and a calcium binding subunit (B subunit), which contains four Ca^{2+} binding regions. Therefore, the effect of Ca^{2+} may either be direct or through the action of calmodulin. One form of the enzyme that has been purified had a molecular weight of approximately 80 kDa, composed of a 60 kDa A subunit and a 20 kDa B subunit. However, the A subunit seems to be quite variable between tissues that may account for reported substrate specificity differences. The subcellular location of the enzyme may be dictated by the B subunit, which may be myristoylated, allowing a tight association with membranes.

PP2C, a 42–45 kDa monomeric protein, on the other hand, appears to be a Mg^{2+}-dependent enzyme, whereas PP5 is primarily found in the nucleus and contains sequences reminiscent of those found with proteins associated with RNA and DNA binding.

Studies of the roles of phosphatases, like those of many other signalling components, have been greatly enhanced by the use of inhibitors. PP1 and PP2A have been shown to be strongly inhibited by okadaic acid, a fatty acid produced by *Dinoflagellates*. It has very little effect on PP2B, and has no effect on PP2C or tyrosine phosphatases. A similar inhibitory pattern is seen with tautomycin from *Streptomyces*.

Tyrosine phosphatases

In 1988, when Tonks and colleagues published the partial sequence of the first tyrosine phosphatase (PTP 1B), it was found that not only did the enzyme lack homology with serine/threonine phosphatases but also that another such enzyme had already been cloned a few years earlier. However, that protein, CD45, had been assigned no function. Obviously such findings have sparked a wave of activity in this area of biochemistry, and it is now known that there are two main classes of tyrosine phosphatases: those that are intracellular and those that are receptor linked, very much as is seen with the tyrosine kinase families.

The intracellular phosphatases are soluble proteins, and reside, as the name suggests, in the inside of the cell, with the activity usually associated with the cytoplasm. PTP 1B can be thought of as the model for this family of phosphatases. The N-terminal end of the single polypeptide contains the catalytic domain that removes the phosphate group from the target protein. It has been found that all of the enzymes in the protein tyrosine phosphatase family contain the consensus sequence:

-[I/V]HCXXGXXR[S/T]-

in their catalytic centre. Of particular importance here is the cysteine, as this cysteine is essential for catalysis. This cysteine is at position 215 in PTP 1B. It is also this cysteine that is able to be modified, so removing catalytic activity, by the inhibitor pervanadate and hydrogen peroxide, as discussed in depth in **Chapter 10**.

The C-terminal end of the polypeptide contains a signal to direct the protein to its intracellular location. PTP 1B appears to be localized to the cytoplasmic side of the endoplasmic reticulum, but removal of the C-terminal end relocates the enzyme into the cytosol. Thirty-five amino acids at the C-terminal end of the protein are particularly hydrophobic, and allow membrane association. Relocation of the enzyme also alters its functional activity and it is thought that regulation of the phosphatase may be associated with a regulation of its location in the cell. Alternative splicing of the mRNA for some PTPs might well be responsible for controlling their ultimate cellular location.

Upon this basic structure for intracellular PTPs, others also contain additional features (**Figure 6.12**). For example, PTP 1C also contains two protein

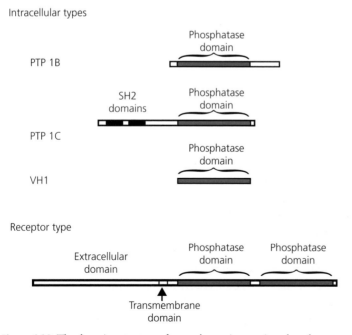

Figure 6.12 The domain structure of several protein tyrosine phosphatases.

binding SH2 domains and, as discussed above for the receptor tyrosine kinases, these domains will be used to locate and bind phosphotyrosine residues. Such localization would aid the catalytic action of the enzyme. The SH2 domains might also be crucial to maintaining the enzyme's intracellular location.

Another addition to the basic PTP 1B structure that has been identified is the addition of a domain which resembles cytoskeletal-associated proteins. For example, PTPMEG1 has an area that shares some homology to erythrocyte protein band 4.1. Again, this is probably important in the maintenance of intracellular location of the phosphatase. Tyrosine phosphorylation is thought to be important in the polymerization of tubulin, a main element of the cytoskeleton, and tyrosine phosphatases would be vital in this regulatory mechanism.

The smallest and simplest PTP identified is one that contains the phosphatase domain and nothing else (Figure 6.12). This protein, VH1, is coded for by the *vaccinia* virus. Interestingly, this phosphatase also shows activity to serine phosphorylated sites as well as tyrosine phosphorylated sites.

Just as PTP 1B can be seen as a model for the intracellular PTPs, CD45 can be used as a model for the receptor linked tyrosine phosphatases. This protein has an extracellular domain, a transmembrane domain and a domain containing phosphatase activity on the other side of the membrane, that is, in the cytoplasm (Figure 6.12). CD45 is actually a glycoprotein, with the N-terminal extracellular domain containing O-linked glycosylation.

Unlike the intracellular PTPs, the receptor linked family nearly all have two phosphatase domains (Figure 6.12), although exceptions to this have been found. However, using mutations of the amino acids thought to be crucial for the catalytic action of this phosphatase, mainly a conserved cysteine residue, it is thought that the second catalytic site is probably inactive in removing the phosphoryl group from tyrosines. However, it could have a function in bringing about the association of the enzyme with its substrate, in a similar manner to the SH2 domains seen elsewhere.

Tyrosine phosphatases, like the tyrosine kinases, are probably regulated by phosphorylation, although the precise mechanisms in many cases have yet to be identified. It may be that some phosphatases are constitutively active, ever ready to switch off pathways activated by kinases. Interestingly, reactive oxygen species, particularly hydrogen peroxide, have been found to inactivate these phosphatases, in particular PTP1B. Oxidation of the active site cysteine to a sulphenyl-amide group is thought to be responsible for the loss of activity of this enzyme. This is further discussed in Chapter 10.

Several inhibitors have been reported, which are useful in identifying the activity of tyrosine phosphatases. Molybdate, and particularly orthovanadate, are inhibitors of all tyrosine phosphatases, whereas other compounds have been used as diagnostic tools for certain PTPs. Such molecules include zinc ions, ethylenediaminetetraacetic acid (EDTA) and spermine.

6.7   **Other covalent modifications**

Although the discussion above has concentrated on the covalent addition of the phosphoryl group and its subsequent removal from a polypeptide as means of regulating activity, it is not the only covalent modification that is seen in the control of enzyme function. Many proteins are proteolytically cleaved, turning them from an inactive precursor into an active enzyme. The classical case here is in the blood clotting cascade, where one factor cleaves and activates another, which then goes on to cleave the next and so on, leading to great amplification of the initial signal. Such action however, unlike phosphorylation, is not readily reversible, as once the polypeptide has been cleaved it is not able to be rejoined. Examples of relevance to cell signalling here are the cleavage activation of protein kinase C (see above), and the cleavage and hence formation of many cytokines (see **Chapter 4**).

A more reversible covalent modification is adenylation. Here the adenylyl group (AMP) from ATP is added to a target tyrosine residue of the polypeptide, releasing inorganic pyrophosphate, the breakdown of which probably helps to drive the reaction. The reversal releases AMP, restoring the enzyme to its former condition. An example of this is seen in *E. coli*, where glutamine synthase in the cell's nitrogen metabolism is controlled in this way. Adenylation inhibits the enzyme's activity.

Many proteins are covalently modified by the addition of a large hydrophobic group, which allows the protein to become associated with a membrane, for example, a myristoyl group. This is a C_{14} fatty acid, which is added to the N-terminal end of the polypeptide. The donor for myristoylation is myristoyl CoA, whereas the enzyme that catalyzes the reaction is N-myristoyl transferase. For this reaction to take place, the target polypeptide must have a glycine residue at its N-terminal with a typical consensus sequence being:

-Gly-X-X-X-[Ser/Thr]-Y-Y-

where X can be a variety of amino acids, but Y is a basic amino acid.

Palmitoylation involves the addition of a palmitoyl group, again often allowing a protein to become membrane associated. The palmitoyl group is a C_{16} fatty acid, donated from palmitoyl CoA. It is usually added to a cysteine residue. An example of a protein that has undergone this type of modification is rhodopsin. Other covalent modifications of large hydrophobic groups include farnesylation, which is the addition of a farnesyl group, a C_{15} fatty acid. The donor in this case is farnesyl pyrophosphate. This process takes place at the C-terminal end of the polypeptide, where a consensus sequence of -Cys-Y-Y-X is found. Cysteine, again, is the target for the modification, Y would be aliphatic in nature, whereas X could be any amino acid. Typically, the -Y-Y-X sequence would be subsequently removed and the end of the polypeptide would also be methylated. An example of a signalling protein that is altered in this way is the monomeric G protein Ras (as discussed in **Chapter 7**).

The last modification of this type to be considered here is the addition of the C_{20} fatty acid geranylgeranyl, added in a process called geranylgeranylation. Again, like farnesylation, the pyrophosphate is the source of the fatty acid, and again it is attached to the C-terminal end of the polypeptide. Here, however, the consensus sequences needed at the C-terminus are -Cys-Cys, -Cys-Cys-X-X or -Cys-X-Cys. One or both cysteine groups could be modified. An example of a protein here is the monomeric G protein Rab (see Chapter 7).

Finally, species chemically referred to as reactive species, such as the reactive oxygen species (ROS) or reactive nitrogen species (RNS), can cause the modification of amino acids in proteins (discussed in more detail in Chapter 9). Oxidation of cysteine groups can lead to the formation of cysteine disulphides, which can then be re-reduced back to reverse the effect. In plants, such reversible reduction is used in the regulation of some enzyme activities, with thioredoxin being used as the reductant in the reaction. The subsequent re-reduction of the thioredoxin is light driven. In animal systems, such mechanisms have been proposed too. Alternatively, ROS might lead to the formation of oxidized derivatives of the cysteine thiol group, such as seen in the formation of the sulphenyl-amide group in the phosphatase PTP1B. Alterations to the amino acids in these ways leads to changes in the conformational shape of the protein, and usually in its activity. Such oxidized forms of the thiol may further react with glutathione, to form glutathionated forms of the protein, where perhaps the activity has been modified. RNS can likewise modify proteins, usually by a process referred to as nitrosylation. Here, the nitric oxide attacks the thiol group to form a –SNO group, again with ramifications for the three-dimensional structure of the protein and its activity. Alternatively, NO can react with tyrosine residues in a process called tyrosine nitration, again with ramifications for the functioning of the protein.

6.8 Ubiquitin-proteasome system

Protein structure

Despite the complex pictures often used to show protein structures, including the ones in this book, proteins do not have a static structure, rather there is a large dynamic nature to their shape. This must be remembered as it impacts on the reactions in which a protein can partake.

Although it sounds almost counter-intuitive, it has now become clear that an extremely important mechanism used for the control of cellular function is the degradation of proteins, that is, they are cleaved into several smaller pieces. Once again, this is a "one-way street" with little chance of being reversed after it has reached completion. Such systems have been studied for three decades, but the exact mechanisms involved are only recently coming to light, and, as they do, their importance is becoming more apparent.

The degradation of proteins involved in cell signalling is not random, but selective and specific, with proteins containing specific degradation signals. These signals are referred to as degrons. Often, the degradation of the proteins involves the mechanism known as the ubiquitin-proteasome system.

It should be remembered that a protein has a three-dimensional structure, and that the functioning of the protein is usually very reliant on the correct three-dimensional structure being maintained. Therefore, if proteins are damaged they may need to be removed and destroyed, as they can no longer carry out their function efficiently. One of the roles of the ubiquitin-proteasome system is to target such proteins, as cells cannot afford the luxury of having dysfunctional proteins present. However, proteins may need to be removed for other reasons too, and, as will be discussed in **Chapter 12**, several proteins are cleaved as part of the process of their control, or even to create a new function using a fragment of a larger protein. Either way, whether simply removal, or as part of a controlled signalling process, the proteins involved have to be identified, targeted and acted on.

Proteins for destruction are tagged with ubiquitin. Usually, the ubiquitin is attached through a lysine residue on the protein. However, the process is quite complex and involves a series of enzymes. Ubiquitin is attached firstly to an enzyme called Ub-activating enzyme (otherwise referred to as E1). This reaction requires ATP, so is energy dependent. The ubiquitin is then transferred to ubiquitin-conjugating enzyme (E2). Final transfer of the ubiquitin to the target protein destined for degradation, relies on the functioning of a third enzyme, called ubiquitin-protein ligase (E3).

There are two main types of E3, but the main difference is whether the ubiquitin is transferred directly from the E2 protein to the target protein, or whether the ubiquitin is transferred to the E3 protein first, and then moved to the target protein. E3 enzymes that contain a domain called a RING (or RING-like) domain will transfer the ubiquitin directly from the E2 to the protein. E3 enzymes that contain a HECT domain have the ubiquitin transferred from a cysteine residue on the E2 protein to a cysteine residue on the E3 protein, before it is transferred to the lysine of the target protein.

The end result of the addition of ubiquitin to proteins is that there is a chain of ubiquitin molecules added, in a process referred to as poly-ubiquitination (**Figure 6.13**), and hence the proteins are now marked for degradation by the next set of proteins in the pathway, as will be discussed below.

Proteins that need to be targeted by the ubiquitin system need to be recognized, and therefore common features of such proteins have been sought. Clearly, if the ubiquitin is added to a lysine group, then the protein has to have a lysine group available for enzymatic modification, and not just buried deep and inaccessible in the protein. But, there must be more than this. One of the determinants of whether a protein undergoes ubiquitin modification is the presence of certain characteristics at the N-terminal end, and this has been termed the N-end rule pathway. The presence of certain amino acids, or modifications of certain amino acids seems to be preferred. For example, protein synthesis starts with a methionine, so many proteins have a methionine at their N-terminal end when they are in a mature state, and this is not a cue for ubiquitin modification, as the protein would be in a functional state and the cell would therefore need to keep it. However, if

Ubiquitin A 76 amino acid peptide that can reside as a soluble monomer in cells, before being attached to other proteins. Addition of ubiquitin to a protein is a process called ubiquitinylation or ubiquitination.

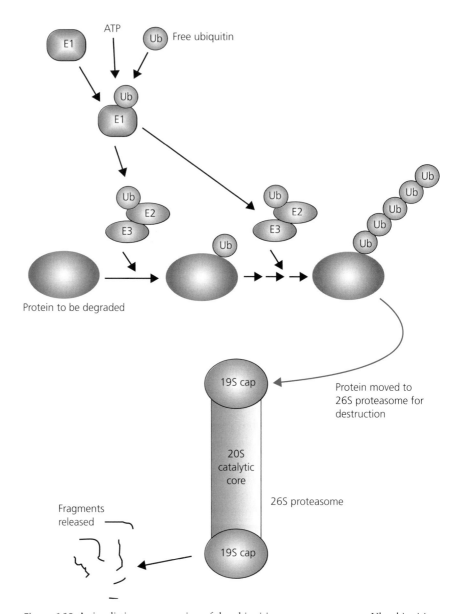

Figure 6.13 A simplistic representation of the ubiquitin-proteasome system. Ub: ubiquitin.

the protein is cleaved, the resulting peptide would have a new N-terminal amino acid, and may be recognized as a signal for ubiquitin modification, and hence ultimate peptide degradation. Some N-terminal end amino acids are modified before they are recognized. Examples here would be asparagine or glutamine, which can be modified and then have an arginine residue added. This results in formation of a degradation signal. In a similar way, cysteine can be oxidized or nitrosylated (see **Chapter 10**) before the addition of an arginine, and thus create a signal for ubiquitin modification of the protein.

However, the N-end rule pathway is not the whole story. Other signals for protein degradation include phosphorylation, as typified by the control of the cyclin and cyclin-dependent kinase (CDK)-dependent pathways during the cell cycle. Other modifications that may mark a protein for degradation via the ubiquitin-proteasome system, include glycosylation and hydroxylation of specific proline residues. Therefore, it can be seen that the identification of individual proteins for degradation by this system has a wide range of mechanisms, but is highly specific and well controlled, as would be expected, and indeed needed.

With the correct proteins tagged and marked out for degradation, now with their multiple ubiquitin peptides added, they are recognizable by the next part of the pathway. This final part of the machinery requires the action of a protein complex known as the 26S proteasome. This is a very large structure of approximately 50 proteins. In the middle is a 20S catalytic core, which has a barrel shape. At one or both ends are 19S structures, which have a role in the regulation of the system. The 19S caps have ATPase activity, and so the process is partly driven by the hydrolysis of ATP. Proteins that are to be cleaved are unfolded, a process which requires such ATP hydrolysis, and then the polypeptide chain is fed into the core of the 20S where it is cleaved into much smaller fragments, which are then released out into the cytoplasm (**Figure 6.13**).

As in most cell signalling pathways, there needs to be the opportunity for reversal of the system. Clearly, once a protein is cleaved and destroyed it is hard to reverse this process, but with ubiquitinylation and the tagging of the protein reversal has been noted. Once a protein has been covalently modified by the addition of ubiquitin, the ubiquitin can be removed by enzymes known as de-ubiquitylating enzymes, or DUBs. Free ubiquitin created can be re-used for another round of ubiquitinylation.

■ As one of the important functions of proteasomes is to control the cell cycle, inhibitors of proteasomes are thought to be good for cancer therapy development.

■ Summary

- Most signalling pathways involve phosphorylation, it is an extremely important biochemical mechanism used in the control of proteins (**Figure 6.14**).

- Phosphorylation is the addition of one or more phosphoryl groups to certain amino acid side groups on a polypeptide chain.

- Dephosphorylation is the removal of one or more phosphoryl groups from certain amino acid side groups on a polypeptide chain.

- Phosphorylation events are either classified as serine/threonine phosphorylations or tyrosine phosphorylations, depending on the target amino acids.

- The enzymes that add the phosphoryl groups are known as kinases, with the reverse reaction being catalyzed by phosphatases.

- The human genome probably contains over 1000 genes for kinases and phosphatases, and other genomes also encode numerous examples.

- Serine/threonine kinases include protein kinase C, cAMP-dependent protein kinase, Ca^{2+}/calmodulin protein kinase and cGMP-dependent protein kinase, which in general have wide substrate specificities.

- cAMP-dependent protein kinase exists in an inactive tetrameric state, which dissociates on activation, that is on binding to cAMP, to release two catalytic subunits.

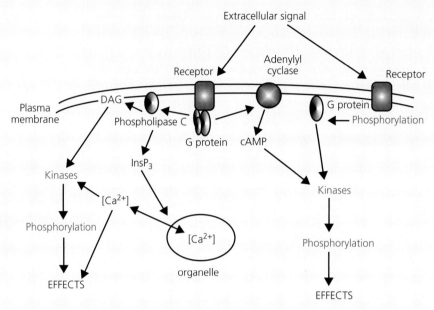

Figure 6.14 A schematic representation of the central role of phosphorylation in cell signalling.

- Protein kinase C, which is activated by the presence of Ca^{2+} ions, is in fact two families of protein kinases, each family containing several members, which probably have subtle differences in both their activation and specificity.

- Another important group of Ca^{2+} controlled kinases are the Ca^{2+}/calmodulin-dependent kinases, again with several isoforms existing.

- The other major class of kinases is the tyrosine kinases, including receptor kinases, which are membrane-bound receptors, which on binding to their ligand result commonly in autophosphorylation of the receptor along with phosphorylation of intracellular proteins.

- Other tyrosine kinases include the soluble Janus kinase (JAKs) family.

- A group of sequential kinases are those of the MAP kinase cascades.

- MAPK pathways lead commonly from receptors that bind growth factors, or indeed insulin, and may involve the G protein Ras and the kinase Raf at the early part of the transduction pathway.

- The two component systems of prokaryotes and some higher organisms are unusual in that they involve the attachment of phosphoryl groups to histidine and aspartate, and are more commonly referred to as the histidine kinases. Histidine kinases fall into two main groups, the hybrid and non-hybrid histidine kinases. Such systems may be more widespread than first thought, although probably not found in mammals.

- The reversal of kinase action is that of phosphatases, and again these are usually either serine/threonine-specific or tyrosine-specific, the latter being either cytosolic or receptor-linked. Again several isoforms of each are known to exist.

- The control of phosphorylation and dephosphorylation has already been identified as a major target for drug development by the pharmaceutical industry, and no doubt many of our future medicines and agro-chemicals will be targeted against the proteins, and their homologues.

- As well as phosphorylation, proteins may undergo other covalent modifications. Some of these add hydrophobic groups to the side chains of proteins and so bestow on the proteins a hydrophobic nature, allowing them to become associated with membranes despite the fact that their amino acid sequence would deem them to be soluble.

- The ubiquitin-proteasome system is an important mechanism based on the covalent modification of a protein, but results in its ultimate destruction and hence usually removal from the signalling pathway. However, such a system can also create small signalling fragments.

→ Further reading

General

Hanks, S.K. and Hunter, T. (1995) Protein kinases 6. The eukaryotic protein kinase superfamily: kinase (catalytic) domain structure and classification. *FASEB Journal*, 9, 576–596.

Hanks, S.K. and Quinn, A.M. (1991) Protein kinase catalytic domain sequence database: identification of conserved features of primary structure and classification of family members. *Methods in Enzymology*, 200, 38–62.

Hunter, T.(1991) Protein kinase classification. *Methods in Enzymology*, 200, 3–37.

Hunter, T. (1995) Protein kinases and phosphatases: the yin and yang of protein phosphorylation and signalling. *Cell*, 80, 225–236.

Kholodenko, B.N. (2003) Four dimensional organisation of protein kinase signaling cascades: The roles of diffusion, endocytosis and molecular motors. *Journal of Experimental Biology*, 206, 2073–2082.

Knighton, D.R., Zheng, J., Ten Eyck, L.F., Ashford, V.A., Xuong, N.-H., Taylor, S.S. and Sowadski, J.M. (1991) Crystal structure of the catalytic subunit of cyclic adenosine monophosphate-dependent protein kinase. *Science*, 253, 40–414.

Mullard, A. (2007) Mechanisms of disease: Kinases out of control - brake line cut! *Nature Reviews Molecular Cell Biology*, 8, 854 [commentary on another paper: Chen *et al.* 2007 – see references at the end of Chapter 4].

Sun, H. and Tonks, N.K. (1994) The co-ordinated action of protein tyrosine phosphatases and kinases in cell signalling. *Trends in Biochemical Sciences*, 19, 480–485.

cAMP kinase

Hanks, S.K., Quinn, A.M. and Hunter, T. (1988) The protein kinase family: conserved features and deduced phylogeny of the catalytic domains. *Science*, 241, 42–52.

Øgreid, D. and Døskeland, S.O. (1981) The kinetics of the interaction between cyclic AMP and the regulatory moiety of protein kinase II. *FEBS Letters*, 129, 282–286.

Smith, S.B., White, H.D., Siegel, J. and Krebs, E.G. (1981) Cyclic AMP-dependent protein kinase I; cyclic nucleotide binding, structural changes and release of the catalytic subunits. *Proceedings of the National Academy of Sciences USA*, 78, 1591–1595.

Taylor, S.S., Buechler, J.A. and Knighton, D. (1990) in *Peptides and Protein Phosphorylation*. (ed. B.E. Kemp), pp.1–41. CRC Press, Florida USA.

cGMP-dependent protein kinase

Lincoln, T.M. and Corbin, J.D. (1983) Characterisation and biological role of the cGMP dependent protein kinase. *Advances Cyclic Nucleotide Research*, 15, 139–192.

Takio, K., Wade, R.D., Smith, S.B., Krebs, E.G., Walsh, K.A. and Titani, K. (1984) Guanosine cyclic 3′-5′-phosphate dependent protein kinase: a dimeric protein homologous with 2 separate protein families. *Biochemistry*, 23, 4207–4218.

Protein kinase C

Coussens, L., Parker, P.J., Rhee, L., Yang-Feng, T.L., Chen, E., Waterfield, M.D., Francke, U. and Ullrich, A. (1986) Multiple, distinct forms of bovine and human protein kinase C suggest diversity in cellular signalling pathways. *Science*, 233, 859–866.

House, C. and Kemp, B.E. (1987) Protein kinase C contains a pseudosubstrate prototype in its regulatory domain. *Science*, 238, 1726–1728.

Nishizuka, Y. (1988) The molecular heterogeneity of protein kinase C and its implication for cellular regulation. *Nature*, 334, 661–665.

Ono, Y., Fujii, T., Ogita, K., Kikkawa, U., Igarashi, K. and Niskizuka, Y.(1988) The structure, expression and properties of additional members of the protein kinase C family. *Journal of Biological Chemistry*, 263, 6927–6932.

Ca²⁺-calmodulin dependent protein kinases

Anderson, K.A. and Kane, C.D. (1998) Ca²⁺/calmodulin-dependent protein kinase IV and calcium signaling. *Biometals*, 11, 331–343.

Hanson, P.I. and Schulman, H. (1992) Neuronal Ca²⁺/calmodulin-dependent protein kinases. *Annual Review of Biochemistry*, 61, 559–601.

Soderling, T.R. (1999) The Ca²⁺-calmodulin protein kinase cascade. *Trends in Biochemical Sciences*, 24, 232–236.

G protein-coupled receptor kinases

Debburman, S.K., Ptasienski, J., Benovic, J.L. and Hosey, M.M. (1996) G protein coupled receptor kinase GRK2 is a phospholipid dependent enzyme that can be conditionally activated by G protein beta/gamma subunits. *Journal of Biological Chemistry*, 271, 22552–22562.

Lefkowitz, R.J. (1993) G-protein-coupled receptor kinases. *Cell*, 74, 409–412.

Palczewski, K. and Benovic, J.L. (1991) G-protein-coupled receptor kinases. *Trends in Biochemical Sciences*, 16, 387–391.

Winstel, R., Freund, S., Krasel, C., Hoppe, E. and Lohse, M.J. (1996) Protein kinase crosstalk: membrane targetting of the beta adrenergic receptor kinase by protein kinase C. *Proceedings of the National Academy of Sciences USA*, 93, 2105–2109.

Protein kinase B

Dummler, B. and Hemmings, B.A. (2007) Physiological roles of PKB/Akt isoforms in development and disease. *Biochemical Society Transactions* 35, 231–235.

Hajduch, E., Litherland, G.J. and Hundal, H.S. (2001) Protein kinase B (PKB/Akt) – a key regulator of glucose transport? *FEBS Letters*, 492, 199–203.

Scheid, M.P. and Woodgett, J.R. (2003) Unravelling the activation mechanisms of protein kinase B/Akt. *FEBS Letters*, 546, 108–112.

AMP-activated protein kinase

Arad, M., Seidman, C.E. and Seidman, J.G. (2007) AMP-activated protein kinase in the heart: role during health and disease. *Circulation Research*, 100, 474–488.

Dyck, J.R.B., Gao, G., Widmer, J., Stapleton, D., Fernandez, C.S., Kemp, B.E. and Witters, L.A. (1996) Regulation of 5'-AMP-activated protein kinase activity by the noncatalytic β and γ subunits. *Journal of Biochemical Chemistry*, 271, 17798–17803.

Gao, G., Fernandez, S., Stapleton, D., Auster, A.S., Widmer, J., Dyck, J.R.B., Kemp, B.E. and Witters, L.A. (1996) Non-catalytic β and γ-subunit isoforms of the 5'-AMP-activated protein kinase. *Journal of Biological Chemistry*, 271, 8675–8681.

Towler, M.C. and Hardie, D.G. (2007) AMP-activated protein kinase in metabolic control and insulin signaling. *Circulation Research*, 100, 328–341.

Haem regulated protein kinase

Kozak, M. (1992) Regulation of translation in eukaryotic systems. *Annual Review Cell Biology*, 8, 197–225.

Proud, C.G. (1992) Protein phosphorylation in translational control. *Current Topics in Cell Regulation*, 32, 243–369.

Plant specific serine/threonine kinases

Coughlan, S.J. and Hind, G. (1986) Purification and characterisation of a membrane-bound protein kinase from spinach thylakoids. *Journal of Biological Chemistry*, 261, 11378–11385.

Stone, J.M. and Walker, J.C. (1995) Plant protein kinase families and signal transduction. *Plant Physiology*, 108, 451–457.

Tyrosine kinases

Bellot, F., Crumley, G., Kaplow, J. M., Schessinger, J., Jaye, M. and Dionne, C.A'. (1991) Ligand-induced transphosphorylation between different FGF receptors. *EMBO Journal*, 10, 2849–2854.

Fantl, W.J., Johnson, D.E. and Williams, L.T. (1993) Signalling by receptor tyrosine kinases. *Annual Review of Biochemistry*, 62, 453–481.

Kawamura, M., McVicar, D.W., Johnston, J.A., Blake, T.B., Chen, Y.-Q, Lal, B.K., Lloyd, A.R., Kelvin, D.J., Staples, J.E., Ortaldo, J.R. and O'Shea, J.J. (1994) Molecular cloning of L-Jak, a Janus family protein-tyrosine kinase expressed in natural killer cells and activated leukocytes. *Proceedings of the National Academy of Sciences USA*, 91, 6374–6378 [known now as JAK3].

Murray, P.J. (2007) The JAK-STAT signaling pathway: input and output integration. *Journal of Immunology*, 178, 2623–2629.

Schindler, C., Levy, D.E. and Decker, T. (2007) JAK-STAT signalling: From interferons to cytokines. *Journal of Biological Chemistry*, 282, 20059–20063.

Schlessinger, J and Ullrich, A. (1992) Growth factor signalling by receptor tyrosine kinases. *Neuron*, 9, 383–391.

Wegrzyn, J. *et al.* (27 authors) (2009) Function of mitochondrial Stat3 in cellular respiration. *Science*, 323, 793–797.

MAP kinases

Blenis, J. (1993) Signal transduction via the MAP kinases: Proceed at your own RSK. *Proceedings of the National Academy of Sciences USA*, 90, 5889–5892.

Blumer, K.J. and Johnson, G.L. (1994) Diversity in function and regulation of MAP kinase pathways. *Trends in Biochemical Sciences*, 19, 236–240.

Herskowitz, I. (1995) MAP kinase pathways in yeast: For mating and more. *Cell*, 80, 187–197.

Marshall, C.J. (1994) MAP kinase kinase kinase, MAP kinase kinase and MAP kinase. *Current Opinion Genetic Development*, 4, 82–89.

Nebreda, A.R. (1994) Inactivation of MAP kinases. *Trends in Biochemical Sciences*, 19, 1–2.

Šamaj, J., Baluška, F. and Hirt, H. (2004) From signal to cell polarity: mitogen-activated protein kinases as sensors and effectors of cytoskeleton dynamicity. *Journal of Experimental Botany*, 55, 189–198 [contains a good illustrated comparison of MAP kinases in mammals, yeast and plants].

Histidine phosphorylation

Besant, P.G., Tan, E. and Attwood, P.V. (2003) Mammalian protein histidine kinases. *International Journal Biochemistry and Cell Biology*, 35, 297–309.

Inouye, M. and Dutta, R. (2003) *Histidine Kinases in Signal Transduction*. Academic Press, London. ISBN 0123724848.

Kofoid, E.C. and Parkinson, J.S. (1992) Communication modules in bacterial signalling proteins. *Annual Review of Genetics*, 26, 71–112.

Volz, K. and Matsumura, P. (1991) Crystal structure of *Escherichia coli* CheY refined at 1.7Å resolution. *Journal of Biological Chemistry*, 266, 15511–15519.

Phosphatases

Barford, D. (2001) The mechanism of protein kinase regulation by protein phosphatases. *Biochemical Society Transactions*, 29, 385–391.

Cohen, P.T.W., Brewis, N.D., Hughes, V. and Mann, D.J. (1990) Protein serine/threonine phosphatases: an expanding family. *FEBS Letters*, 268, 355–359.

Gang, L. (2003) Protein tyrosine phosphatase 1B inhibition: opportunities and challenges. *Current Medicinal Chemistry*, 10, 1407–1421.

Guan, K.L., Broyles, S.S. and Dixon, J.E. (1991) A Tyr/Ser protein phosphatase encoded by vaccinia virus. *Nature*, 350, 359–362.

Moorhead, G. (Ed.) (2007) *Protein Phosphatase Protocols: Methods in Molecular Biology Series*, Vol. 365, Humana Press, Totowa, New Jersey, ISBN 978-1-58829-711-2.

Tonks, N.K. (2003) PTP1B: from the sidelines to the front lines! *FEBS Letters*, 546, 140–148.

Virshup, D.M. (2000) Protein phosphatase 2A: A panoply of enzymes. *Current Opinion in Cell Biology*, 12, 180–185.

Ubiquitin-proteasome system

Adams, J. (2004) The development of proteasome inhibitors as anticancer drugs. *Cancer Cell*, 5, 417–421.

Ang, X.L. and Harper, J.W. (2004) Interwoven ubiquitination oscillators and control of cell cycle transitions. *Science STKE* 2004, pe31.

Grabbe, C. and Dikic, I. (2008) Going global on ubiquitin. *Science*, 322, 872–873.

Ravid, T. and Hochstrasser, M. (2008) Diversity of degradation signals in the ubiquitin-proteasome system. *Nature Reviews Molecular Cell Biology*, 9, 679–690.

7 Cyclic nucleotides, cyclases and G proteins

Once signalling molecules have been perceived by a cell, intracellular cascades of events usually ensue. One of the most common ways for the signal to be propagated in a cell is by the alteration of the concentration of other molecules. In this chapter, alteration of a class of molecules called cyclic nucleotides is discussed. These are crucial intermediates between the receptor and phosphorylation events in many cases, and control many important cellular processes; for example, glycogen metabolism and hence the availability of glucose. They are also mediators of NO signalling, discussed further in **Chapter 10**, and importantly allow for a large amplification of a signal inside a cell.

In some cases, production of cyclic nucleotides is controlled by a class of proteins known as G proteins. These have been dubbed "molecular switches", which in many ways succinctly sums up their action: they can exist in an "on" state or an "off" state. However, they control far more than nucleotide metabolism, and are found in cascades between receptor activation and phosphorylation cascades, for example. Many isoforms of both classes of these proteins have been identified in a wide range of organisms, highlighting their integral signalling role. Furthermore, some forms of cancerous cell growth have been attributed to G proteins being stuck in the "on" state.

7.1 Introduction

The reception of a signal at the cell surface is a crucial event in many cell signalling cascades, as discussed in the chapters above. However, once the cell has received an external signal, such as the binding of a ligand to a cell surface receptor, a response inside the cell needs to rapidly follow. Such responses often involve molecules that have been referred to as "second messengers", as they are, or were thought to be, the second signalling component in the chain of action. However, the term second messenger, although widely used, is somewhat misleading. Many of the second messengers are not in fact the second component at all, and the term is rather loosely used to refer to small molecules that are formed inside the cell and diffuse to their point of action, where they relay their message to the next part of the signalling cascade.

In this chapter, the role of one group of messenger molecules, the cyclic nucleotides, is discussed, along with the mechanisms that are involved in the control of their production.

7.2 cAMP

One of the major groups of "second messengers" is the cyclic nucleotides, and in particular cyclic adenosine monophosphate or adenosine 3′,5′-cyclic monophosphate to give it its more chemical name. It is understandable that it is usually simply referred to as cyclic AMP, or more commonly cAMP. It is, in fact, the intracellular concentration of cAMP that constitutes the signal here. Some of the hormones that have their cellular effects mediated by cAMP are extremely important and widely found, and include epinephrine and glucagon along with those listed in Table 7.1

Table 7.1 Some signalling molecules that act via cAMP.

Hormone	Major tissues affected
Adrenocorticotropic hormone	Adrenal cortex, fat
Epinephrine (adrenaline)	Muscle, fat, heart
Glucagon	Liver, fat
Luteinising	Ovaries
Parathormone	Bone
Thyroid-stimulating hormone	Thyroid gland, fat
Vasopressin	Kidney

cAMP concentrations

Often it is assumed that there is no cAMP in cells until it is produced, and then after the signal has been terminated that the concentration of cAMP returns to zero. In reality, the concentration probably fluctuates from a low level to a higher level, and back down again, in a transient manner.

cAMP is produced at the plasma membrane of the cell by the enzyme adenylyl cyclase, alternatively known as adenylate cyclase (originally known as adenyl cyclase). The cAMP is produced on the inner side of the membrane and is released into the cytosol, where it can diffuse and act on the next part of the signal transduction pathway. For example, the next response might be the binding of cAMP to cAMP-dependent protein kinase, so increasing the phosphorylation of certain proteins, with the concomitant alteration of their activity (for a discussion of the kinase see Chapter 6). An excellent example of the role of cAMP in a signalling pathway is the control of glycogen metabolism, as illustrated in Figure 7.1. Here, an extracellular signal such as epinephrine binds to a membrane receptor, which in turn activates a G protein (the action and role of G proteins will be discussed below). The G protein activates adenylyl cyclase to produce cAMP. The increase in intracellular cAMP subsequently activates cAMP-dependent protein kinase, which phosphorylates phosphorylase kinase. This latter enzyme phosphorylases, and in doing so activates, phosphorylase, which breaks down glycogen, releasing glucose for energy metabolism. This, of course, was the action within the cell that the arrival of the epinephrine at the cell surface was sent to provoke.

Futile cycle

As discussed in Chapter 5, if the product of a reaction is simply converted back again to its original form, for example, glycogen being broken down to form glucose, and then the glucose being converted back to glycogen, then such cycling of chemicals is referred to as a futile cycle. Clearly, an event like this is a waste of energy for the cell and needs to be avoided.

However, if the epinephrine was simply to result in the increase in the glucose level then this new glucose would be rapidly converted to glycogen again, and a futile cycle would be created. To prevent this, cAMP-dependent protein kinase also phosphorylates and inhibits glycogen synthase, reducing the synthesis of glycogen in the same cell, so allowing the glucose to be used for the purpose for which the epinephrine was sent in the first place.

As with all signals, the cAMP message needs to be reversed. Cyclic nucleotides like cAMP are rapidly broken down by a family of enzymes called the phosphodiesterases, and once removed from the cytosol of the cell, no further

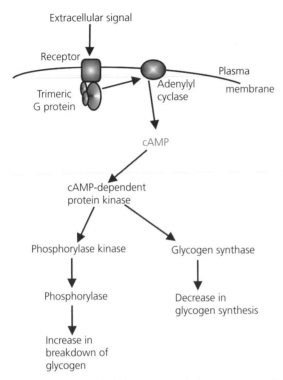

Figure 7.1 The coordinated control of glycogen metabolism. Hormones binding to receptors activate a G protein (G$_s$), which activates adenylyl cyclase. The rise in cAMP activates cAMP-dependent protein kinase, leading to the phosphorylation of phosphorylase kinase and glycogen synthase, causing inhibition of the latter. Phosphorylase kinase phosphorylation of phosphorylase causes an increase in glycogen breakdown. Therefore, the overall result is an inhibition of glycogen synthesis and an increase in glycogen breakdown.

signalling will take place. Therefore, an extremely simplified scheme of how cAMP, and other related chemicals might signal is shown in **Figure 7.2**.

With this brief discussion of the role of cAMP, it can be seen that it is ideal as a signalling molecule. cAMP is rapidly made by an enzyme, is small and readily diffusible, and is readily and quickly broken down by another enzyme, phosphodiesterase. Therefore, the signal can be rapid and reversed. Furthermore, the production and use of cAMP in cells, like most signalling routes, may lead to massive amplification of the signal. The activation of one receptor can cause the activation of several adenylyl cyclase enzymes each leading to the production of hundreds or thousands of cAMP molecules, each of which can go on to have an effect.

Although the majority of the effects of cAMP are mediated through the kinase cAMP-dependent protein kinase (PKA) as seen above, cAMP can also have more direct effects. For example, in *E. coli* it can act through cAMP receptor protein, or CRP, encoded by the *crp* gene. This is alternatively known as catabolite gene activator protein (CAP); a dimer of 22 kDa subunits. On an

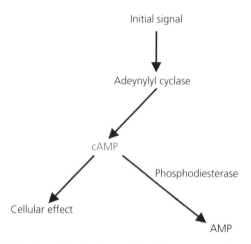

Figure 7.2 A simplified scheme showing how cAMP fits into a signalling pathway.

increase in intracellular cAMP concentration, cAMP binds to CRP to form a cAMP–CRP (or cAMP–CAP) complex, which binds to DNA and alters gene expression of several genes, including those of the lactose and arabinose operons.

In the majority of cases, cAMP is seen as an extremely important intracellular signal. However, cAMP has also been seen as an extracellular signal, acting as a type of hormone in the slime mould *Dictyostelium*, where cAMP is used as a signal between cells, controlling cellular aggregation and differentiation.

7.3 Adenylyl cyclase

The enzyme that produces cAMP is adenylyl cyclase. It is also referred to as adenylate cyclase, and in the past as adenyl cyclase. Therefore, to understand how cAMP fits into signalling pathways, it is necessary to discuss the structure, function and control of this important enzyme.

In mammals, adenylyl cyclase is a single polypeptide that resides in the plasma membranes of cells. It catalyzes production of cAMP from the ever present ATP. The 3′-OH ribose group of the ATP attacks the α-phosphoryl group resulting in a cyclization of the molecule. The by-product from the reaction is inorganic pyrophosphate, which is itself broken down by an enzyme pyrophosphatase. The energy released from this latter reaction probably helps to drive the cyclization reaction (Figure 7.3).

As with many of the proteins involved in signal transduction, there are various isoforms of the enzyme adenylyl cyclase. Their cellular location is usually in the plasma membrane, as with the mammalian ones which are integral in that membrane, but other forms have been found to be peripheral plasma membrane proteins, for example in the yeast *S. cerevisiae* or in *E. coli*,

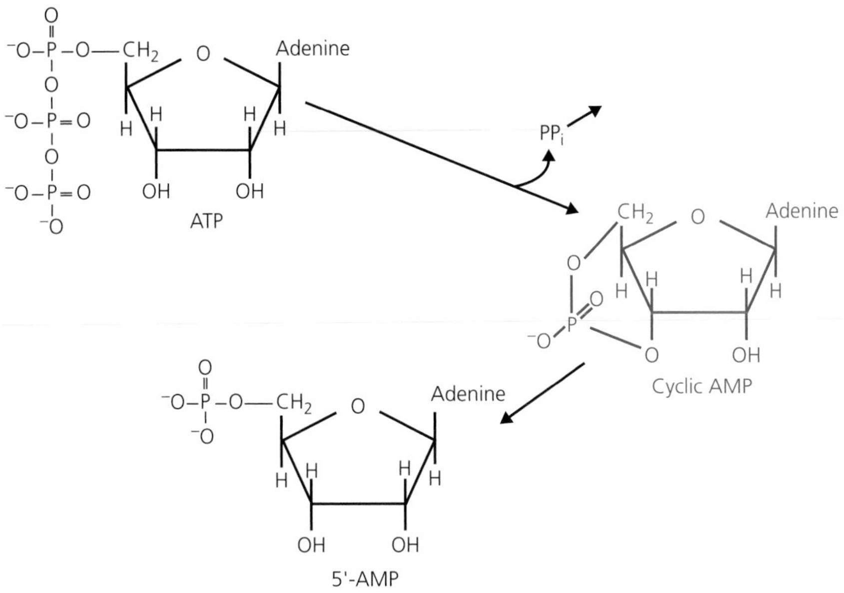

Figure 7.3 The production and hydrolysis of cAMP.

and even soluble forms have been reported in some bacteria. A soluble form also exists in mammalian sperm, and probably in many other cells and tissues in mammals.

At least nine isoforms of the plasma membrane adenylyl cyclase have been identified in mammals (types I–IX: see **Table 7.2**), but they all share a similar structural topology. In general, the proteins contain two clusters of six transmembrane spanning highly hydrophobic domains, which separate two catalytic domains on the cytoplasmic side of the membrane (**Figure 7.4**).

It, of course, makes sense to have the catalytic areas on the cytoplasmic side, as it is here that the product, cAMP, needs to be created to have its desired effect. Some areas of the two catalytic domains have been shown to be well conserved across the cyclases studied, and interestingly they also show

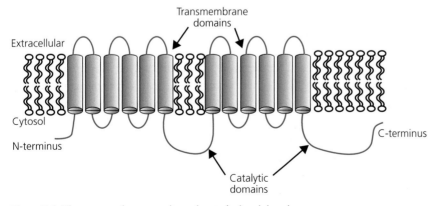

Figure 7.4 The proposed structural topology of adenylyl cyclase.

homology to regions of membrane bound guanylyl cyclases. Of course, in this latter enzyme the catalysis is very similar, so a similar catalytic mechanism and structure might be expected. It is probable that both these conserved regions in the catalytic domains of adenylyl cyclase, called C_{1a} and C_{2a}, are responsible for the activity of the enzyme, as it has been found that point mutations in either region are extremely inhibitory.

Studies have concluded that the enzyme has a molecular weight of 200–250 kDa, but the actual molecular weight is approximately 120 kDa, begging the question as to whether the enzyme exists as a dimer in the membrane. The extra apparent molecular weight may come from its association with other proteins, such as G proteins or even receptors.

All enzymes that are involved in signal transduction need to be carefully controlled, and adenylyl cyclase is no different. One of the most important controlling mechanisms is through the action of G proteins, and the next section will be dedicated to that. However, for experimental purposes, adenylyl cyclases can be non-competitively inhibited by Mg^{2+}/ATP analogues such as 3′-AMP and it can be activated by a lipid soluble diterpene, forskolin. It is presumed that the hydrophobic domains of the enzyme are the target for forskolin. In fact, isoforms without the six transmembrane spans seem to be insensitive to forskolin.

In *Dictyostelium*, the mould that uses cAMP as an extracellular signal, one of the forms of adenylyl cyclase conforms to the mammalian model, although another form differs in only having one transmembrane spanning region. A hypothesis has also been put forward that the hydrophobic domains might act as a channel across the membrane, and this is seen as important in cells that excrete cAMP. However, little evidence for the adenylyl cyclase acting as a transporter has been found and the heterogeneity of the sequences in these regions between different isoforms would suggest that this is not a role of the enzyme.

7.4 Adenylyl cyclase control and the role of G proteins

When a hormone binds to the relevant receptor on the cell surface, activation of the enzyme adenylyl cyclase is not a direct process, but rather, requires the use of other proteins, and furthermore, it was discovered that breakdown of GTP was also involved. The other proteins here are in fact guanyl nucleotide binding proteins or G proteins, and receptors that use this mechanism are referred to as G protein-coupled receptors. In this case, activation of the receptor causes the activation of the G protein, which regulates the activity of adenylyl cyclase. Some extracellular signalling molecules that bind to G protein-coupled receptors are discussed in Chapter 5 (see Table 5.1).

An example of a signalling pathway involving the receptor/G protein/adenylyl cyclase pathway was discussed above (see Figure 7.1), in particular the control of glycogen metabolism. However, the action of a G protein on the activity of adenylyl cyclase can be either stimulatory or inhibitory, depending on the G protein involved. If the result of the G protein activation is a stimulation of adenylyl cyclase activity, the G protein is known as stimulatory G protein or G_s, but if it is inhibitory the G protein is referred to as G_i.

G proteins at their simplest can be seen as molecular switches, that is, they can exist in an "on" state and in an "off" state, and can toggle between the two forms. Here, with their role in controlling adenylyl cyclase, they associate with the receptor in their "off", inactive, state, and then conformational changes induced in the receptor by ligand binding induce changes in the G protein so that it adopts the "on" state. In this new state, it can therefore transmit its message onto the next component in the signalling cascades, here, the adenylyl cyclase.

G proteins can be grouped into two main classes, those that consist of a single polypeptide, the monomeric family, and those that have three subunits, the heterotrimeric G proteins. To consider the control of adenylyl cyclase, the heterotrimeric G proteins will be discussed. In the inactive state, these G proteins exist as a complex of three polypeptides, α, β and γ. α has a molecular weight of approximately 45 kDa, β approximately 35 kDa, whereas γ has a molecular weight of only 7 kDa.

As discussed above, the G proteins are guanyl nucleotide binding proteins, and in the inactive state the G protein has GDP bound to a single GDP binding site on the $G\alpha$ subunit. On activation of the receptor, the GDP on the G protein is released in exchange for GTP. This causes the breakdown of the G protein complex into a free α subunit (which has bound to it a GTP), and a β/γ complex that does not dissociate further. The free $G\alpha$-GTP subunit diffuses to the adenylyl cyclase, where it binds and causes activation (in the case of G_s), with the result of the subsequent release of cAMP (**Figure 7.5**).

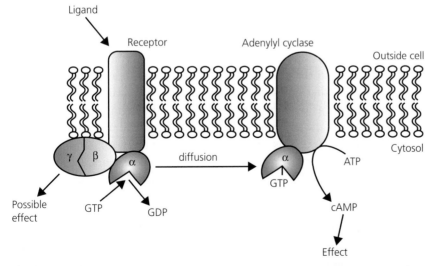

Figure 7.5 The regulation of adenylyl cyclase by the activation of a trimeric G protein by a G protein-coupled receptor.

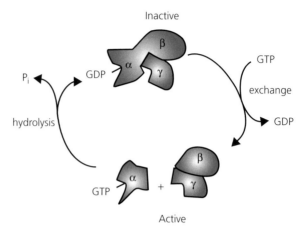

Figure 7.6 The G protein cycle: activation by nucleotide exchange and subunit dissociation with deactivation by nucleotide hydrolysis and subunit re-association.

Any signal that is turned on needs to be turned off again. Here, the deactivation of the G protein is achieved by the breakdown of GTP back to GDP on the α subunit. This breakdown of GTP is catalyzed by an intrinsic GTPase activity, which resides on the α subunit itself. This GTPase activity is activated by the complexing of the α subunit to adenylyl cyclase, and, once it has converted the GTP back to GDP, the α subunit subsequently releases from the adenylyl cyclase and reforms the original complex with the β/γ subunits. Therefore, it can be seen that the G protein can undergo a cycle of activation and deactivation, with the former involving the exchange of GTP and GDP along with dissociation of the complex, and the latter involving the catalytic cleavage of GTP back to GDP, and the re-association of the complex (Figure 7.6).

Certain oncogenes have been shown to code for G proteins that contain a defect in their intrinsic GTPase activity. Such proteins once bound to GTP are unable to return to the inactive state, causing the cell to continue to receive the "on" signal, even in the absence of receptor binding to a ligand. A heterotrimeric G protein that had a lack of intrinsic GTPase activity would only be able to go halfway around the G protein cycle as shown in Figure 7.7, and be left permanently in the active state. A similar effect can be emulated *in vitro* by the addition of GTPγ-S. This analogue of GTP contains a sulphur in the bond that is usually broken down by the GTPase when converting GTP to GDP. Hence, GTPγ-S is not able to be broken down by the GTPase, and again the α subunit will remain in an active state. The importance of turning off the G protein signal is emphasized by the disease cholera. The toxin released from the bacterium *Vibrio cholerae* inactivates the α subunit of these G proteins, in particular G_s. Cholera toxin, otherwise known as choleragen, is an oligomeric enzyme, consisting of a complex of an A_1 subunit of 25 kDa, linked to an A_2 subunit of 5.5 kDa by a disulphide bond, along with five B subunits of 16 kDa.

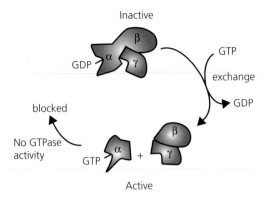

Figure 7.7 Blockage of the G protein cycle causes permanent activation of the G protein.

The A_1 subunit of the enzyme uses NAD^+ as a substrate, and catalyzes the transfer of ADP-ribose to a specific arginine residue of the α subunit of G_s, destroying the α subunit's intrinsic GTPase activity. The efficient modification of the $G_s\alpha$ requires the presence of another G protein called ADP-ribosylation factor (ARF), which is a subgroup of the Ras family of proteins (see below for discussions of Ras). Once modified, $G_s\alpha$ has no GTPase activity, cannot convert its GTP back to GDP, and is permanently active. Therefore, it continues to activate adenylyl cyclase, which is its target in the signalling cascade, and the subsequent loss of cAMP control in the gut epithelial cells leads to movement of Na^+ and water into the intestine causing resultant diarrhoea. It is worth noting here that the $G_s\alpha$ subunit can also be activated by the presence of aluminium tetrafluoride (AlF_4^-) if together with Mg^{2+}.

Turning off the α subunit of the G proteins may also be accelerated, in fact by the action of another group of proteins, the regulators of G protein signalling (RGS). These proteins contain a domain that promotes the GTPase activity of the $G\alpha$, so ensuring that the signal is terminated quickly. These can, therefore, be referred to as GTPase activating proteins (GAPs), as will be discussed for the monomeric G proteins below (see Section 7.7). All these RGS proteins seem to share a common domain of around 120 amino acids that bestows on them this function.

As mentioned above, G proteins are not only able to cause activation of adenylyl cyclase, as seen with G_s, but also another form can lead to inhibition of the enzyme. This G protein, termed G_i is analogous to G_s described above, except that on activation and diffusion to adenylyl cyclase, the release of cAMP from adenylyl cyclase is depressed. The G protein contains the same β/γ subunit complex as the G_s, but the $G\alpha$ subunit differs. It has been suggested that there are two possible ways in which this G_i protein might be causing adenylyl cyclase inhibition. Firstly, once released, the $G_i\alpha$ subunit may be able to cause direct inhibition of adenylyl cyclase, in a similar manner to the way G_s causes activation. Secondly, the $G_i\alpha$ subunit appears to be less inhibitory than the β/γ subunit released by the G protein's activation, and,

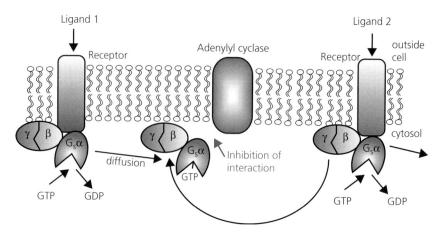

Figure 7.8 A diagram of subunit exchange: activation of G_i prevents stimulation by G_s through the action of the β/γ subunits.

interestingly, the β/γ subunit is much more inhibitory in the presence of G_s. This suggests that the β/γ subunit from the G_i is able to bind to the $G_s\alpha$ that emanates from the dissociation of G_s, and stop activation of the adenylyl cyclase by the G_s route. As the β/γ complexes of both G_i and G_s are proposed to be the same, this would be possible. This interaction of the β/γ complex with different Gα subunits has been termed the subunit exchange (**Figure 7.8**), and further suggests that the exact activation profile of adenylyl cyclase relies on a balance of the activation of the G_s and G_i subunits in some cases.

In a similar way to the permanent activation of the $G_s\alpha$ by cholera toxin, the $G_i\alpha$ subunit is susceptible to modification mediated by another toxin, this time the pertussis toxin. This is released by the bacterium that causes pertussis or whooping cough. Here, an ADP-ribose moiety is added to a specific cysteine residue at the C-terminal end of the $G_i\alpha$ polypeptide, destroying its interaction with the receptor, and this prevents $G_i\alpha$ activation. The lack of activation of the G_i protein will stop the inhibition of the adenylyl cyclase. As this is a double negative, the end result may be similar to the permanent activation of the G_s protein, that is the adenylyl cyclase will be able to be turned off. However, this might depend on whether the adenylyl cyclase had been turned on by some other mechanism. Again, it is the balance of activation and inhibition that might be crucial here, and toxins will sway that balance one way or the other with profound effects.

The interplay of the G_s and G_i proteins is illustrated by the action of epinephrine on different receptors. On binding to the cell's β-adrenergic receptors, the activity of adenylyl cyclase is increased through a route that uses G_s. On the other hand, epinephrine binding to the α_2-adrenergic receptors causes a decrease in the activity of the cell's adenylyl cyclase through a route that uses G_i.

The exact nature of the binding of the adenylyl cyclase to the G proteins has been investigated by the use of chimeric G proteins, that is, proteins made up of different parts of different G proteins. For example, part of the α chain may be derived from an α subunit that is known not to stimulate adenylyl cyclase, such as a chimera made of segments from G_{i2} and G_t. G_t is a G protein known as transducin and has a defined role in the photoreception of the rod cells of the eye (discussed in **Chapter 11**). By the use of various combinations in such chimeras, the active parts of the α subunits can be deciphered. It was found that if a protein contained 40% of the C-terminal end of the G_s subunit, adenylyl cyclase was activated, thus indicating that this was a crucial part of the polypeptide for this interaction.

Therefore, it can be seen that G proteins are crucial for the activation and control of adenylyl cyclase, but there are other mechanisms too. Ca^{2+}/calmodulin can also activate some of the isoforms of this enzyme, particularly type I but also types III and VIII, as well as some non-mammalian forms. Types V and VI seem to show inhibition by Ca^{2+}, whereas types II and VII are stimulated by protein kinase C (see **Table 7.2**).

The heterotrimeric G protein family

Although cholera toxin and pertussis toxin have profound effects on G proteins, there are some G proteins that are immune to their effects. This suggests that there are many isoforms of these important proteins, and that is exactly what has been found. For example, there is a subfamily known as G_q, which is very important in the control of phospholipase C activity (see also Chapter 8), but are not affected by either toxin.

This is no surprise, for as with other signalling proteins, heterotrimeric G proteins form a family of related proteins. For example, there is not a single G_s polypeptide, but rather a family of G protein subunits, with the $G_s\alpha$ arising from the alternative splicing of the same mRNA. This leads to at least four forms of the $G_s\alpha$ subunit. Since the discovery of heterotrimeric G proteins in the early 1980s, numerous forms have been isolated, and the number of different receptors that have their effects through the action of G proteins must run into the hundreds. The different α subunits of the trimer are used in general to define the G protein, but, as well as the different α subunits, at least five isoforms of the β subunit exist and more than 10 γ subunits have been identified. If every type of α, β and γ subunit could randomly mix in the trimer, there could theoretically be around 1000 differing forms of heterotrimeric G proteins in mammals such as humans. However, reconstitution studies show that not all combinations are possible. For example, the isoform Gγ1 can complex with Gβ1 but not with Gβ2. It is probable, in fact, that the stability, and therefore existence, in the cell of some of the subunits may be dependent on the presence and complexing with the other relevant partners. If a Gβ polypeptide is formed in the absence of a Gγ polypeptide, which it is normally able to complex with, it appears to be aggregated and degraded. Therefore, not all combinations will exist, some will be reliant on coordinated protein expression and many combinations expressed will be tissue specific. It is the specific mixture of G proteins in a cell that will enable it to tailor its response to receptor activation and react appropriately to the initial signal.

Table 7.2 The more common members of the G protein α subunit families.

G_s				
α_{olf}	α_s			
Gi				
α_{i1}	α_{i2}	α_{i3}		
α_{oA}	α_{oB}			
α_{t1}	α_{t2}			
α_g				
α_z				
Gq				
α_q	α_{11}	α_{14}	α_{15}	α_{16}
G12				
α_{12}	α_{13}			

Along with G_s, G_i and G_q, other heterotrimeric G proteins that have been characterized include the G_o of brain neurons, G_{olf} and G_t of the olfactory and light sensitive cells of the eye respectively. These do not all of course have an effect on adenylyl cyclase, for example G_q regulates the action of phospholipase C, which leads to the release of other intracellular messengers through the inositol phosphate pathway, further leading to the release of calcium and the activation of phosphorylation, whereas G_t controls the activity of cGMP phosphodiesterase. Some of these are listed in Table 7.2, whereas the functions of G proteins will be considered again in Chapters 8 and 12.

As well as being able to use the inhibitory effects of toxins to elucidate the role of such G proteins in pathways, the ability to stimulate heterotrimeric G proteins artificially in the laboratory will also help greatly. For example, a tetradecapeptide isolated from wasp venom, mastoparan, mimics the action of G protein-coupled receptors and causes the activation of G proteins, particularly G_i and G_o.

Although not directly related to this discussion, it is interesting to note that signalling is never obvious in cells. We may assume that G proteins control cAMP levels in many cells, but can it be the other way around, where the levels of cAMP are regulating the activation of the G protein? This is the mechanism of the slime mould *Dictyostelium*, where the level of extracellular cAMP is detected by receptors that lead to the activation of the G protein subunit Gα2.

The roles of the β/γ complex

When looking at the roles of G proteins in signal transduction, the active part of the heterotrimeric G proteins has traditionally been thought to be the α subunit, with the β/γ complex only being involved in its association with the α subunit when the complex is inactive. However, the β/γ complex has also been shown to be important in its own right. As many isoforms of the β and γ subunits exist, in studies similar to those carried out with the chimeric α subunits, defined β/γ complexes with known β and γ subunits were formed by the use of expression of the subunits in insect Sf9 cells. These could be subsequently purified by subunit affinity chromatography and used to study whether there was any interaction of the β/γ complexes with, for example, the adenylyl cyclase enzyme. It transpired that different isoforms of adenylyl cyclase are controlled differently by the subunits of the G protein trimer. The action of β/γ can be stimulatory, as with types II or IV adenylyl cyclase, or inhibitory, as seen with type I. Furthermore, some adenylyl cyclase isoforms do not appear to be affected by β/γ at all. It was also found that the interactions of the different isoforms were dependent on prenylation of the γ subunit of the G protein. Of course, a discussion of subunit exchange above suggests that the β/γ subunits have a role in the modulation of the action of the G proteins themselves. In fact, it was noted that the concentration of β/γ subunits needed to regulate adenylyl cyclase is greater than that of the $G_s\alpha$ subunit which is normally released. As the heterotrimeric G proteins undergo a stoichiometric dissociation, that is, one $G_s\alpha$ gives rise to one β/γ subunit when the G protein is activated, then the β/γ that can give rise to the effect probably does not arise solely from the breakdown of the G_s trimer. The β/γ complex may come from G_i or another G protein, for example G_o, which in some tissues are more abundant, such as in the brain. This sort of interaction may be commonplace, enabling different G protein trimers to have different effects on adenylyl cyclase at the same time, as well as effecting the activity of each other through subunit exchange.

However, the β/γ subunits may have far wider roles in signal transduction. Besides the possible regulation of adenylyl cyclase, other roles in which the β/γ subunits have been found to be involved include regulation of K^+ channels in myocytes and the pheromone signalling pathway of yeasts, here, through the activation of MAP kinase cascades. β/γ has also been implicated in the control of the kinase activity and phosphorylation of β-adrenergic receptors, as discussed in Chapter 6, whereas βγ subunits activate GRK. A similar interaction is with a protein called phosducin, which may be involved in the regulation of the concentration of β/γ subunits by forming complexes with them. In the photoreceptor cells of the eye, phosducin is found to be a phosphoprotein whose phosphorylated state is light sensitive, mediated by light induced changes in cyclic nucleotide concentrations. In the dark, the protein is phosphorylated by cAMP-dependent protein kinase, whereas it is dephosphorylated in the light by phosphatase 2A. The role of the phosducin seems to be to complex with the β/γ subunits, thereby stopping their interaction with Gα subunits and preventing the Gα subunits from recycling (discussed again in Chapter 12). Similar proteins to phosducin have been reported to be present in the cytosol of bovine brain cells, and this type of control might be quite ubiquitous.

The β/γ subunit has also been shown to activate one of the forms of phospholipase C, that is the PLCβ, leading to the activation of the inositol phosphate pathway. The isoforms of PLC that appear to be most susceptible are β2 and β3 (see Chapter 8). Therefore, depending on the PLC isoform, members of the family of this enzyme can be controlled by the α or β/γ subunit of heterotrimeric G proteins.

7.5 Other roles of the heterotrimeric G proteins

As can be seen from the discussion above, there are many roles of the heterotrimeric G proteins, not just the activation of adenylyl cyclase. As well as those mentioned, the α subunit released from the activation of G_i in muscle tissue, that is the $G_i\alpha$ subunit, is also able to activate K^+ membrane channels, allowing K^+ to exit the cells, thus inhibiting the contraction of the muscle. So, the exact signalling process that transpires from the dissociation of any G protein trimer may be very complex, as subunits may have more than one effect, the exact nature of the response being likely to be dependent on the tissue in which the G protein resides. Similarly, $G_s\alpha$ subunits have been seen to inhibit cardiac Na^+ channels and an effect of $G_s\alpha$ on voltage-gated Ca^{2+} channels has also been reported.

In most cases, the G proteins are activated at the membrane, and have their effect there too. If these subunits from the G proteins are readily diffusible along the membrane from receptor to enzyme, what is it that keeps them attached to the membrane when dissociation takes place and stops them from diffusing out into the cytosol? For the γ subunit it has been proposed that the subunit is bound to an isoprenoid lipid, which being a hydrophobic molecule will attach the subunit to the membrane. For the α subunits the attachment is probably through another lipid, this time myristic acid (see Chapter 6 for a discussion on protein modifications).

Although, classically, these G proteins have been thought of as activators of enzymes on the plasma membrane, particularly adenylyl cyclase, it is thought that their action may be far more widespread throughout the cell. For example, an isoform of the subunit $G_i\alpha$ has been localized to the Golgi apparatus, where it is probably involved in the control of the packaging of proteins and in the regulation of vesicular activity.

It is likely that the roles of G proteins in disease and disorders will be far more widespread than those reported to date. Defects of the G_s and G_i G proteins have been implicated in the excessive proliferation of pituitary cells that leads to pituitary cancer, and it is reasonably certain that the role of these G proteins will be implicated in many disorders and be the target for many therapies. Of course, most of the discussion here is based on data obtained from mammalian tissue, but the existence of G proteins is not limited to these organisms.

7.6 **Guanylyl cyclase**

In an analogous way to the production of cAMP, cGMP is also used by cells as a signalling molecule. Such nucleotides were identified from urine in 1963 by Ashman and his colleagues, and much research has been targeted on these molecules since. cGMP is produced by the enzyme guanylyl cyclase, otherwise known as guanylate cyclase. Although a little like adenylyl cyclase, the cAMP producing enzyme, guanylyl cyclase, is found in two forms, a soluble type that resides in the cytoplasm of the cell and a membrane bound form, located in the plasma membrane. These forms are distinct proteins that have their own modes of regulation, as discussed below. The catalysis of the enzymes is similar to that of adenylyl cyclase, however, as the cGMP is produced from GTP, as cAMP is produced from ATP, and so the chemistry shown in Figure 7.3 would be pertinent (with guanine and not adenine as the base).

Therefore, cGMP can fit into pathways as a diffusible messenger, in a similar manner to cAMP (Figure 7.9). One of the most discussed roles for cGMP is in the regulation of cGMP-gated ion channels, as in the photosensitive cells of the retina, as discussed further in Chapter 12 (Figure 12.4). However, cGMP has also been found to control cGMP-dependent phosphodiesterases and cGMP-dependent protein kinase (cGPK).

Soluble guanylyl cyclase

A very important version of the guanylyl cyclase enzyme is the soluble form, which is characterized by the presence of haem at the catalytic site. The haem prosthetic group is an iron-containing protoporphyrin IX, as it is in most cytochrome molecules in cells. It is the haem group within the enzyme that is

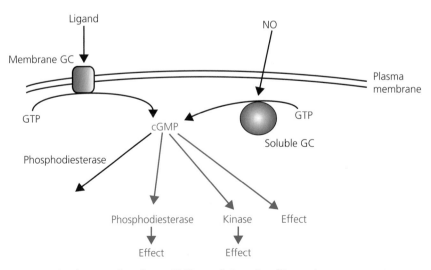

Figure 7.9 A scheme to show how cGMP may fit into signalling pathways.

the target of nitric oxide (NO), one of the main activators of the enzyme, as discussed in Chapter 10. In this signalling scheme, NO may emanate from another cell, cross the plasma membrane as it is non-charged, and activate soluble guanylyl cyclase with the concomitant rise in intracellular cGMP levels (Figure 7.9).

The topology of the guanylyl cyclase enzyme shows it to be a heterodimer of α and β subunits of approximately 70 kDa and 85 kDa. Each of these subunits contains a region that is homologous to the C_{1a} and C_{2a} catalytic regions of adenylyl cyclase, and both are required for catalysis. The presence of Mn^{2+} ions as opposed to Mg^{2+} increases the activity of the enzyme, which also is slightly stimulated by Ca^{2+} ions. However, ATP acts as an inhibitor. Haem binding has, not surprisingly, been shown to involve a histidine residue, which is on the β subunit of the enzyme.

However, the enzyme appears to exist in tissue-specific isoforms, for example, two isoforms have been reported to exist in bovine lung. The two forms contain one subunit in common, but the second subunit differs, one being 85 kDa whereas the other is a little smaller, being 73 kDa. Interestingly, as it is so important for function, the attachment of the haem group may vary, as seen by differences in the haem spectra for the enzymes. It has also been suggested that the haem group of some enzymes is penta-coordinate, whereas that of others is hexa-coordinate, which suggests that the two enzymes may have subtle differences in their catalytic action and in their control. Such subtle differences, as discussed above, are not uncommon amongst families of proteins and enzymes involved in cell signalling.

Membrane bound guanylyl cyclase

As well as soluble forms of guanylyl cyclase, there are also membrane bound forms. However, except for relatively small regions, these show little topological similarity to adenylyl cyclase, or indeed the soluble forms of guanylyl cyclase. The membrane guanylyl cyclase contains a single membrane spanning domain, which crosses the plasma membrane, and therefore has domains on the outside and inside of the cell. On the cytoplasmic side along with a protein kinase homology domain, is an intracellular catalytic domain (Figure 7.10). This makes sense, as any cGMP produced needs to be in the cytoplasm to enable it to pass its "message" down the signalling pathway. Some areas of homology to the mammalian adenylyl cyclases have been reported for this catalytic area, but as was noted above, this is no great surprise as the chemistry that needs to be catalyzed is very similar, i.e. the triphosphate converted to cyclic monophosphate forms. On the outside of the membrane is the extracellular domain, which is in fact a ligand binding site, and so these enzymes are receptors. In the classification given in Chapter 5 (Section 5.1, and see Section 5.2), they would be referred to as receptors "containing intrinsic enzymatic activity". They act as receptors for two main classes of molecules: signalling peptides or the so-called naturietic peptides and the heat stable enterotoxins including guanylins.

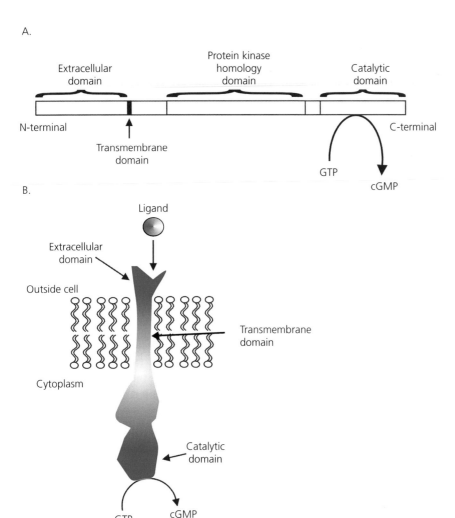

Figure 7.10 The domain structure and schematic topology of membrane associated guanylyl cyclase. A. Depicted as a simple linear representation. B. Schematic of how the protein will arrange in a membrane, with the ligand binding site on one side and the catalytic site on the other.

These membrane cyclase forms, like many receptor proteins, probably function as dimers or oligomers. ATP has been shown to stimulate the activity of these cyclases, but, despite the presence of a protein kinase homology domain, non-hydrolyzable forms of ATP will have the same effect, and neither ATPase activity nor phosphorylation activity seem to be required for the functioning of these receptors. However, it appears that the cyclase normally resides in a highly phosphorylated state. Furthermore, a novel protein phosphatase has been found associated with one of the forms of the receptor and it is thought that the state of phosphorylation of the cyclase may be crucial to its activity, dephosphorylation causing a desensitization of the receptor.

Calcium ion concentrations are also important in the regulation of the membrane forms of guanylyl cyclase. In the retina, Ca^{2+} sensitive proteins have been found, which interact with guanylyl cyclase. One such protein, of 26 kDa, is recoverin. Other such proteins have molecular weights of 20 kDa and 24 kDa: known as p20 and p24 respectively. These proteins are cytoplasmic where they respond to the changes in the Ca^{2+} concentration. Increases in Ca^{2+} will lead to a complex of the calcium sensitive protein and the cyclase, and the production of cGMP will be decreased. It is possible that Ca^{2+} might also have a direct interaction with the cyclase but this has never been demonstrated.

Interestingly, one particulate form of the enzyme, that from bovine rod outer segments, is reported to be activated up to 20-fold by nitric oxide, suggesting that it might not just be the soluble forms of guanylyl cyclase that are targets of this intercellular signal. This would mean that nitric oxide has two distinct enzyme targets in cells that impinge on cGMP levels.

7.7 Phosphodiesterases

All signals need to have the potential to be reversed, and cAMP and cGMP are no different in this respect. However, despite the fact that they are produced by distinct and surprisingly different enzymes, the destruction of both of them is down to a single family of enzymes. The breakdown of cAMP and cGMP, and hence the reduction of their intracellular concentrations, usually leading to a cessation of the message, is catalyzed by the enzymes called cyclic nucleotide phosphodiesterases (PDEs), usually referred to simply as phosphodiesterases. The reaction can be written as follows:

$$cAMP + H_2O \xrightarrow{Mg^{2+}} AMP + H^+$$

Although cAMP and cGMP are relatively stable in the cell, removal of the cyclic nature of the molecules and hence their breakdown overall is exergonic, that is, the change in Gibbs free energy (ΔG) is negative, the enzyme overcoming the activation energy barrier, which stops the reaction becoming spontaneous.

In mammals, at least 11 different types of phosphodiesterase have been distinguished, (see Table 7.3), encoded for by at least 20 genes, with the existence of more classes suggested by some. As would be expected, some of the types have more than one form, for example, type I has three forms, type III has at least two forms, and type IV as at least four forms. They all seem to be highly homologous in their C-terminal regions, but show greater divergence towards the N-terminal end of the polypeptides. The expression of different forms of the same family of PDEs is also tissue-specific, suggesting subtle differences in their functionality.

The structures of phosphodiesterases are known (for example, see Figure 7.11), and if the cDNAs for all the PDEs are compared the 3' ends

Table 7.3 The classes of phosphodiesterases.

Class of enzyme	Characteristics	Comments
I	Ca^{2+}/calmodulin-dependent	Two calmodulin (CaM) domains
II	cGMP stimulated	Two GAF domains, which are cGMP binding domains
III	cGMP inhibited	Transmembrane domains (6?)
IV	cAMP specific	At least four members in this family
V	cGMP specific	Two GAF domains
VI	Photoreceptor type	Associated with control proteins
VII	High affinity cAMP specific	Appears simple, only having a catalytic domain
VIII	cAMP specific	Ligand binding domain?
IX	High affinity cGMP specific	Appears simple, only having a catalytic domain
X	Hydrolyzes cAMP and cGMP	Might act as a cAMP inhibited cGMP enzyme
XI	Hydrolyzes cAMP and cGMP	Contains a GAF domain

show a great deal of homology, over a stretch of approximately 800 base pairs, as would be expected from the homology of the C-terminal ends of the polypeptides. These regions code for a common catalytic core of the molecules. As with the enzymes that produce the cyclic compounds, similarity here would be expected in the enzymes that break them down, as the chemistry is the same, regardless of whether the base is adenine or guanine. However, comparisons of the 5' ends show low homology and account for the differences that make up the eleven family types. The N-terminal end of the polypeptides may also influence their subcellular location, either by containing signal peptides that direct their post-translational route within the cell or allowing their attachment to membranes.

The genes for at least five cAMP-specific phosphodiesterases have been located in both humans and mice. In humans, two of the genes are found on chromosome 19, with others on chromosomes 5, 1 and 8, whereas in mice phosphodiesterase genes were found on chromosomes 4, 8, 9 and 13. However, multiple forms of a PDE within a family may also arise from the same gene. In some cases multiple promoters within one gene have been found, enabling transcription to start at different points and therefore resulting in polypeptides of varying length. Even if the same promoter is used, the gene transcript may undergo alternative splicing, and again varying polypeptides will result. Such differences in expression and post-transcriptional modification may be

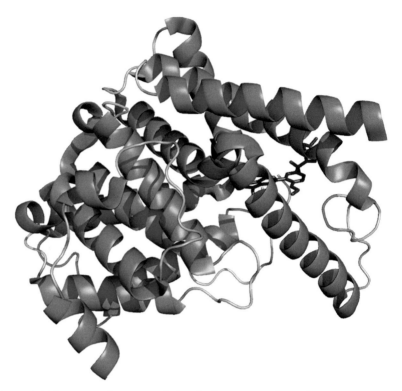

Figure 7.11 The proposed structure of phosphodiesterase 5, obtained using X-ray diffraction; structure was obtained from the RCSB Protein Data Bank (www.rcsb.org/pdb/) PDB ID: 1udt. (Sung, B.-J., Hwang, K.Y., Jeon, Y.H., Lee, J.I., Heo, Y.-S., Kim, J.H., Moon, J., Yoon, J.M., Hyun, Y.-L., Kim, E., Eum, S.J., Park, S.-Y., Lee, J.-O., Lee, T.G., Ro, S. and Cho, J.M. (2003) Structure of the catalytic domain of human phosphodiesterase 5 with bound drug molecules. *Nature*, 425, 98–102).

tissue-specific, with each of the PDEs retaining the catalytic core but having alternative modules that alter their function and cellular location. However, some PDEs are particularly susceptible to proteolysis, and, therefore, several reports in the literature of the presence of new isoenzyme forms may only be the result of breakdown products from larger PDEs. Despite this, it is clear that there are many forms of this enzyme, and subtle differences will no doubt be required in different tissues to maintain the balance of intracellular cyclic nucleotides used in signalling.

Although cAMP and cGMP are often broken down by different phosphodiesterases, the presence of one cyclic nucleotide may still influence the intracellular concentration of the other. Some forms of cAMP PDE are influenced by the presence of cGMP. For example, type III has been shown to be inhibited by the presence of cGMP. Such phosphodiesterases are therefore referred to as cGMP inhibited phosphodiesterases, or cGI-PDEs. These PDEs have been shown to have a molecular weight of approximately 110 kDa, but proteolytic products between 30 and 80 kDa have also been reported.

Other cAMP phosphodiesterases have been reported to contain allosteric binding sites for cGMP, which cause a stimulation. These PDEs also have an apparent molecular weight of approximately 105 kDa, but further analysis suggests that the native structure is as non-spherical dimers, although a tetrameric form is reported to have been isolated from rabbit brain tissue.

Other factors besides the cyclic nucleotides might influence the activity of these enzymes. One family of PDEs (type I) are controlled by the intracellular concentration of Ca^{2+} in association with calmodulin, and a look at the structure of these enzymes shows that they have two calmodulin binding domains in their N-terminal end. This family has been further subdivided into four groups, which have molecular weights of 60 kDa, 63 kDa, 58 kDa and 67 kDa.

One of the most well-characterized cGMP PDEs is that from the rod cells of the eye (type VI). Here, as discussed in **Chapter 12**, cGMP acts as a regulator of ion channels. The membrane bound PDE from bovine rods exists as a $\alpha\beta\gamma_2$ complex, where the α subunit has a molecular weight of 88 kDa, the β subunit has a molecular weight of 84 kDa, and the γ subunit has an apparent weight of only 11 kDa. However, other complexes may exist, such as $\alpha\alpha\gamma_2$ and $\beta\beta\gamma_2$. Here, as well as the membrane form, there is also a soluble form where the α, β and γ subunits appear to be of the same sizes as the membrane PDE but they are joined by a further δ subunit of 15 kDa. The analogous PDE found in the cone cells, responsible for colour vision, has a large subunit of 94 kDa, which is joined by three small subunits of 11 kDa, 13 kDa and 15 kDa. The overall weight of this complex has been estimated to be approximately 230 kDa.

Of significant importance are the type V phosphodiesterases, the cGMP-specific members of the family. It has been known for many years that the blood supply though the vessels in mammals can be influenced by the signalling molecule nitric oxide (NO). However, as we have seen above, NO targets the soluble form of guanylyl cyclase, causing its activation. Therefore, NO causes a rise in intracellular cGMP. But, if type V phosphodiesterase is active, the rise in cGMP will be very transient, and in some individuals the rise is too transient. The obvious phenotype here is the lack of a sustained penile erection. Therefore, a drug, commonly known as Viagra™, has been developed to enhance the longevity of the raised cGMP levels. It does this by inhibiting type V phosphodiesterase. Therefore, despite some reports of Viagra™ causing NO release, it does not, but it is a phosphodiesterase inhibitor. However, as mentioned, the chemistry and catalytic domains of the phosphodiesterases are very similar across the family, and therefore it is no surprise to learn that the drug also has inhibitory effects on other members of the family in some individuals, in particular phosphodiesterase type VI. This manifests itself as an alteration in eyesight, a blue haze is reported, and this is because phosphodiesterase type VI is so important in the signalling that takes place in the rods and cones of the eye in mammals. This is also discussed in **Chapters 10 and 12**.

■ Chemically Viagra™ is sildenafil citrate or more fully 1-[[3-(6, 7-dihydro-1-methyl-7-oxo-3-propyl-1*H*-pyrazolo [4, 3-*d*]pyrimidin-5-yl)-4-ethoxyphenyl]sulphonyl]-4-methylpiperazine citrate.

It is also possible that binding of cyclic nucleotides to phosphodiesterases takes place in which there is no catalytic turnover, and this might serve to buffer the intracellular concentration of such signalling molecules, which means that production of a cyclic nucleotide such as cGMP may not automatically mean that the molecule will be free in solution to have a further effect on other enzymes.

Some phosphodiesterases may in fact be excreted, and are then used to control the levels of extracellular cyclic nucleotides. This would be of importance to organisms such as *Dictyostelium*, which uses the concentrations of extracellular cAMP as a signal for aggregation.

The isolation of a phosphodiesterase that appears to be specific for cCMP suggests that this third cyclic nucleotide may also be found to have a role in signalling, and no doubt such data will lead to further research in this area.

7.8 Compartmentalization of nucleotide signalling

Although cAMP appears to be produced in the cytoplasm by the enzyme adenylyl cyclase, the concentration in the cytoplasm is not uniform, but rather should be considered as patchy. Later, in Chapter 9, the idea of calcium ion concentrations appearing as "hotspots" in the cytoplasm will be discussed. Calcium ions are small, and they should readily diffuse, so why is the concentration not the same throughout the cytoplasm? The same should be able to be asked about cAMP, which is also small and can readily diffuse. However, detailed analysis of the cell reveals that is it not.

The answer comes not so much from the localization of the signalling molecule per se, but rather from the uneven distribution of the enzymes that both make it and remove it. It is clear that the adenylyl cyclase proteins will not be evenly distributed around the plasma membrane, or that not all the adenylyl cyclase will be activated. Certain regions of the cell will be active in producing cAMP at a particular moment in time. Therefore, in the locale of the active cyclase enzymes the cAMP is likely not to be higher than elsewhere. But it should diffuse. However, it will only diffuse until it is removed from the cytoplasm by any active phosphodiesterase enzymes able to act on it. Therefore, if the active adenylyl cyclase is at one end of the cell, in the region where the receptor has been activated by its ligand and the G protein has activated the cyclase, the cAMP at that end of the cell will be high. If there is then a significant phosphodiesterase activity between that end of the cell and the other end, the cAMP will never get the chance to diffuse to the other end of the cell, as it will be converted to AMP before it gets there. A researcher then observing the cell would see high cAMP only in one region of the cell. This is not in this case a static concentration, but rather a steady-state concentration, where the cAMP is being both produced, and removed, with the enzymes being very busy.

The other regions of the cell will appear, and are in this respect, inactive. Such regionalization of signals is called compartmentalization, although there are no real compartments involved. However, such mechanisms allow great subtlety of action of signals, allowing the same compound to have different effects in different parts of the cell. Proteins and downstream components may also have regional localization, and perhaps their binding constants and activation thresholds will reflect the levels of signal that they are likely to be exposed to. Such complications need to be considered and added to any computer modelling or systems biology approach to signalling research.

7.9   The GTPase superfamily: functions of monomeric G proteins

The role of G proteins, that is, proteins that bind GDP and GTP, have an exchange of the two nucleotides leading to an activated state, and hydrolysis of the nucleotide to restore the protein back to an inactive state, is far more common than that outlined above for the control of adenylyl cyclase. As well as the plasma membrane associated heterotrimeric G proteins discussed above, there exists a group of small monomeric G proteins that control such diverse cellular functions as proliferation, gene expression and protein synthesis, differentiation and regulation of movement of proteins through the cytoplasm. All these G proteins possess intrinsic GTPase activity and the phrase "GTPase superfamily" has been coined to encapsulate this group of proteins. In a manner that is very similar to that seen with the heterotrimeric G proteins, the monomeric G proteins can act as molecular switches, in a mechanism that allows for a large amount of amplification of the signal.

As with the heterotrimeric G proteins, the monomeric G proteins are inactive when bound to GDP, which on activation is exchanged for GTP causing a conformational change in the structure of the proteins and so allowing the signal to be transmitted along the pathway. The intrinsic GTPase activity converts the GTP back to GDP, turning off the signal and allowing the protein to await the next signal (**Figure 7.12**).

It has also been postulated that a third state exists in which GDP has been removed but GTP has yet to bind. This is a transient state in which either GTP can bind to turn the protein on, or GDP can bind to prevent the system turning on. Normally in the cell, the GTP binding step would be favoured. An analogous empty state of the guanine nucleotide binding site probably also exists in the cycle of the α subunit of the heterotrimeric G proteins. This might allow the cell to "reflect" on its health as GTP levels would tend to reflect those of ATP. Therefore, if ATP was low, it might not be beneficial to turn on what may be an energy requiring pathway or response.

A.

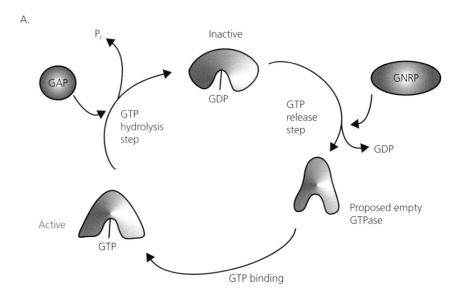

B.

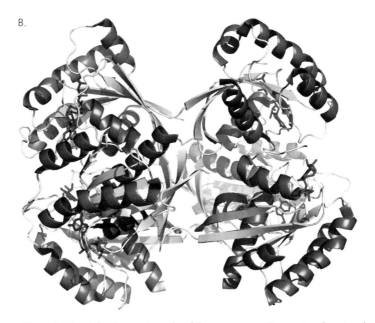

Figure 7.12 A. The G protein cycle of the monomeric G proteins showing the actions of GAP and GNRP. B. The structure of RAS protein, in the presence of a GTP analogue, as solved using X-ray diffraction. Structure was obtained from the RCSB Protein Data Bank (www.rcsb.org/pdb/) PDB ID: 2pmx. (Tong, Y., Tempel, W., Shen, L., Arrowsmith, C.H., Edwards, A.M., Sundstrom, M., Weigelt, J., Bochkarev, A. and Park, H. Human K-Ras in complex with a GTP analogue. To be published).

However, there are significant differences between this system and that of the heterotrimeric G proteins described above. These proteins are monomeric, and in general act as monomers, but in no sense do they act alone, but rather associate and are aided in their function by other polypeptides. Both the dissociation step resulting in the loss of the GDP, and the GTPase step converting the GTP back to GDP, require the assistance of other proteins. The first GDP releasing step is aided by the presence of guanine nucleotide releasing proteins (GNRP), whereas the GTP hydrolysis is aided by the presence of GTPase activating proteins (GAPs; Figure 7.12A).

It is probable that the α subunit of heterotrimeric G proteins, although acting as a monomer when associated with GTP, does not need the assistance of a GAP protein for activating the GTPase activity because it contains a region on the polypeptide that has GAP-like activity, that is a GAP-like domain.

So, how do these G proteins fit into a signalling pathway? An overall, but very simplified scheme is shown in Figure 7.13. A receptor, often a tyrosine kinase receptor, will bind its ligand, and an autophosphorylation event will lead to the formation of binding sites for an intracellular adaptor molecule (as discussed in Chapter 6). This will activate a GNRP, which will facilitate the exchange of the GDP for GTP on the monomeric G protein. Once active, the G protein can activate its downstream effector, often a kinase cascade, which will lead to the cellular effect required by the ligand perception. More detailed and characterized pathways are shown in Figures 6.8 and 11.4.

The most well-characterized type of monomeric G protein is the product of the *ras* oncogene, p21^{ras}, or known simply as Ras. This is a plasma membrane associated protein of molecular weight 21 kDa. Like many such proteins, it was first identified as encoded for by a gene from acute transforming retroviruses, but now the Ras protein has been crystallized and a structure resolved. A representation of the structure is shown in Figure 7.12B.

The role of Ras was originally hinted at by the discovery that transforming genes were, in fact, alterations of normal genes, activation of which was due to

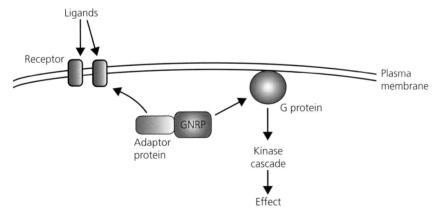

Figure 7.13 An extremely simplified scheme to show how a monomeric G protein might fit into a signalling pathway. GNRP, guanine nucleotide releasing protein.

the presence of a point mutation. Their involvement has been suggested in the formation of several forms of cancer, including the formation of tumours induced by physical, and some chemical, reagents. Most of the assays for the presence of Ras depend on its ability to induce cell proliferation. Bioassays have been based on DNA synthesis, transformation of established cultured cells and the formation of tumours. Other assays have exploited the ability of Ras to induce differentiation of cells. However, it is interesting that full transformation of cells *in vitro* often also depends on the presence of another oncogene, for example *myc*.

The *ras* gene has been well characterized. In humans and some other mammals there are three functional Ras genes present in the genome. These are H-*ras*, K-*ras* and N-*ras*. H-*ras* and K-*ras* are related to the Harvey (Ha) and Kirsten (Ki) sarcoma viruses of mice. The structure of the *ras* gene is unusual in that the first 5' exon appears to be non-coding. The actual coding region is split into four exons, although the K-*ras* gene contains two alternative fourth exons, which can give rise by alternative splicing to two separate gene products. The introns are very different between the various genes, leading to very different lengths of mRNA for each.

The *ras* products have been detected in all tissues explored, including foetal tissue. However, histochemical staining was more intense in cells that were proliferating, as opposed to those which were fully differentiated. Northern blot analysis, to reveal the levels of gene expression, showed that the genes did show some tissue-specificity. Studies of the transcriptional control for these genes suggest that the first intron may contain an enhancer, with another small weak enhancer downstream of the gene, whereas a short alternative exon within another intron also appears to be important, although the exact nature of its role remains obscure. However, the presence of serum or some growth factors can enhance the expression of Ras.

With the gene relatively well characterized, what is known about the protein itself? Most Ras proteins contain 189 amino acids, except the K-*ras*B product, which has 188 amino acids. A striking feature is that the first 164 amino acids are homologous across all the *ras* products, with the C-terminal region being referred to as the heterogeneous region, being more divergent. However, sequence patterns are conserved even here between species, suggesting that certain C-terminal sequences confer functional specificity to the proteins. Obviously, if a region of many proteins is so highly conserved, it suggests that there is a function of great importance harbouring there, and, indeed, the first 164 amino acids contain the GTPase activity of the proteins. Also, in all Ras proteins a cysteine residue is conserved at a position four residues before the C-terminus, again suggesting importance attached to this amino acid. The protein is made as a pro-p21^{ras} product and this C-terminal region also undergoes a great deal of post-translational modification, including farnesylation at the conserved cysteine amino acid, cleavage of the end amino acids and methylation (also see Chapter 6). These modifications are involved in the increase in the hydrophobicity of the protein, which encourages its association with the plasma membrane. Mutant proteins without this conserved cysteine are cytosolic and have no transforming activity.

The general structure of the p21^{ras} protein, Ras, shows that it has a hydrophobic core consisting of six β-sheet strands connected by hydrophilic loops and α-helices. Five regions of the proteins, which have been designated the names G1 to G5, lie on one side of the protein and are associated with the hydrophilic loops. G1 is probably involved in the binding of the first two phosphate groups, α and β, of either GDP or GTP. The catalytic step, that is the GTPase activity, of the cycle requires the presence of Mg^{2+}, and this is probably associated with regions G2 and G3. These regions, G2 and G3, have also been implicated as the sites of conformational changes induced by GTP binding, and so activation of the protein. They have also been implicated in the association with GAP proteins. The other association needed for the functioning of the GTPase is an interaction with the protein next in line down the pathway that the GTPase needs to act on, that is, the effector. Here again, the GTPase region G2 has been implicated.

Ras oncogenes, which have the ability to promote tumour formation, can be used to determine the active residues in the protein. Commonly, substitutions of residues at positions 12, a glycine, and 61, a glutamine, reduces the intrinsic GTPase activity of the protein. It is now believed that the catalytic cycle involves a glutamine residue on the protein, which activates a water molecule, which then attacks the γ-phosphoryl group of the GTP in a nucleophilic fashion. However, different GTPases also certainly vary in the precise mechanism of GTP hydrolysis.

Some viral *ras* genes encode a threonine at position 59, with autophosphorylation taking place at this threonine, but the role of this phosphoryl addition is unclear. Another, rather fascinating, mutant form of Ras has no ability to bind to guanine nucleotides, and therefore its conformation appears to be permanently locked into an active form.

Another good model for the monomeric G protein family is EF-Tu, elongation factor Tu, which is involved in protein synthesis. However, this protein is much larger, with a molecular weight of approximately 43 kDa, but like Ras, it has been crystallized. Other members of the GTPase family include the proteins Rab, Rho and Rac. At least four Rho polypeptides have been identified along with two Rac polypeptides, as well as other related proteins such as TC10 and CDC42H. Like Ras, such proteins also work in conjunction with GAP-like proteins, such as Rho-GAP and GNRP-like proteins such as Dbl and Rho-GDS.

Let us now turn to the proteins that interact with the monomeric G proteins, and so are instrumental in the functioning of the monomeric G proteins. Several proteins have been found that contain GAP activity. One of these is a protein with an approximate molecular weight of 120 kDa (p120GAP). It consists of 1047 amino acids, and appears to be expressed everywhere. When in its active state, it has the ability to increase up to fivefold the GTPase activity of the Ras, and this function is due to a catalytic domain in the C-terminal end of the GAP protein. The polypeptide also contains several protein binding domains: two SH2 domains, a SH3 domain and a PH domain (pleckstrin homology domain), suggesting that it has great ability to interact with other polypeptides. Two such polypeptides, which are themselves phosphorylated, include one of 190 kDa, p190, and one of 62 kDa, p62. If p120GAP has formed a complex with p190 it has a reduced ability to stimulate GTPase activity of Ras. p120GAP is also phosphorylated on tyrosine residues by tyrosine kinase receptors, which therefore probably have a measure of control on its activity. It is also inhibited by some lipids, including arachidonic acid (see Chapter 8).

A second group of GAP proteins is a family known as GAP1. The mammals contain at least two homologues, GAP1^{m} and GAP1^{IP4BP} (Superscript ^{IP4BP} here denotes inositol 1,3,4,5-tetrakisphosphate binding: see Chapter 8). These proteins, of approximately 850 amino acids, contain a PH domain towards the C-terminal end of the polypeptide, a GAP related domain, known as the GRD domain in the middle and two domains that are homologous to the C2 regulatory domains of protein kinase C towards the N-terminus. Interestingly, as the name suggests, these proteins have been shown to bind to inositol 1,3,4,5-tetrakisphosphate, one of the metabolites derived from InsP$_3$, which suggests exciting possibilities in its control. As discussed below in Chapter 8, several inositol derivatives have been thought to have signalling roles, and, clearly, here is an example. Like the other GAP proteins, these proteins also show inhibition by phospholipids. Expression of this protein is not so widespread as for p120GAP, with the highest expression seen in the placenta, brain and kidneys.

GAP is not the only protein that has been found to stimulate GTPase activity in Ras. A rather large protein of approximately 250 kDa known as NF1, otherwise called neurofibromin, is responsible for Recklinghausen's neurofibromatosis (NF1 disease). Out of its 2818 amino acids, a region of 350 amino acids found in the centre section of the polypeptide is analogous to GAP (see Figure 7.14 for a structural representation). The protein appears to be expressed primarily in the nervous system, particularly in neurons and Schwann cells, where it exists in at least three isoforms, the different forms arising from alternative splicing. The role of neurofibromin also certainly involves its interaction with other proteins in a complex and again, like other GAPs, its activity is inhibited by lipids.

The GAP1 family also shows GAP activity against other G proteins, such as Rap, although this activity is not affected by lipids or by the presence of InsP$_4$. However, other members of the Ras superfamily are not necessarily acted on by GAP itself, but have their own proteins that contain GAP-like activity.

The other G protein regulatory proteins are the GNRPs. One of the most important, and certainly the most well studied is the protein known as Son of sevenless or Sos. It was originally found to be involved in signalling in the sevenless system in the fly *Drosophila*. It is often found associated with an adaptor protein, such as GRB2 in mammals, or Drk in *Drosophila*, the interaction being possibly through the adaptor's SH3 domains. As discussed above, on binding of a ligand to its respective receptor protein kinase, the receptor usually can autophosphorylate and the formation of the new phosphotyrosine groups allows an interaction with adaptor protein through its SH2 domains. This ensures that the adaptor protein and its associated GNRP, for example Sos, are now in close association with the membrane

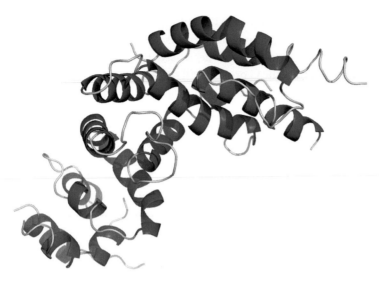

Figure 7.14 The proposed structure of a GAP domain of neurofibromin, obtained by X-ray diffraction. Structure was obtained from the RCSB Protein Data Bank (www.rcsb.org/pdb/) PDB ID: 1nf1. (Scheffzek, K., Ahmadian, M.R., Wiesmuller, L., Kabsch, W., Stege, P., Schmitz, F. and Wittinghofer, A. (1998) Structural analysis of the GAP-related domain from neurofibromin and its implications. *EMBO Journal*, 17, 4313–4327).

where the latter leads to nucleotide exchange and activation of the monomeric G protein. Such a system is further discussed in **Chapter 11** (**Figure 11.7**) and previously in **Chapter 6** (**Figure 6.8**). Mutations of Sos in humans have been implicated in hereditary gingival fibromatosis, which manifests itself as a gingival enlargement that may cover the crowns of teeth.

Sos is not, as expected, the only GNRP to be found and characterized. Other GNRPs include CDC25 in *S. cerevisiae*, a 1545 amino acid polypeptide. Mutants of *S. cerevisiae* lacking the CDC25 protein, *cdc25⁻* mutants, have been useful in the identification of similar proteins. An analogous gene in mammals, *cdc25^{Mm}*, encodes for a protein of 140 kDa, which is found in brain tissue. The gene, however, is quite complex as it can give rise to at least four distinct proteins of very disparate molecular weights. Other GNRPs include a protein of 35 kDa, which appears to have the role of stimulating guanine nucleotide exchange, whereas another cytosolic protein, encoded for by *smgp21GDS*, has been found to be active only if the Ras has undergone its post-translational modifications. To further complicate the issue, inhibitor proteins of guanine nucleotide exchange have been reported that act on several members of the Ras superfamily, particularly members of the Rab, Rho and Rac families, for example Rho-GDI.

Although schemes can be drawn to show how these monomeric G proteins fit into transduction pathways, it is not always clear exactly how the signalling mechanism works. In particular, the activity of Ras can be increased in one of two ways. It is usually assumed that the G protein is activated by an increase in GTP/GDP exchange, through the increase of GNRP activity, perhaps involving Sos (as shown in **Figures 6.8 and 11.7**). However, if the GTPase activity is decreased, for example through a decrease of GAP activity, then the G protein will remain longer in its active state, and signalling will continue, the result being similar to an increase in GNRP activity.

As for the pathways in which monomeric G proteins are involved, they are numerous. Ras and its family members are often seen to be involved in pathways leading from the activation of growth factor receptors for example, and in the pathways leading from insulin perception. It has also been shown that the transformation abilities of several oncogenes encoding protein tyrosine kinases, for example, *src* and *fms*, require the activation of Ras. Ras itself is often found to activate the serine/threonine kinase, Raf, also an oncogene product, and activation of Raf can lead to the activation of the MAP kinase cascades. Ras can also activate other serine/threonine kinases, and it has been reported to activate MAP kinase kinases as well as MAP kinases. Activation of ribosomal protein S6 kinase and protein kinase C have also been reported. As with the heterotrimeric G proteins, in *S. cerevisiae* Ras has also been shown to activate the enzyme adenylyl cyclase.

However, the effector role of Ras may be not necessarily be directly related to changes in the Ras protein. If Ras is in interaction with its associated proteins, for example GAP, then conformational changes are possible in these proteins too, and that in itself might constitute a signal. Binding of Ras to GAPs is thought to lead to exposure of protein interacting domains, such as SH2 and SH3 domains. Once exposed, these domains would enable GAP proteins to interact with other proteins and so propagate a signal within the cell, without the direct involvement of Ras. In this case, this might mean that the binding of GAP and Ras allows the divergence of the signal, with two distinct signals being propagated into the cell.

Other Ras related proteins

A subgroup of the Ras family of proteins is the family of ADP-ribosylation factors (ARF). These were first identified as factors needed for the efficient modification of $G_s\alpha$ of the heterotrimeric G proteins by cholera toxin (see above). These proteins have been shown to be involved in the vesicle-mediated protein traffic of cells as well as in the control of some lipid signalling by the regulation of the activity of phospholipase D, an enzyme which yields phosphatidic acid and choline from membrane lipids.

Lastly here, as mentioned above, an elongation factor involved in protein synthesis, EF-Tu is also very similar to Ras in its mode of action. However, the GNRP involved in the GDP/GTP exchange is another elongation factor, EF-Ts. This protein contains constitutive activity, unlike the GDP/GTP exchange on many other G proteins, which involves the activation of a receptor or receptor-associated proteins.

■ Summary

- Cyclic nucleotides, in particular cAMP and cGMP, are central signalling molecules in many transduction pathways (**Figure 7.15**).

- cAMP is produced from ATP by the enzyme adenylyl cyclase.

- In mammals, adenylyl cyclase is a single polypeptide integral to the plasma membrane, but several isoforms are seen in nature.

- The main control of adenylyl cyclase activity is through an interaction with heterotrimeric G proteins, particularly G_s and G_i, which are stimulatory and inhibitory respectively.

- Heterotrimeric G proteins are constructed of an α subunit, which binds to GDP in its inactive state and GTP in its active state, along with a β/γ subunit complex.

- Although most of the signalling by heterotrimeric G proteins has been reported to be through the action of the α subunit, it has now been found that the β/γ complex has many signalling roles.

- Heterotrimeric G proteins are, in fact, a wide family of signalling proteins and include members that interact with phosphodiesterases, G_t, and regulate phospholipase C, G_q.

- cGMP is produced by an enzyme analogous to that producing cAMP: guanylyl cyclase. This enzyme has two forms: one soluble and one membrane bound.

- Cyclic nucleotide signalling is ended by the hydrolysis of the cyclic forms to the monophosphate forms by the enzyme phosphodiesterases.

- Eleven classes of phosphodiesterases have been reported in mammals, including ones that are controlled by $Ca^{2+}/$calmodulin, stimulated by cGMP and inhibited by cGMP.

- cGMP-specific phosphodiesterases (type V) are the target of drugs such as Viagra™.

- Monomeric G proteins are an extremely important group of signalling molecules, which amongst their number include the oncogene products, Ras and Rac as well as the protein synthesis elongation factor EF-Tu.

- Like the heterotrimeric forms, monomeric G proteins are active when bound to GTP, and intrinsic GTPase activity hydrolyzes the GTP to GDP accompanied by a conformational change to the protein and its return to an inactive state.

- GTPase activity in monomeric proteins is enhanced by the protein GAP, whereas the GDP release is enhanced by the proteins called GNRPs, such as Sos.

- Monomeric G proteins have been shown to be instrumental in the transduction of the signal from many receptors, particularly those for growth factors, on the plasma membrane to kinase cascades such as the MAP kinases, and hence often on to alterations of activity in the nuclei of cells.

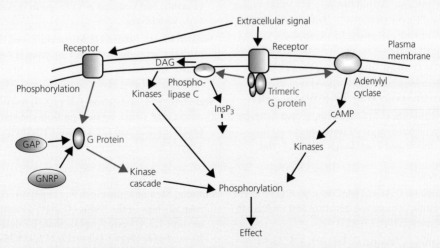

Figure 7.15 A simplified diagram showing the role of cAMP and G proteins in the transduction of cellular signals.

→ Further reading

Adenylyl cyclase

Alexander, S.P.H., Mathie, A. and Peters, J.A. (2008). Guide to Receptors and Channels (GRAC), 3rd edition (2008). British Journal of Pharmacology, 153 (Suppl. 2), S175–S176 [short reference on adenylyl cyclase types].

Cooper, D.M.F., Mons, N. and Karpen, J.W. (1995) Adenylyl cyclases and the interaction between calcium and cAMP signalling. Nature, 374, 421–424.

Tang, W-.J. and Gilman, A.G. (1991) Type specific regulation of adenylyl cyclase by G protein beta/gamma subunits. Science, 254, 1500–1503.

Tang, W-.J. and Gilman, A.G. (1992) Adenylyl cyclases. Cell, 70, 869–872.

Adenylyl cyclase control and the role of G proteins

Birnbaumer, L. (1992) Receptor-to-effector signalling through G proteins: roles of βγ dimers as well as α subunits. Cell, 71, 1069–1072.

Hepler, J.R. and Gilman, A.G. (1992) G proteins. Trends in Biochemical Sciences, 17, 383–387.

Iñiguz-Lluhi, J., Kleuss, C. and Gilman A.G. (1993) The importance of G-protein β/γ subunits. Trends in Cell Biology, 3, 230–235.

Hao, N., Behar, M., Elston, T.C. and Dohlman, H.G. (2007) Systems biology analysis of G protein and MAP kinase signaling in yeast. Oncogene, 26, 3254–3266.

Smrcka, A.V. (ed.) (2004) G Protein Signaling: Methods and Protocols. Humana Press, Totowa, New Jersey. ISBN 1588291375.

Guanylyl cyclase

Ashman, D.F., Lipton, R., Melicow, M.M. and Price, T.D. (1963) Isolation of adenosine 3',5'-monophosphate and guanosine 3',5'-monophosphate from rat urine. Biochemical Biophysical Research Communication, 11, 330–334.

Garbers, D.L. and Lowe, D.G. (1994) Guanylyl cyclase receptors. Journal Biological Chemistry, 269, 30741–30744.

Goraczniak, R.M., Duda, T., Sitaramayya, A and Sharma, R.K. (1994) Structural and functional characterisation of the rod outer segment membrane guanylyl cyclase. Biochemical Journal, 302, 455–461.

Phosphodiesterase

Alexander, S.P.H., Mathie, A. and Peters, J.A. (2008). Guide to Receptors and Channels (GRAC), 3rd edition (2008). British Journal of Pharmacology, 153 (Suppl. 2), S186–S188 [short reference on phosphodiesterase types].

Gijsbers, R. (2003) Domain Structure and Function of Nucleotide Pyrophosphatases/Phosphodiesterases (Npps). Leuven University Press. ISBN 9058673006.

Manganiello, V.C., Smith, C.J., Degerman, E. and Belfrage, P. (1990) in Cyclic Nucleotide Phosphodiesterases: Structure, regulation and drug design. Beavo, J. and

Houslay, M.D. eds, pp 87-116. John Wiley and Sons, Ltd., Chichester.

Milatovich, A., Bolger, G., Michaeli, T. and Francke, U. (1994) Chromosome localizations of genes for five cAMP-specific phosphodiesterases in man and mouse. Somatic Cell and Molecular Genetics, 20, 75–86.

Soderling, S.H. and Beavo, J.A. (2000) Regulation of cAMP and cGMP signaling: new phosphodiesterases and new functions. Current Opinion in Cell Biology, 12, 174–179.

The GTPase superfamily

Bourne, H.R., Sanders, D.A. and McCormick, F. (1991) The GTPase superfamily: conserved structure and molecular mechanism. Nature, 349, 117–127.

Cullen, P.J., Hauan, J.J., Truong, O. Letcher, A.J., Jackson, T.R., Dawson, A.P. and Irvine, R.F. (1995) Identification of a specific Ins(1,3,4,5)P$_4$ binding protein as a member of the GAP1 family. Nature, 376, 527–530.

Fukuda, M. and Mikoshiba, K. (1996) Structure-function relationships of the Mouse Gap1^m: determination of the inositol 1,3,4,5-tetrakisphosphate binding domain. Journal of Biology Chemistry, 271, 18838–18842.

Lowy, D.R. and Willumsen, B.M. (1993) Function and regulation of Ras. Annual Review Biochemistry, 62, 851–891.

Pai, E.F., Krengel, U., Petsko, G.A., Goody, R.S., Kabsch, W. and Wittinghofer, A. (1990) Refined crystal structure of the triphosphate conformation of H-Ras P21 at 1.35Å resolution: implications for the mechanism of GTP hydrolysis. *EMBO Journal*, 9, 2351–2359.

Schlessinger, J. (1993) How receptor tyrosine kinases activate Ras. *Trends in Biochemical Sciences*, 18, 273–275.

Schlichting, I., Almo, S.C., Rapp, G., Wilson, K., Petratos, K., Lentfer, A., Wittinghofer, A., Kabsch, W., Pai, E.F., Petsko, G.A. and Goody, R.S. (1990) Time resolved X-ray crystallographic study of the conformational change in Ha-Ras P21 protein on GTP hydrolysis. *Nature*, 345, 309–315.

Symons, M. and Settleman, J. (2000) Rho family GTPases: more than simple switches. *Trends in Cell Biology*, 10, 415–419.

Vojtek, A.B., Hollenberg, S.M. and Cooper, J.A. (1993) Mammalian Ras interacts directly with the serine threonine kinase Raf. *Cell*, 74, 205–214.

Other Ras related proteins

Brown, H.A., Gutowki, S., Moomaw, C.R., Slaughter, C. and Sternweis, P.C. (1993) ADP-ribosylation factor, a small GTP-dependent regulatory protein, stimulates phospholipase D activity. *Cell*, 75, 1137–1144.

8 Inositol phosphate metabolism and roles of membrane lipids

Inositol lipids and compounds derived from these lipids are used by cells to transmit messages into their interior. One of the main pathways here can be controlled by G proteins (as discussed in Chapter 7), which transmit the signal from a cell surface receptor, leading to the production of compounds that control phosphorylation events and calcium signalling. The role of calcium is discussed in the next chapter. As with other intracellular signals, mechanisms using lipids and their derivatives allow for amplification of the signal, and allow for divergence, potentially regulating more than one pathway. The inositol compounds used as signals can also be further modified, giving the potential for much more signalling. Hence, mechanisms discussed in this chapter are central in the control of many cellular events.

As with all signalling events, dysfunction of these pathways can also lead to disease, and such pathways are targets for manipulation by drugs.

8.1 Introduction

One of the key events in signal transduction in cells that takes place on the plasma membrane is the perception of the signal, facilitated by a receptor. But this is not the only role that the plasma membrane has in signalling. An important mechanism in transducing the signal into the cell involves the modification of some of the lipids that comprise the membrane. Such modifications might involve their phosphorylation, or their breakdown to yield further signalling molecules.

Between 2 and 8% of the lipids of eukaryotic membranes contain inositol. The three main forms of these lipid structures are phosphatidylinositol (PtdIns), phosphatidylinositol 4-phosphate (PtdIns 4-P) and phosphatidyl-inositol 4,5-bisphosphate (PtdInsP$_2$: otherwise abbreviated to PIP$_2$), with the non-phosphorylated form, phosphatidylinositol accounting for more than 80% of the total inositol lipid content. Although in relatively small concentrations, these lipids are nonetheless extremely important, not just as part of the membrane structure, but as molecules that can be used in signalling.

A major route in signalling can occur as key proteins on the membrane are responsible for cleavage of certain inositol lipid molecules into smaller, diffusible molecules, which then convey the signal to other parts of the cell. Primarily, PtdInsP$_2$ is broken down by the enzyme phospholipase C to give the primary products, inositol 1,4,5-trisphosphate (InsP$_3$) and diacylglycerol (DAG), both of which can diffuse and signal into the cell. This pathway, like others, gives a great deal of amplification to the signal, as one activated phospholipase C molecule will produce many InsP$_3$ and DAG molecules, but further to this, importantly, the system leads to great divergence of the signal. InsP$_3$ is responsible for the release of Ca^{2+} ions from the intracellular stores, which will lead to the activation of the calcium arm of the signalling pathways, including the turning on of calmodulin and its associated effects, as discussed in the next chapter, whereas DAG leads to the activation of protein kinase C and the associated phosphorylation of a host of proteins along with modulation of their activity, as discussed in **Chapter 6**. A typical scheme is depicted in **Figure 8.1**. Therefore, inositol phosphate metabolism can be

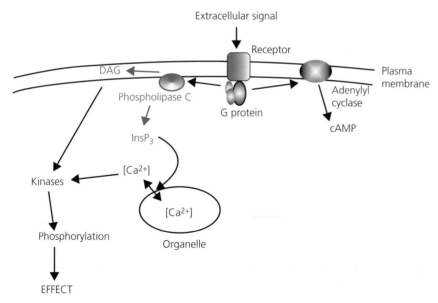

Figure 8.1 The control and role of phospholipase C in inositol phosphate metabolism and its influence on kinase activity and intracellular calcium release. The role of phospholipase C is highlighted in blue.

seen as a keystone pathway linking events at the membrane with other transduction pathways deep within a cell.

Once this system was discovered, it was thought that this was probably the end to the story. It appeared to answer many questions. However, a plethora of reports since have indicated that it is just not that simple, perhaps as would be expected in cell signalling. For example, more than one type of membrane associated lipid can be cleaved, whereas $InsP_3$ can lead to a host of secondary products, and gradually many of these have been assigned signalling roles within the cell.

At this point it is probably worth mentioning nomenclature. Here, inositol 1,4,5-trisphosphate is referred to as $InsP_3$, but many reports now find that a more full abbreviation is needed to clarify which trisphosphate is being referred to, and a similar story goes for other phosphorylated derivatives. For example, phosphatidylinositol bisphosphate could be phosphorylated at positions 3 and 4, or at positions 4 and 5. However, to try to simplify the situation, here I have used a minimalist approach and the abbreviation refers to the most important isomers recorded, that is, $InsP_3$ for inositol 1,4,5-trisphosphate and $PtdInsP_2$ for phosphatidylinositol 4,5-bisphosphate, whereas other derivatives that have a less well-defined role within the cell are given a more full name and the numbers as well for clarity. It should also be noted that some researchers and journals use another type of nomenclature, the "Chilton" forms, that is, $InsP_3$ is known as IP_3 and $PtdInsP_2$ as PIP_2. Conventions have been published that give guidelines to which nomenclature to use, but many journals still accept publication of both forms, that is, $InsP_3$ or IP_3.

Numbering for inositols

The numbering for the inositols is derived from the position of the carbon around the inositol to which the extra phosphates are attached, i.e. not the one used as part of the backbone of the structure. An example can be seen in Figure 8.2. Thus, when Ptds 4,5-bisphosphate is cleaved, the numbering of the derivative includes the phosphate on carbon number 1, and hence the product is a trisphosphate 1,4,5.

8.2   Events at the membrane

The main inositol containing lipid cleaved in the membrane, which leads to the production of $InsP_3$ and DAG, is $PtdInsP_2$. The hydrophobic tails of the lipid comprise part of the inner leaflet of the membrane bilayer, whereas the inositol group and phosphates stick out to the cytoplasm of the cell (Figure 8.2). Therefore, on cleavage, the lipid part, that is the DAG is left in the membrane, whereas the inositol phosphate which, as we shall see, needs to diffuse to the endoplasmic reticulum, is produced in the cytoplasm where it can have its action.

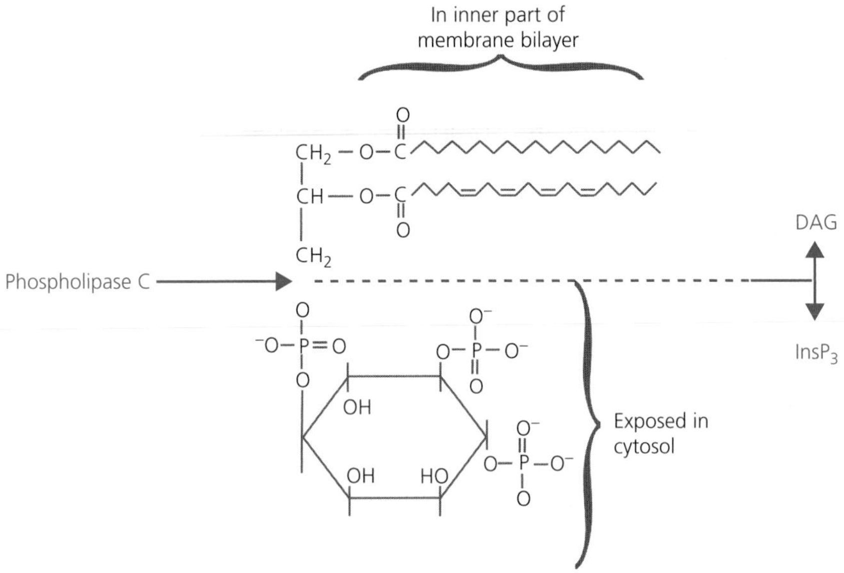

Figure 8.2 The molecular structure of phosphatidylinositol 4,5-bisphosphate and its breakdown by phospholipase C to form InsP$_3$ and DAG.

PtdInsP$_2$ is itself produced by the sequential phosphorylation of phosphatidylinositol (PtdIns). PtdIns is first phosphorylated to phosphatidylinositol 4-phosphate (PtdIns 4-P) by PtdIns 4-kinase and then further by the addition of a phosphoryl group to the 5 position of the inositol ring, catalyzed by PtdIns 4-P 5-kinase, to produce PtdInsP$_2$. The former enzyme, PtdIns 4-kinase has been purified from several sources, including liver and brain, and in most tissues has been found to be associated with the membranes. PtdIns 4-P 5-kinase, on the other hand, has been found to be in both the particulate and soluble fractions of cells.

Therefore, it can be seen that PtdInsP$_2$ is extremely important for signalling in cells. However, as we shall see when discussing the inositol lipids and their derivatives, the story is never quite as simple as first thought. PtdInsP$_2$ is phosphorylated at positions 4 and 5 of the inositol ring, but several phosphatidylinositol derivatives that have also been phosphorylated on the 3 position of the inositol ring have been found to exist (**Figure 8.3**). These were first identified in fibroblasts by Whitman and colleagues in 1988, and it was suggested that they were involved in the control of cell proliferation. However, their presence in cells that are differentiated and do not undergo proliferation suggests that other functions also need to be assigned to these lipids. More lately, they have been shown to be involved in events that lead from the perception of insulin for example, as discussed further in **Chapter 11**.

These inositol lipids are interconvertible by simple phosphorylation and dephosphorylation steps, as shown in **Figure 8.3**, and this suggests that some of these lipids are themselves signalling molecules. For example, phosphatidylinositol 3,4,5-trisphosphate (PtdIns(3,4,5)P$_3$), derived mainly from the

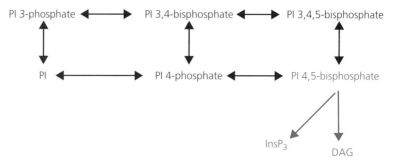

Figure 8.3 The phosphorylated forms of phosphatidylinositol which may be formed and their interconversion. The substrate and products of phospholipase C activity are highlighted in blue.

phosphorylation of PtdInsP$_2$, may be involved in the activation of the small G protein Rac, and has been shown to activate a specific kinase known as 3-phosphoinositide-dependent kinase (PDK1; although there are other forms too). PDK1, as discussed for insulin signalling (**Chapter 11**), can activate PKB and possibly PKC, for continuation of signal transduction.

The kinase involved in the production of these 3 phospho-derivatives of the inositol lipids, PtdIns 3-kinase, is probably activated though several routes, depending on the type, including tyrosine kinase activity, such as receptor tyrosine kinases, action of monomeric G proteins such as Ras, or insulin signalling via the multiphosphorylated protein insulin receptor substrate 1 (IRS-1). Other adaptor proteins might also be instrumental in the control of PtdIns 3-kinase. Research into the role of PtdIns 3-kinase has been greatly facilitated by the finding that it can be reasonably specifically inhibited by the fungal metabolite, wortmannin, as well as other inhibitors (such as LY294002).

PtdIns 3-kinase is not a single entity, but a family of proteins that can be grouped in classes: IA, IB, II and III. They all contain a kinase domain as expected for their activity, and they all appear to contain a kinase accessory domain (PI3Ka domain: N terminal to the kinase domain), and most contain towards the N-terminus a protein kinase C homology domain 2 (C2 domain: Class II as an extra C2 domain at the C terminal end). Class IA, IB and II also contain Ras binding sites, whereas class IB also has binding sites for βγ subunits from trimeric G proteins. Therefore, by a close look at the domain structures here, the spectrum of interactions that may be used for the control of each class of kinase can be predicted (**Figure 8.4**).

As discussed in previous chapters, the reversal of signalling must take place, and the signalling through the PtdIns 3-kinase pathway is no exception. The PtdIns(3,4,5)P$_3$ produced has to be dephosphorylated to remove the signal, and this is catalyzed by a phosphatase. The phosphatase that seems to be most involved here is the product of the tumour suppressor gene *PTEN*. PTEN is expressed in all eukaryotic cells. The human enzyme structure shows that it has a phosphatase domain and C2 domain, although in yeast the C2

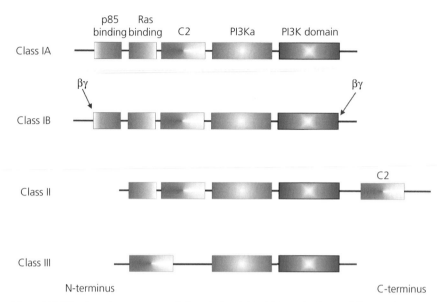

Figure 8.4 The domain structures of phosphoinositide 3-kinase isoforms. C2; protein kinase C homology domain 2: PI3Ka; PtdIns 3-kinase accessory domain: βγ; indicates binding site for β/γ subunits of trimeric G proteins.

domain is missing. PTEN is not the only phosphatase that has this activity, other phosphatases are also involved in the removal of PtdIns$(3,4,5)$P$_3$, including ones called SHIP1 and SHIP2.

The lipid PtdInsP$_2$, which is used as a starting point for making other phosphorylated forms, as well as being the main source of DAG and InsP$_3$, has been shown to have roles in its own right. For example, it has been found to be involved in interactions with proteins associated with the cytoskeleton, particularly with gelsolin and profilin, and such function has been shown to be important in the control of cytoskeletal formation.

> **PTEN** The name PTEN comes from "phosphatase and tensin homolog on chromosome 10", but it is also known as MMAC1 or TEP1.

8.3   The breakdown of the inositol phosphate lipids

Phospholipase C

The hydrolysis of phosphatidylinositol 4,5-bisphosphate (PtdInsP$_2$) in the lipid bilayer of the membrane is catalyzed by the enzyme phospholipase C, often referred to as PLC. This enzyme also hydrolyzes the other inositol lipids PtdIns and PtdIns 4-P. This reaction releases the inositol phosphates such as InsP$_3$ and DAG (**Figure 8.2**).

Protein based studies, and genomic and cloning work, have revealed the existence of many isoforms of phospholipase C, and therefore again we see that what was once thought of as a single protein entity is in fact a family of polypeptides, each with the same overall function, but with subtleties in their actions and control, which would account for their roles in signalling in different tissues. In mammals, for example, nine isoforms have been identified, which can be classified into four main groups, α, β, δ and γ. At the N-terminal end of most, for example forms β, δ and γ, there is a PH domain that probably aids in the enzymes' association with the lipids of the membrane. Next, comes an EF hand region (see Chapter 9), allowing calcium ion chelation, and hence control by the intracellular calcium ion concentration. Then we find the catalytic core, which breaks down the lipids and creates the signalling molecules. At, or towards, the C-terminal end, we see a protein kinase C homology domain 2 (C2 domain). Members of the β class of phospholipases also have a G protein interaction domain at their extreme C-terminal ends, whereas the γ class have extra domains associated with protein/protein interactions, including two SH2 domains, an SH3 domain and a slit PH domain.

Therefore, activation of PLCs can again be predicted from a study of the structures of the isoforms (as above with PtdIns 3-kinase). One of the major mechanisms for turning on PLC is through the interaction with components of the heterotrimeric G proteins (which were discussed in more detail in Chapter 7). PLCβ1 is activated by the α subunit of the trimeric G protein G_q, that is $G_q\alpha$ (Figure 8.5). As discussed above, the β class of phospholipases contain a G protein interaction domain, and two sites, referred to as P and G, at the C-terminal end of PLCβ1 have been found to be important for this interaction. As predicted from the structures, this G protein subunit has no stimulatory activity on any of the other forms of PLC. However, it should be noted that it is not only the α subunit of the trimeric G proteins which is important for the control of the β isoforms, but also the β/γ subunit complex of the trimeric G protein family has been seen to activate the β2 and β3 isoforms of phospholipase C in some cell types.

Phosphorylation has also been seen to be crucial in the activation of other isoforms. PLCγ is phosphorylated on some tyrosine residues, usually three at positions 771, 783 and 1254. This may be catalyzed by a tyrosine kinase-linked receptor, for example, by the epidermal growth factor receptor (Figure 8.5B). Treatments with protein tyrosine phosphatases suggest that this is an important step in the activation of this enzyme. When the cell is in the unstimulated state, PLCγ is mainly found in the cytosol. However, a sequence of events similar to that seen with the activation of the monomeric G proteins (Chapter 7) may occur. On binding to its ligand, the tyrosine kinase-linked receptor firstly has to dimerize and phosphorylate itself. The phosphoryl groups added to the receptor create binding sites for the SH2 domains of PLCγ, and so allow the translocation of PLCγ to the membrane and its association with the receptor, and subsequent phosphorylation, and activation, of the PLC enzyme. Therefore, such a mechanism not only activates this enzyme, but also brings the PLCγ to the membrane where its substrate resides. Hence, it is only active when, and where, needed.

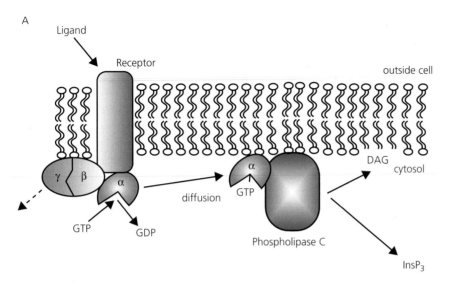

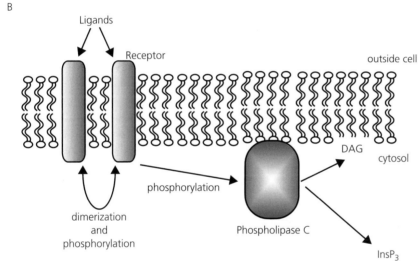

Figure 8.5 The activation of phospholipase C may be by different mechansims depending on the isoform. A: activation may be through a receptor coupled to a trimeric G protein, where the α subunit acitvates phospholipase, as with PLCβ1. Alternatively, B: activation may be through a tyrosine kinase-linked receptor, leading to phosphorylation of phospholipase C on tyrosine residues as see which PLCγ.

Interestingly, phosphorylation on a serine residue by both protein kinase C and cAMP-dependent protein kinase has been reported to be inhibitory, and therefore may reflect a negative control on the activity of PLC and a reduction in the formation of subsequent intracellular signals.

It has also been reported that Ca^{2+} may well be involved in PLC control. PLCδ contains one EF hand motif, which will confer Ca^{2+} binding potential to this protein, and therefore PLCδ may represent a Ca^{2+} controlled form of the PLC family. Therefore, taken together, it can be seen that various PLC isoforms can be controlled in a variety of ways, depending on the isoform. Ca^{2+}, G proteins and kinases are all involved in controlling $InsP_3$ and DAG formation.

As mentioned earlier, inositol containing lipids have been found associated with the cytoskeleton and it has been shown that if the lipid is associated with profilin, an actin binding protein, then the activity of PLCγ is severely reduced, suggesting that this too might be a regulatory mechanism.

A PLC has also been found that hydrolyzes phosphatidylcholine (PC). This enzyme is a heterodimer and preferentially hydrolyzes PC rather than $PtdInsP_2$ as a substrate.

8.4 Inositol 1,4,5 trisphosphate and its fate

Although, originally, the only inositol phosphate formed from the hydrolysis of the inositol lipids to be assigned a role in cell signalling was $InsP_3$, it is clear that the hydrolysis of the membrane lipids is only the first step in a very convoluted and complicated pathway. $PtdInsP_2$ is not the only lipid to be hydrolyzed. The cleavage of PtdIns will lead to the formation of inositol 1-phosphate, whereas the hydrolysis of PtdIns 4-P leads to the formation of inositol 1,4-bisphosphate, with $InsP_3$ itself arising from the hydrolysis of $PtdInsP_2$. So, it can be seen that different lipids will give rise to different inositol phosphate products. Furthermore, as alluded to, inositol has the capacity to contain a variable number of phosphate groups, ranging from none, that is inositol, right through the spectrum of compounds up to the hexaphosphate form, $InsP_6$. $InsP_3$ is poised in the middle of this jungle of compounds and therefore can lead to the formation of lower inositol phosphates, that is, the inositol compounds containing one or two phosphate groups. Alternatively, $InsP_3$ can lead to the formation of the higher phosphate forms, that is, the four, five or six phosphate containing compounds. Some, but not all of the compounds that can be formed, along with their interconversions, are shown in **Figure 8.6**. The addition of phosphate groups to the inositol ring is catalyzed, as expected, by kinase enzymes, whereas the removal is catalyzed by phosphatases.

So, what function does this interconversion of the phosphate forms serve? Firstly, it deactivates inositol phosphates, that is, turns off the signal that they are relaying. For example, if $InsP_3$ is converted to another form, either with more phosphates or fewer phosphates, then the concentration of $InsP_3$ itself is lower, and therefore it has a reduced capacity to cause the release of Ca^{2+} from the intracellular stores, and so the end result is in effect a turning off of that signal. Secondly, the new inositol phosphate formed might itself have a signalling role. $InsP_3$ is not the only inositol phosphate to have been shown to be a signal. More and more of these inositol phosphates are being assigned roles in the control of cellular functions, so the removal of one form, reducing

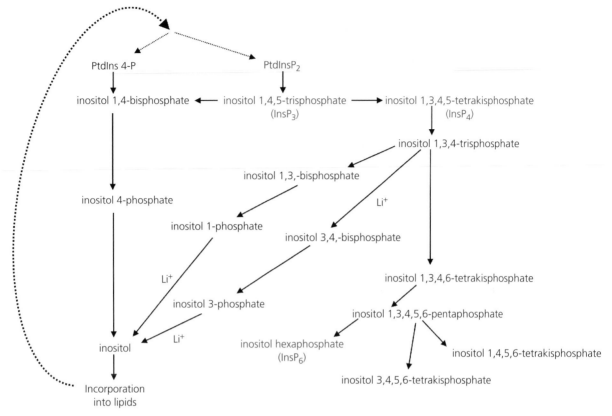

Figure 8.6 Some of the most important inositol phosphate metabolism pathways, with the influence of lithium ions included. The most highly studied compounds are highlighted in blue.

the signal associated with the rise in its concentration, may well be leading to the rise in the concentration of another form and the formation of a new signal. It is beyond the scope of this book to attempt to look at all the isoforms involved and their postulated roles, but some of the more prominent ones and the enzymes involved will be discussed.

As mentioned, $InsP_3$ lies in the middle of the inositol phosphate phosphorylation states, where phosphates can be added or removed to form other compounds. Addition of a phosphate can be carried out by inositol 1,4,5, trisphosphate 3-kinase, yielding a new inositol, inositol 1,3,4,5-tetrakisphosphate, commonly known as $InsP_4$ (or IP_4). The enzyme involved is a soluble protein, which, like most kinases, uses ATP as a substrate and requires Mg^{2+}. The K_m for its substrate is lower than that of the phosphatases that can remove phosphate groups from $InsP_3$, and hence this might be the favoured reaction, that is, $InsP_4$ may be more likely to be made than $InsP_2$. The kinase contains two catalytic subunits and also contains calmodulin allowing stimulation via the Ca^{2+} pathway, whereas additional regulation of the kinase is almost certainly via other kinases and phosphorylation, either by protein kinase C or cAMP-dependent protein kinase.

As for the role of the $InsP_4$ that is formed from the phosphorylation of $InsP_3$, some debate has occurred. It is thought that it is involved like $InsP_3$ in the immobilization of Ca^{2+} from internal stores, such as seen with histamine receptor stimulation, but some papers refute the role of $InsP_4$ in Ca^{2+} release here. Some studies suggest that $InsP_4$ stimulates an ATP-independent uptake of Ca^{2+} into the sarcoplasmic reticulum, whereas others suggest that it may be involved in the control of Ca^{2+} back into the cell through the plasma membrane. Certainly, proteins that are capable of binding inositol tetrakisphosphate have been reported. Putative plasma membrane associated receptors for $InsP_4$ in human platelets have been reported. One protein, isolated from pig cerebellar tissue was found to bind inositol 1,3,4,5-tetrakisphosphate 100 times better than $InsP_3$. Other groups have reported two $InsP_4$ binding sites associated with the nuclear envelope. One of these is in the outer nuclear membrane, and binds to $InsP_4$ with relatively high affinity and is involved in the uptake of nuclear Ca^{2+}. The other binding protein is on the inner membrane and has much lower $InsP_4$ binding affinity. Interestingly, binding of $InsP_4$, and $InsP_3$, to putative receptor sites in rat cortical membranes appears to be enhanced by peptides derived from β-amyloid protein, which is a protein associated with Alzheimer's disease.

Of course, one of the key proteins seen to be able to bind to $InsP_4$ is a protein with GAP (GTPase activating protein) activity, which was discussed in Chapter 5. This protein, a member of the GAP1 family, GAP1^{IP4BP}, affects the activity of monomeric G proteins, and therefore $InsP_4$ may not only be involved in the control of intracellular Ca^{2+} but also can control G protein function. Often, such G proteins are involved in the pathways that control gene expression, so $InsP_4$ may be involved in long-term, as well as short-term, effects in the cell.

As well as the tetrakis forms of the inositol phosphates, further phosphate groups can be added to create the penta- and hexaphosphate derivatives. There are theoretically six possible forms of inositol pentaphosphate, but in animals the most favoured form is the one referred to as inositol 1,3,4,5,6-pentaphosphate. Such compounds are also found in plants and in *Dictyostelium*, suggesting that inositol pentaphosphate also might have a defined role. Inositol pentaphosphates, at least in animals, appear to be produced by the route outlined in Figure 8.4, that is, $InsP_3$ is converted to inositol 1,3,4,5-tetrakisphosphate, which, by the removal of a phosphate from position 5, creates the trisphosphate, inositol 1,3,4-trisphosphate. Addition of a phosphate to position 6 will lead to the production of 1,3,4,6-tetrakisphosphate and a further addition of a phosphate to position 5 will produce inositol 1,3,4,5,6-pentaphosphate. Once produced, this new inositol compound can lead to the production of lower inositols, such as inositol 3,4,5,6-tetrakisphosphate or inositol 1,4,5,6-tetrakisphosphate, or by the further addition of a phosphate group, can lead to the production of inositol hexaphosphate, known as $InsP_6$. Inositol 3,4,5,6-tetrakisphosphate has been shown to accumulate in a dose-response related manner. Its exact roles remain to be determined, although it has been shown to selectively block epithelial Ca^{2+}

activated chloride channels. In the meantime, such inositol compounds, which have no defined function, have been dubbed "orphan signals".

Upon stimulation of some cells, for example, the premyeloid cell line HL-60, the concentrations of the penta- and hexa-inositol phosphates rise rapidly, certainly within minutes, although in other systems the peak of accumulation of these compounds may take hours if not days. However, the exact function of these higher inositol phosphates is unclear, and it has been proposed that the pentaphosphate form may just be a precursor of the four phosphate derivative inositol 3,4,5,6-tetrakisphosphate. It has even been proposed that the penta- and hexa- forms might have an extracellular role. Using cerebellar membranes, an inositol hexaphosphate binding protein has been found, but a full understanding of the functions of such proteins has yet to be fully uncovered.

As well as the addition of phosphates to the inositol ring, their removal is also crucially important, not only to create even more derivatives of the inositol phosphates but also to liberate free inositol, which will be used to make new inositol lipids. The removal of the phosphate from position 5 of several inositol phosphates is carried out by inositol polyphosphate 5-phosphatase. This enzyme can use $InsP_3$ as a substrate, so creating inositol 1,4-bisphosphate. This inositol is also formed by the hydrolysis of the lipid PtdIns 4-phosphate. Removal of the 5 phosphate from $InsP_3$ will remove the $InsP_3$ from the signalling pathway and so effectively turn off its message. However, this enzyme can also use inositol 1,3,4,5-tetrakisphosphate and cyclic inositol 1,2,4,5-trisphosphate as substrates. The phosphatase enzyme exists in many isoforms, which have been found located both in the soluble and particulate fractions of cells. In human platelets, for example, two immunologically distinct forms have been isolated (Type 1 and Type II). The Type I enzyme can be phosphorylated by protein kinase C with a concomitant increase in its activity, suggesting that protein kinase C may be instrumental in the reduction of $InsP_3$ levels in cells, and may be a trigger for the termination of the $InsP_3$ signal. In other tissues, other isoforms have been isolated, with diverse molecular weights, and in some cases the substrate specificities vary. The existence of a variety of forms of the phosphatases, and the potential reactions that such enzymes can catalyze suggests that they are not there merely to remove $InsP_3$ from the cell, and suggests that their products have defined roles in their own rights.

The removal of the 5 phosphate from inositol 1,3,4,5-tetrakisphosphate leads to the production of another trisphosphate, that is, inositol 1,3,4-trisphosphate. This molecule itself can undergo dephosphorylation to produce inositol 3,4-bisphosphate. The enzyme involved here is inositol polyphosphate 1-phosphatase. Its only other substrate recorded is another bisphosphate, inositol 1,4-bisphosphate, and here inositol 4-phosphate is formed. The enzyme is monomeric and it is inhibited by the presence of Ca^{2+}. Of particular significance here is the fact that this reaction is also inhibited by lithium ions, as lithium ions have been used for years as a treatment for manic depression, and experimentally in the laboratory lithium ions have been used to stop inositol metabolism. They inhibit the activity of this enzyme and another phosphatase, inositol monophosphatase, and hence prevent the recycling of inositol, ultimately preventing its re-incorporation back into phosphatidylinositol lipids. The lithium ion concentrations found in patients undergoing treatment are consistent with the K_i of the inhibition by these ions of these enzymes, adding weight to the hypothesis that it is through this action that lithium is having its therapeutic effect. However, lithium ions have effects on other signalling systems too, so care needs to be exercised here in the exact interpretation of data obtained following lithium ion addition.

The recycling of the inositol may also proceed via a different route. As well as the conversion of inositol 1,3,4-trisphosphate to the bisphosphate, inositol 3,4-bisphosphate, an enzyme called inositol polyphosphate 4-phosphatase can convert the 1,3,4-trisphosphate form to a different bisphosphate, inositol 1,3-bisphosphate. It is also responsible for the conversion of inositol 3,4-bisphosphate to inositol 3-phosphate.

The inositol 1,3-bisphosphate formed is converted to inositol 1-phosphate by the enzyme inositol polyphosphate 3-phosphatase. This enzyme exists in more than one isoform, for example, there are two in rat brain. Data from electrophoresis studies suggest that both isoforms exist as dimers. Type I appears as a dimer, while Type II is a heterodimer. The larger subunit of the Type II isoform probably has a regulatory role, as it seems to decrease the efficiency of the catalysis. Interestingly, the substrate specificity of this enzyme is not restricted to the soluble forms of the inositols, and it has even been found that this enzyme can catalyze the removal of a phosphate group from the lipid phosphatidylinositol 3-phosphate to produce phosphatidylinositol. Therefore, this enzyme has the potential to be involved in two very disparate arms of the inositol signalling pathways.

Once the inositol phosphate has undergone dephosphorylation by the above array of enzymes and has been reduced to the monophosphate form, the final dephosphorylation can be carried out by the enzyme inositol monophosphatase. This enzyme will remove phosphates from all the positions around the inositol ring except those attached at the 2 position. The enzyme exists as a dimer of identical subunits. This enzyme is the second one in the inositol phosphate pathway that is inhibited by lithium ions, and here again, the lack of production of free, non-phosphorylated inositol will prevent its re-incorporation back into phosphatidylinositol and hence prevent further production of $InsP_3$ and continual cycling, and therefore signalling.

With the complete removal of all the phosphate groups, the resultant inositol is re-used in the formation of phosphatidylinositol lipids in the endoplasmic reticulum, which will then be taken by vesicular transport to, and re-incorporated back into, the plasma membrane, ready for another round of signalling. Such re-incorporation into the plasma membrane will, of course, also help to restore the integrity of the membrane too.

As well as the phosphorylated forms of the inositols described above and shown in **Figure 8.6** there are also cyclic forms (**Figure 8.7**), that is, where one of the phosphates bridges back across to a second carbon in the inositol ring.

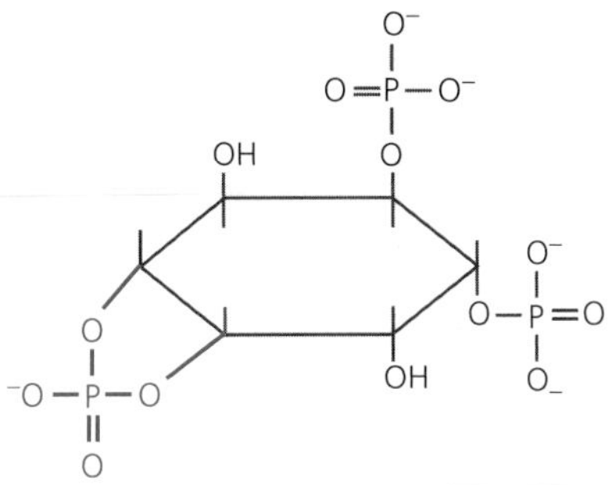

Figure 8.7 The molecular structure of a cyclic inositol compound.

If some cells are stimulated for long periods of time, the cyclic forms of the inositol phosphates tend to increase. Like their non-cyclic counterparts they are produced by the action of phospholipase C, but they are, in general, not substrates for the enzymes that add and remove phosphates from the non-cyclic forms of inositol phosphates. In the normal phospholipase C reaction, the hydrolysis of the bond between the phosphate of the inositol ring and the glycerol backbone of the lipid involves the supply of a hydroxyl group from water. However, this hydroxyl group can also be supplied from the inositol ring itself, and if this happens then the phosphate is cyclized (**Figure 8.7**).

The enzyme that breaks the cyclic phosphate bonds is inositol 1,2-phosphate 2-phosphohydrolase or cyclic hydrolase. However, the only substrate for this enzyme appears to be inositol cyclic 1,2-phosphate. Therefore, presumably all of the cyclic inositol phosphates are finally metabolized through this route, and hence their concentrations in the cell are probably controlled by this enzyme, along with the rate of their production by PLC. The activity of cyclic hydrolase is inhibited by the metal ion Zn^{2+} but activated by Mn^{2+}. Inhibition also occurs by inositol 2-phosphate, suggesting that the levels of cyclic inositols can be controlled by the non-cyclic forms. Interestingly, when cyclic hydrolase was isolated from human placenta it was found to be the same as another protein, lipocortin III. This protein, belonging to a group of eight related polypeptides, was known to be able to bind lipids and also calcium ions. By the study of cDNAs and the use of over-expression techniques, it has been shown that the activity of cyclic hydrolase is correlated to the levels of proliferation of cell cultures. It was suggested that this enzyme was antiproliferative and could be an example of an anti-oncogene product, but cyclic hydrolase was not shown to prevent transformation of cells.

Other forms of inositol phosphates are the inositol pyrophosphates, where two phosphates are joined together and then attached to the inositol ring. Such compounds have been mainly found in *Dictyostelium*. The pyrophosphate groups are usually at the 1 or 3 positions of the ring, but pyrophosphate attached to the 4 and 6 positions has also been seen. Similarly, mammalian cells have been shown to contain pyrophosphate inositols, which are formed by an ATP-dependent phosphorylation of inositol 1,3,4,5,6-pentaphosphate and InsP$_6$, designated InsP$_5$P and InsP$_6$P respectively. However, once again, the significance of these findings has yet to be fully determined.

8.5 The role of diacylglycerol

Any discussion of the role of DAG must emphasize that DAG is not, in fact, a single chemical, but is a family of related compounds, the structures of which are determined by the acyl groups that were present in the original lipid that was hydrolyzed by the phospholipase (refer to **Figure 8.2**). Therefore, it is likely that different DAGs will have subtly different roles in the cell. The main role of DAG has always been seen as an activator of protein kinase C, and there is clear evidence that activation of phospholipase C leads to production of InsP$_3$ and DAG, and that this is accompanied by a rise in protein kinase C activity. Such increases in PKC activity can also be induced by the use of DAG emulators such as phorbol esters.

Phorbol esters Tumour-promoting compounds used for the experimental activation of PKC. They were discussed previously in Chapter 6.

However, DAG may be further metabolized to other compounds, which themselves may have a signalling role. One such compound is phosphatidic acid (PA). This chemical has been seen to stimulate inositol 4,5-bisphosphate formation, activate phospholipase C and act as a cell mitogen. Its production involves the phosphorylation of DAG by diacylglycerol kinase (DGK). Isoforms of this enzyme have been found in both the membrane and cytoplasmic fractions of mammalian cells, with the molecular weights of reported enzymes varying enormously. Interestingly, when two of the isoforms were cloned they were found to contain protein folding regions known as zinc finger domains, suggesting the binding to DNA and possible control of gene expression. It has also been seen that receptor activation can lead to regulation of DGK activity and some forms are phosphorylated by protein kinase C or cAMP-dependent protein kinase.

The breakdown of DAGs, and hence the termination of the signal they carry, is an open question, with the exact route for their metabolism depending on the acyl groups involved. However, like all signals, their removal from transduction pathways is important, otherwise the signals would perpetuate beyond the period needed, with the obvious consequences.

8.6 Inositol phosphate metabolism at the nucleus

When the inositol pathways were first unravelled, metabolism of inositol lipids was thought of as an event at the plasma membrane. However, the same biochemical pathways may also take place at the nucleus, that is PtdInsP$_2$, and possibly PtdIns 4-phosphate, are broken down, yielding InsP$_3$ and DAG, which can then carry the signal to their relevant effectors. Interestingly, many of the studies involving the nucleus have been carried out following the removal of the nuclear membranes, suggesting that it is the lipids and enzymes associated with the skeletal structures within the nucleus which are involved here and not the lipids that constitute the nuclear membranes. This has obvious analogies to the association of the inositol lipids with the cytoplasmic cytoskeleton. Activation of this nuclear pathway is probably involved in the control of the cell cycle, with the likely involvement of protein kinase C activated by the released DAG. DNA synthesis may also be under the control of inositol derivatives such as PtdIns 4-phosphate or inositol 1,4-bisphosphate. Clearly, such involvement of inositol metabolism will be of interest to many cell biologists.

8.7 Other lipids involved in signalling

Most of the work on the involvement of lipids and lipid derived products in cell signalling over the years has been centred on the role of the inositols, especially as it can be seen from the discussion above that there is such a plethora of these compounds, and no doubt a plethora of effects and roles. However, it is not just these inositol-based lipids that are active in cell signalling. Here, the roles of some of the others are briefly discussed.

Phosphatidylcholine and arachidonic acid metabolism

Lipids other than the phosphatidylinositols are used for signalling from the plasma membrane to the inside of a cell. For example, it is well documented that phosphatidylcholine (PC) is also the precursor of signalling molecules. PC can be hydrolyzed by several phospholipase enzymes, including phospholipase A$_2$ (PLA$_2$); phospholipase C (PLC), or phospholipase D (PLD, **Figure 8.8**).

As a general rule, agonists that lead to the increase in hydrolysis of PC also cause an increase in the breakdown of PtdIns in the same cell. This can lead to the rise in DAG being biphasic. Firstly, the breakdown of PtdIns leads to a rapid increase in the concentration of DAG, but slightly later, the hydrolysis of PC leads to a more prolonged accumulation of DAG, which is accompanied by production of choline and phosphorylcholine.

Importantly, however, PC, along with the breakdown of PtdIns and PE, is the major source of intracellular arachidonic acid, which itself has been implicated as a major signalling molecule, as well as leading to the production of eicosanoids. As arachidonic acid is most commonly found in the second, or middle, position of the glycerol backbone of the phospholipid, its release is catalyzed by the enzyme PLA$_2$ (**Figure 8.9**).

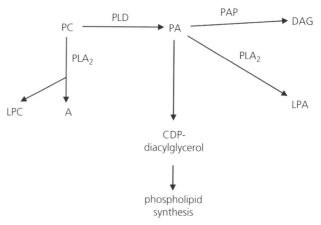

Figure 8.8 The breakdown of phosphatidylcholine (PC) by phospholipase A$_2$ (PLA$_2$) and phospholipase D (PLD). AA, arachidonic acid; DAG, diacylglycerol: LPA, lysophosphatidic acid; LPC lysophosphatidylcholine; PA, phosphatidic acid; PAP, phosphatidate phosphohydrolase.

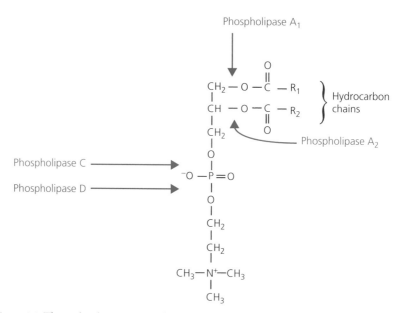

Figure 8.9 The molecular structure of phosphatidylcholine and its breakdown by phospholipase A$_2$, phospholipase A$_1$, phospholipase D and phospholipase C. R denotes the fatty acid chains where R$_2$ is likely to be arachidonic acid.

PLA$_2$ is commonly found associated with the plasma membrane of cells, but it can also reside in the cytoplasm, or even be associated with intracellular membranes. Purification has shown that the enzymes are usually approximately 80–100 kDa in size and show activation by Ca^{2+}. The enzyme contains in its N-terminal region a Ca^{2+}-dependent phospholipid binding sequence, which mediates its interaction with the membrane. Activation in this way is associated with a concomitant translocation of the enzyme to the membrane, and therefore it is activated and moved at the same time, ensuring that it is only active when and where needed. This type of activation/translocation has been discussed above for other enzymes involved in cell signalling, such as PLCγ. PLA_2 will also hydrolyze PtdIns and PE at this second carbon of the glycerol backbone, and, although arachidonic acid is the usual fatty acid there, PLA_2 is not specific for this substrate.

The activation of PLA_2 activity is certain to be complex. Besides its activation by Ca^{2+}, it has also been reported that PLA_2 can be activated by phosphorylation by MAP kinases, as well as by specific isoforms of protein kinase C. Protein kinase C can also lead to the activation of the MAP kinases, which may subsequently cause the phosphorylation and activation of PLA_2. As well as phosphorylation of the enzyme, it is thought that G proteins have a role in PLA_2 regulation. Although effects have been seen on the addition of GTPγ-S, a non-hydrolyzable form of GTP used to lock G proteins in their active state, a direct interaction between a G protein and the enzyme has to be demonstrated, and it is possible that the G proteins are themselves having an indirect effect through a kinase.

So what do the breakdown products created by PLA_2 actually do? Arachidonic acid, produced by hydrolysis of PC, is a 20 carbon unsaturated fatty acid, containing four double bonds. It can act as a signalling molecule, but it can also lead to the production of prostaglandins, thromboxanes, leukotrienes and other eicosanoids. Some of the enzymes that act on arachidonic acid or are involved in its further metabolism include cyclooxygenase, cytochrome P450 and lipoxygenases (Figure 8.10). Cyclooxygenase, otherwise known as prostaglandin G/H synthase, or PGHS, leads to the production of prostaglandins G_2 and H_2 (PGG_2 and PGH_2), which further lead to the formation of other prostaglandins, prostacyclins and thromboxanes. Cytochrome P450 leads to the production of epoxyeicosatrienoic acids (EET), which can be acted on by epoxide hydrolase to release diols, whereas lipoxygenases convert arachidonic acid to hydroperoxyeicosatetraenoic acids (HPETE) leading to the formation of hydroxyeicosatetraenoic acids (HETE). Alternatively, HPETE can lead to the production of leukotrienes and epoxyhydroxides. Added confusion arises when it is realized that some of these enzymes, for example lipoxygenases, can utilize other polyunsaturated fatty acids as well as arachidonic acid, for example linoleic acid, released by PLA_2, and this may also be leading to the production of even more molecules with signalling potential. Furthermore, arachidonic acid can arise from the hydrolysis of DAG via the action of DAG-lipase. DAG, as discussed, comes from the hydrolysis of phospholipids by PLC, which normally would be used in the activation of protein kinase C. However, it can, through DAG-lipase, also lead to arachidonic acid and its further metabolism.

Many of the products of the metabolism of arachidonic acid have regulatory roles within the cell. Ion channels, Na^+/K^+ ATPase activity and cell proliferation are amongst a whole host of functions under such control. Receptors for some of the eicosanoid family have recently been cloned. These, not

■ The enzyme system cytochrome P450, although discussed here as having a signalling role, is more usually associated with the removal of xenobiotics (foreign molecules) such as toxins from an organism.

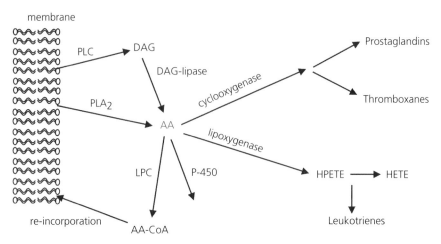

Figure 8.10 The central role of arachidonic acid (AA): its production and further metabolism. DAG, diacylglycerol; HETE, hydroxyeicosatetraenoic acid; HPETE, hydroperoxygeicosatetraenoic acid; LPC, lysophophatidylcholine; PLA$_2$, phospholipase A$_2$; PLC, phospholipase C.

surprisingly, contain seven putative membrane spanning regions and are thought to act through trimeric G proteins in the cell. As well as its further products, arachidonic acid itself can lead to the activation of some isoforms of protein kinase C and to the activation of PLD. No doubt more and more regulatory functions for arachidonic acid and the eicosanoids will come to light.

Arachidonic acid can also be used in the reformation of the membrane lipids. This is through a route that includes firstly its conversion to arachidonoyl-coenzyme A by the enzyme arachidonoyl-CoA synthetase, and then an esterification is catalyzed by the enzyme arachidonoyl-lysophospholipid transferase. This route will obviously reduce the arachidonic acid concentration in the cell available for the production of the host of signalling molecules previously mentioned, as well as reduce the signal that arises directly from the presence of the arachidonic acid.

The formation of arachidonic acid is also seen to be modulated by the presence of a 37 kDa protein called lipocortin, otherwise known as annexin, possibly through the inhibition of PLA$_2$. Its identity was first discovered by studies of the suppression of eicosanoid production by glucocorticoid treatment. However, not surprisingly, lipocortin is, in fact, a family of at least 12 proteins. Different members of the family are characterized by their variable N-terminal regions, but lipocortins appear to contain between four and eight highly conserved repeating domains. These domains are capable of binding to phospholipids in the presence of Ca^{2+} ions. However, studies including the use of peptides designed from the known sequences of the lipocortins have revealed that the proteins may have far wider roles in cellular regulation, and their presence has been implicated in the control of neutrophil migration, differentiation and growth of cells, and in the protection of neuronal tissues.

Hydrolysis of PC by PLA$_2$ not only produces AA, but also leaves behind a by-product, lysophosphatidylcholine (LPC). This too is thought to have a signalling role, and, although its exact action has not been fully defined, it probably acts through protein kinase C.

Some PLA$_2$ enzymes have also been found to be extracellular, for example in snake venom, synovial fluid and pancreatic secretions. These enzymes are of very low molecular weight, less than 20 kDa, and fall into one of two groups, PLA$_2$-I or PLA$_2$-II. Although generally not thought to be involved in cell signalling cascades, their products might have a signalling role on cells in the vicinity of their production.

Phospholipase D

Another enzyme that has an important role in the hydrolysis of PC is phospholipase D (**Figure 8.8**). Like all phospholipase enzymes, it has a specificity for its cleavage site within the lipid, and here this cleavage of the phospholipid yields phosphatidic acid (PA) as a product, and it is thought that a significant amount of DAG in a cell arises from the further breakdown of PA by the enzyme phosphatidate phosphohydrolase (PAP). This, of course, is contrary to the production of PA from DAG by DAG kinase as discussed earlier. This breakdown of PA to DAG, however, has been shown to be increased by several agonists in a wide range of cells. PAP exists in at least two forms, which differ in their subcellular locations as well as their enzymatic properties. PLD, like many enzymes, appears to exist in several isoforms with varying subcellular locations.

Early work was based on the observation that PLD activity was modulated by the addition of stable, non-hydrolyzable GTP analogues such as GTPγ-S. However, the factor that was shown to be the activator of PLD activity is a low molecular weight GTP-binding protein known as ARF (see also **Chapter 7**). This protein was originally found to be required for the ADP-ribosylation of the G protein subunit G$_s\alpha$ in the presence of cholera toxin. ARF has also been shown to have a role in the control of vesicle protein trafficking. Other researchers have also suggested the involvement of a monomeric G protein, Rho, in the regulation of PLD, whereas further control of the activities of PLD in the hydrolysis of the PC also comes from protein kinase C. As well as serine/threonine phosphorylation, tyrosine phosphorylation has been shown to be involved here. Interestingly, hydrogen peroxide causes an increase in PLD activity, and this effect could be mediated by a tyrosine phosphorylation step, but evidence of a direct tyrosine phosphorylation of the PLD polypeptide leaves it open to debate what the exact mechanism of activation is.

The main role of PLD catalyzed breakdown of lipids may be to result in the increase of DAG concentrations in the cell, but also the product PA itself may have signalling properties. It is thought that it may have regulatory roles on other signalling molecules such as adenylyl cyclase (with the concomitant production of cAMP), kinases and phosphatases. PA has also been implicated in the control of superoxide release of neutrophils, which can signal itself or lead to the generation of other ROS signals (see **Chapter 10**), and in thrombin-induced actin polymerization of fibroblasts. Furthermore, the action of PLA$_2$ on PA leads to the production of lysophosphatidic acid (LPA), itself an extra-cellular signal. Receptors for LPA have been found on some cells by the use of photo-reactive analogues of LPA, although, once again, much work is needed to elucidate fully the exact mechanisms involved.

As mentioned above, the concentration of DAG in the nucleus has been shown to be of crucial importance for the regulation of many nuclear events, and a substantial proportion of this DAG may be a result of the breakdown of PC.

An important point that should be borne in mind when studying the hydrolysis of membrane lipids, be it PtdIns, PC or PE, is that the lipids are being removed from the membrane, and therefore the structure and integrity of that membrane may be altered. Such changes in the membrane may well be acting as a control mechanism directly, besides the obvious effects of the products being made. The alteration of lipid membrane fluidity caused by the removal of certain lipids may be involved in the regulation of secretion and neurotransmission, as well as in the control of the activity of some membrane proteins. For example, detergents have been shown to activate the superoxide ion release of neutrophils, the enzyme involved probably responding to the alteration of the lipid structures of the membrane.

Sphingolipid pathways

A further signalling pathway involving the use of lipids, independent of the phospholipase C, phosphatidylinositol and phosphatidylcholine pathways, also came to light a few years ago, and has reached more prominence recently. Sphingosine was reported to inhibit protein kinase C activity, and this has led to a flurry of research that has revealed what is referred to as the sphingomyelin cycle. Essentially, extracellular signals such as γ-interferon, interleukin 1 and complement can lead to the activation of an enzyme, sphingomyelinase, which hydrolyses sphingomyelin (SM) releasing ceramide. Unlike other phospholipids, sphingomyelin does not have a backbone of glycerol, but rather its backbone is sphingosine (**Figure 8.11**). Sphingosine is an amino alcohol, which contains a long unsaturated hydrocarbon chain. To produce SM, a fatty acid is added to sphingosine by an amide bond, creating ceramide, whereas its primary hydroxyl group is esterified to phosphoryl choline. In fact, in the biosynthesis, SM is made by an exchange of the phosphoryl choline head group from phosphatidylcholine to ceramide, catalyzed by the enzyme phosphatidylcholine:ceramide choline phosphotransferase. Breakdown by sphingomyelinase will once again yield ceramide, which can enter a catabolic cascade.

Ceramide has been shown to be an important regulator in protein trafficking, cell growth, differentiation and in the viability of cells, having an important role in the orchestration of apoptosis in some cases. At a molecular level, ceramide has been shown to activate a nuclear transcription factor, NF-κB, in

Complement The complement cascade is a series of proteins involved in the immune system of animals.

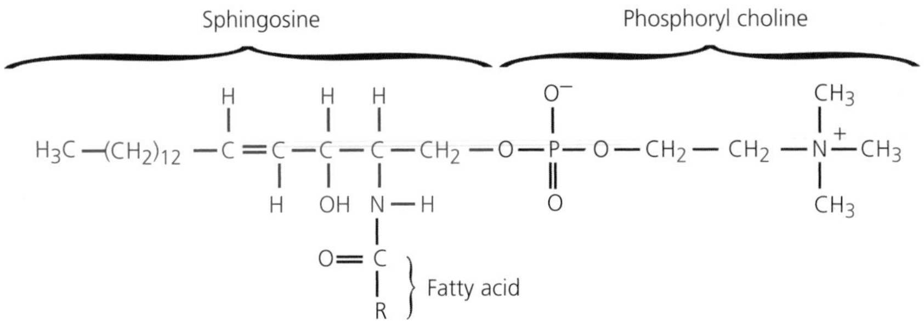

Sphingosine   Phosphoryl choline

$$H_3C-(CH_2)_{12}-C=C-C-C-CH_2-O-P-O-CH_2-CH_2-N^+-CH_3$$

Figure 8.11 The molecular structure of sphingomyelin showing its backbone of sphingosine (in blue).

permeabilized cells, and to be important, not surprisingly, in the control of transcription of several genes. Ceramide also has an effect through phosphorylation, with an activation of MAP kinase cascades, which again can lead to alterations in gene expression. However, it has been shown to have a major control route through the use of a phosphatase. A serine/threonine phosphatase has been identified that is specifically and directly activated by the presence of ceramide and this enzyme has been called ceramide-activated protein phosphatase (CAPP; see **Figure 8.12**). This phosphatase belongs to the 2A group of serine/threonine protein phosphatases, and, accordingly, has a trimeric structure. Therefore, this seems like an elegant signalling cascade with ceramide as a central player, but the overall result being opposite to the

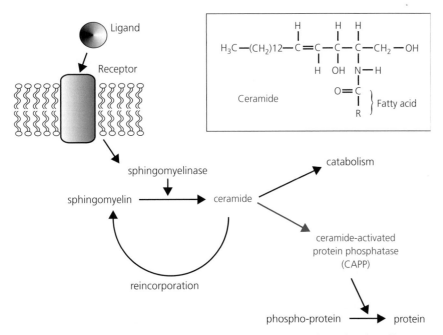

Figure 8.12 The production of ceramide and one of its possible roles within the cell.

promotion of protein phosphorylation seen with the activation of protein kinase C by DAG.

Sphingosine itself has been shown to have a signalling role, and can stimulate the hydrolysis of phosphatidylinositol resulting in the production of InsP$_3$ and the subsequent release of Ca^{2+} ions. Furthermore, another metabolite, sphingosine 1-phosphate, has been shown to have second messenger activity, where it is involved in the inositol-independent release of calcium ions from intracellular stores. Such calcium signalling that does not involve inositols is creating much interest. This is further discussed in **Chapter 9**.

8.8 Related lipid derived signalling molecules

It is now evident that many of the developmental stages of plant growth, as well as many cellular functions of plants are under the control of complex compounds which can be derived from lipids. Some of these compounds were first shown to be of interest as they could induce tuber formation in plants, such as potatoes and yams. One of the major compounds is jasmonic acid (JA). Jasmonic acid is produced by a short metabolic pathway, which starts by the action of lipoxygenase and catalyzes the incorporation of O$_2$ into linolenic acid. Most of the linolenic acid in the cell is found associated with lipids of the plasma membrane and can be released by the action of phospholipase A$_2$. The 13-hydroperoxylinolenic acid produced by the action of phospholipase A$_2$ is further acted upon sequentially by hydroperoxide dehydrase, allene oxide cyclase and a reductase. Three further β-oxidation steps finally yields the 12 carbon jasmonic acid (**Figure 8.13**).

Other derivatives of this compound include methyl jasmonic acid, curcurbic acid and tuberonic acid, the latter named after its tuber-inducing capacity. Just as with the inositol pathway, there are further derivatives of these compounds that have been isolated, and, because of the presence of chiral carbon atoms, several stereoisomers can exist for these molecules, although there is probably a favoured conformation that occurs naturally.

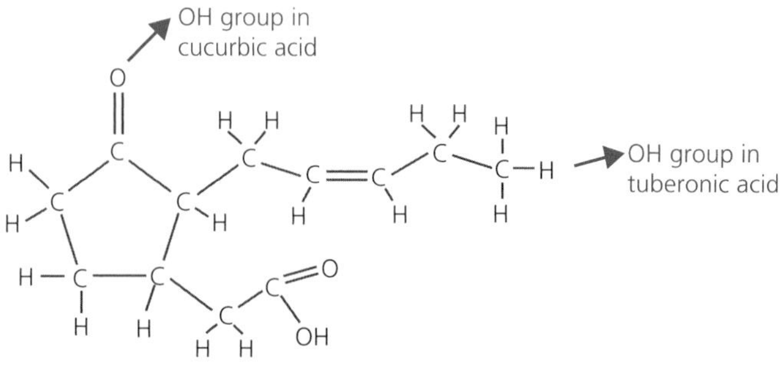

Chiral compounds Ones where there are two forms that are non-superimposable mirror images of each other, and therefore the two forms often have different biological activity.

Figure 8.13 The molecular structure of jasmonic acid, with derivatives shown.

Jasmonic acid and methyl jasmonic acid have been shown to induce senescence in leaves as well as the promotion of tuber formation. If leaves were floated on a solution of JA for 4 days, it was found that the polypeptide profile of the leaves changed as analyzed by SDS-polyacrylamide gel electrophoresis. These new proteins were termed jasmonate-induced proteins or JIPS. Further analysis of the regulation of the synthesis of these proteins indicated that the control was probably at both the levels of transcription and translation. JA has also been shown to induce the expression of other proteins with a known specific function, including storage proteins of leaves, known as vegetative storage protein (VSP). These proteins accumulate in vacuoles of the mesophyllic cells of the leaf and the bundle sheath cells before flowering. The mRNA coding for VSP is increased on the addition of JA. Cell division has also been suggested as a target for the control by JA or its related compounds.

■ Summary

- A central part of many cell signalling pathways is the breakdown of lipids in the plasma membrane, with the subsequent formation of messenger molecules that can transmit the signal(s) into the cell.

- The formation of new inositol lipids by phosphatidylinositol 3-kinase has also been shown to be important in many transduction events, including insulin signalling (see Chapter 11).

- Products of phosphatidylinositol 3-kinase can activate kinases such as protein kinase B.

- Two important signalling molecules are produced by breakdown of PtdInsP$_2$ by phospholipase C: inositol 1,4,5-trisphosphate (InsP$_3$) and diacylglycerol (DAG).

- DAG is known to act as an activator of protein kinase C.

- InsP$_3$ releases Ca^{2+} from intracellular stores (Figure 9.1, and further discussions in Chapter 9).

- Phospholipase C may be activated by the heterotrimeric G protein G$_q$ (see Chapter 7), as well as by other means.

- Inositol trisphosphate (InsP$_3$) may be the precursor of both higher and lower inositol phosphates, that is, it can be phosphorylated or dephosphorylated to create new metabolites, each of which may have a signalling role within the cell.

- Several inositol phosphates are thought to have defined signalling functions, including inositol tetrakisphosphate (InsP$_4$), and binding proteins for several of these compounds have now been reported.

- Several of the enzymes involved in this metabolism are the target for lithium ion inhibition, which might account for its therapeutic action in treating manic depression.

- Other inositols such as cyclic phosphate isomers and pyrophosphates have also been reported.

- Lipid hydrolyzing enzymes such as phospholipase D and phospholipase A$_2$ can generate other lipid-derived signalling molecules.

- Arachidonic acid, generated from lipid breakdown, can be a substrate for cyclooxygenase, cytochrome P450 and lipoxygenase, leading to the formation of prostaglandins, leukotrienes and thromboxanes.

- Other lipid metabolism also has a clear influence on cellular functioning including sphingosine metabolites such as ceramide and sphingosine 1-phosphate, whereas, in plants, jasmonic acid has been shown to have a diverse range of actions, including the induction of senescence of leaves and tuber formation.

→ Further reading

Berridge, M.J. (1993) Inositol trisphosphate and calcium signaling. *Nature*, 361, 315–326.

Michell, R.H. (2008) Inositol derivatives: evolution and functions. *Nature Reviews Molecular Cell Biology*, 9, 151–161.

Events at the membrane

Anderson, K.E. and Jackson, S.P. (2003) Class I phosphoinositides 3-kinases. *International Journal of Biochemistry and Cell Biology*, 35, 1028–1033.

Kodaki, T., Worchoski, R., Hallberg, R., Rodriguez-Viciana, P., Downward, J. and Parker, P.J. (1994) The activation of phosphatidylinositol 3-kinase by Ras. *Current Biology*, 4, 798–806.

Koyasu, S. (2003) The role of PI3K in immune cells. *Nature Immunology*, 4, 313–319.

Smrcka, A.V., Hepler, J.R., Brown, K.O. and Sternweis, P.C. (1991) Regulation of polyphosphoinositide specific

phospholipase C activity by purified G_q. *Science*, 251, 804–807.

Stein, R.C. and Waterfield, M.D. (2000) PI3-kinase inhibition: A target for drug development? *Molecular Medicine Today*, 6, 347–357.

Sulis, M.L. and Parsons, R. (2003) PTEN: from pathology to biology. *Trends in Cell Biology*, 13, 478–483.

Wymann, M.P., Zvelebil, M. and Laffargue, M. (2003) Phosphoinositide 3-kinase signalling: Which way to target? *Trends in Pharmacological Sciences*, 24, 366–376.

InsP$_3$ and its fate

Cowburn, R.F., Wiehager, B. and Sundstrom, E. (1995) Beta-amyloid peptides enhance binding of the calcium mobilising 2nd messengers, inositol (1,4,5)trisphosphate and inositol (1,3,4,5)tetrakisphosphate to their receptor sites in rat cortical membranes. *Neuroscience Letters*, 191, 31–34.

Cullen, P.J., Patel, Y., Kakkar, V.V., Irvine, R.F. and Authi, K.S. (1994) Specific binding-sites for inositol 1,3,4,5-tetrakisphosphate are located predominantly in the plasma membranes of human platelets. *Biochemical Journal*, 298, 739–742.

Malviya, A.N. (1994) The nuclear inositol 1,4,5-trisphosphate and inositol 1,3,4,5-tetrakisphosphate receptors. *Cell Calcium*, 16, 301–313.

Menniti, F.S., Oliver, K.G., Putney, J.W. and Shears, S.B. (1993) Inositol phosphates and cell signalling: new views of InsP$_5$ and InsP$_6$. *Trends in Biochemical Sciences*, 18, 53–56.

Miller, A.T., Chamberlain, P.P. and Cooke, M.P. (2008) Beyond IP$_3$: roles for higher order inositol phosphates in immune cell signalling. *Cell Cycle*, 7, 463–467.

Stephens, L.R., Hawkins, P.T., Stanley, A.F., Moore, T., Poyner, D.R., Morris, P.J., Hanley, M.R., Kay, R.R. and Irvine, R.F. (1991) Myoinositol pentakisphosphate: structure, biological occurrence and phosphorylation to myoinositol hexakisphosphate. *Biochemical Journal*, 275, 485–499.

The role of DAG

Florin-Christersen, J., Florin-Christersen, M., Delfino, J.M. and Rasmusssen, H. (1993) New pattern of diacylglycerol metabolism in intact cells. *Biochemical Journal*, 289, 783–788.

Fujikawa, K., Imai, S., Sakane, F. and Kanoh., H. (1993) Isolation and characterisation of the human diacylglycerol kinase gene. *Biochemical Journal*, 294, 443–449.

Inositol phosphate metabolism at the nucleus

Banfic, H., Zizak, M., Divecha, N. and Irvine, R. (1993) Nuclear diacylglycerol is increased during cell proliferation *in vivo*. *Biochemical Journal*, 290, 633–636.

Divecha, N., Banfic, H. and Irvine, R.F. (1993) Inositides and the nucleus and inositides in the nucleus. *Cell*, 74, 405–407.

Irvine, R.F. (2006) Nuclear inositide signalling—expansion, structures and clarification. *Biochimica et Biophysica Acta (BBA) - Molecular and Cell Biology of Lipids*, 1761, 505–508.

Phosphatidylcholine and arachidonic acid metabolism

Brown, H.A., Gutowski, S., Moonaw, C.R., Slaughter, C. and Sternweis, P.C. (1993) ADP-ribosylation factor, a small GTP-dependent regulating protein stimulates phospholipase D activity. *Cell*, 75, 1137–1144.

Durieux, M.E. and Lynch, K.R. (1993) Signalling properties of lysophosphatidic acid. *Trends Pharmacological Sciences*, 14, 249–254.

Exton, J.H. (1994) Phosphatidylcholine breakdown and signal transduction. *Biochimica Biophysica Acta*, 1212, 26–42.

Piomelli, D. (1993) Arachidonic acid in cell signalling. *Current Opinion Cell Biology*, 5, 274–280.

Sphingolipid pathways

Hannum, Y.A. (1994) The sphingomyelin cycle and the second messenger function of ceramide. *Journal Biological Chemistry*, 269, 3125–3128.

Spiegel, S. and Milstein, S. (1995) Sphingolipid metabolites: members of a new class of lipid second messengers. *Journal Membrane Biology*, 146, 225–237.

Related lipid derived signalling molecules

Anderson, J.M. (1991) Jasmonic acid dependent increase in vegetative storage protein in soybean tissue cultures. *Journal Plant Growth Regulation*, 10, 5–10.

Koda, Y., Kikuta, Y., Tazaki, H., Tsujino, Y., Sakamura, S. and Yoshihara, T. (1991) Potato tuber inducing activities of jasmonic acid and related compounds. *Phytochemistry*, 30, 1435–1438.

9 Intracellular calcium: its control and role as an intracellular signal

Calcium ions have widespread use as signals in cells. Cells expend large amounts of energy expelling calcium ions from the cytoplasm, either to the outside or into organelles. Such ions are subsequently allowed to return to the cytoplasm, constituting a signal. The movement of calcium ions may be controlled through the production of inositol phosphate compounds as discussed in Chapter 8, but many other mechanisms also exist - this is discussed in this chapter.

Calcium ions themselves control a myriad of functions in cells, their action often being mediated by a protein known as calmodulin. The production of other signals, phosphorylation events, and even calcium ion movements themselves, may be controlled.

The movement of calcium ions is so central to cell signalling events that it will have been mentioned in previous chapters, and will appear again in later chapters. Calcium ions control many metabolic and physiological functions in cells, and are even involved in the process of apoptosis; that is, cell suicide. Therefore, the discussion in this chapter is pivotal in gaining an understanding of cell signalling.

Calcium signalling and disease

There are numerous diseases or pathologies in which calcium signalling has been suggested to be involved. Amongst these are Down's syndrome, Brody's disease, diabetes, Alzheimer's disease, hypertension and cardiac hypertrophy.

9.1 **Introduction**

The signalling in which calcium ions are involved is both extremely important and central to cellular communication. Numerous events and activities within the cell, as well as the release of products, perhaps even other signals, from the cell are under direct or indirect control by calcium ions. As will be discussed in this chapter, there are many mechanisms for the control of intracellular calcium, and many effectors that are regulated by the fluctuations that take place in its concentration within the cell.

However, despite its ubiquity, and importance, one respect of the use of calcium as a signal is reasonably unique. Most intracellular signalling molecules are produced by the cell, they have an effect and then they are destroyed, for example, with the production of cAMP, its role in the activation of a kinase and its destruction by phosphodiesterase (see **Chapter 7**). However, with calcium ions this rationale is not followed. The calcium ions are not made when needed, or destroyed after they have served their role. It is the variation in the concentration of calcium ions inside the cell, $[Ca^{2+}]_i$, which constitutes the signal. Calcium ions are simply moved around, not produced or destroyed, but it is the concentration of the ion that is experienced by the effector molecule in a particular compartment in which that effector acts as the message or signal. However, calcium signalling is ubiquitous, being found in bacteria, and right across the diversity of higher organisms, including plants and mammals: it is especially important in the neuronal pathways of higher animals.

The intracellular concentration of Ca^{2+} ions can be controlled in many ways, each of which have to work in a coordinated manner to allow the amount of calcium to be an effective signal. The steady-state level of calcium in the cytoplasm of a cell is usually less than 10^{-7} M. If the cell is activated by an influx of calcium, this level might rise to approximately 10^{-6} M, this higher concentration being the signal to trigger further events in the cell. Putting these concentrations into perspective, the extracellular Ca^{2+} concentration may be in the order of 2×10^{-3} M. Therefore, to maintain this very low steady-state level, with a relatively high extracellular Ca^{2+} concentration, requires calcium ions to be actively pumped out of the cell, a process that uses proteins in the plasma membrane. Cells also sequester Ca^{2+} in organelles within the cell, for example the endoplasmic reticulum, or sarcoplasmic reticulum in muscle tissue, or in the mitochondria. Again, this is an active process. Therefore, a considerable amount of a cell's energy is devoted to control of the intracellular calcium concentration. A signal may be propagated in the cytoplasm for example, when previously excluded calcium is released back into the cytoplasm, so rapidly increasing the Ca^{2+} concentration in the desired area. The calcium may be released from the organelles, or a subpopulation of the organelles, or it may be released back into the cell from the outside. Such release of calcium ions is usually under the control of other second signalling molecules, and in many cases several signalling molecules might be attempting to control calcium movements at the same time.

One of the earliest examples of a signalling molecule involved in the control of calcium ion movements, and perhaps the one which has been most well characterized, is the use of inositol trisphosphate (InsP$_3$), the production of which was further discussed in **Chapter 6**. InsP$_3$ is released from the plasma membrane by the breakdown of phosphatidylinositol bisphosphate (PtdInsP$_2$) and will bind to InsP$_3$ receptors (InsP$_3$R) on the membranes of the intracellular stores. These receptors are also Ca^{2+} channels, which will allow the rapid release of Ca^{2+} back into the cytoplasm of the cell. An example of such a signal transduction pathway is shown in Figure 9.1.

The third, and important, way in which calcium concentrations are controlled in cells is by the presence of calcium binding proteins. Such proteins actually have two roles. Firstly, they remove the free calcium ions from the solution, and therefore effectively reduce [Ca^{2+}]$_i$, and, secondly, the binding of the calcium ions to such proteins is usually the mechanism by which the signal is propagated to the next component of the signal transduction cascade. For example, the action of calcium on an effector molecule, such as a kinase, may be through such a binding protein. One of the most important, and ubiquitous, calcium binding proteins is a protein called calmodulin, which is discussed in more detail below. For calcium ions to bind to proteins, there has to be a recognition of the ion and some form of physical attraction. For this, oxygen atoms are often involved. Calcium ions can coordinate up to 12 oxygen atoms, but commonly coordination numbers range between six and eight. This is exploited by many binding proteins, which bind calcium ions through

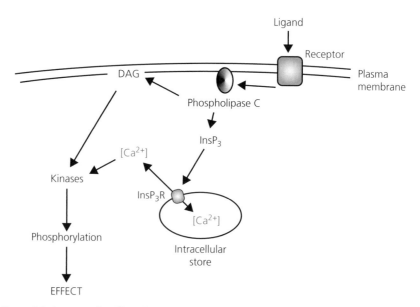

Figure 9.1 An example of how ligand binding can lead to release of Ca^{2+} from intracellular stores. The activation of phospholipase C gives rise to the messenger InsP$_3$, which binds to InsP$_3$ receptors of the endoplasmic reticulum membranes. Such receptors are also channels and open to release calcium ions.

six or seven oxygen atoms, usually provided by glutamate or aspartate residues, which are normally charged at physiological pH and therefore able to partake in this role.

A common feature of many calcium ion binding proteins is the presence of a structure known as an EF hand. This structure gained its name from the E and F regions of the protein parvalbumin, which serves as a calcium buffer in muscle. Parvalbumin has an amino acid secondary structure, which comprises a series of α-helices labelled by letter. The α-helices E and F are positioned in such a way as to point like the forefinger and thumb of a right hand, hence the name, EF hand (**Figure 9.2**). A calcium binding loop containing active calcium binding glutamate and aspartate residues lies between these α-helices. Many calcium binding proteins contain more than one binding site, and it is not unusual to see cooperative binding. It is interesting to note that the presence of EF hands in proteins can be predicted by the use of bioinformatics, and therefore, the control by calcium of the protein activity might be assumed, and therefore tested. An example of this will be met in **Chapter 10**, that is, when isoforms of the NADPH oxidase large subunit were identified from their cDNA sequences, bioinformatic analysis suggested that some variants had EF hands towards their N-terminal ends. Calcium control of the oxidase had already been suggested, and further studies will now enable researchers to confirm which isoforms are involved.

Besides the immensely important calcium binding protein called calmodulin, which is discussed further below, a family of proteins that have been identified as important calcium sensors and also contain EF hands are the S100 Ca^{2+}-sensing proteins. There are more than 25 genes for such proteins in humans, and it is thought that they may control the functioning of nearly 100 other proteins. The S100 proteins appear to have individual cellular locations, which can be modified on calcium binding, and some are even secreted from cells, so extending the range of signalling. Interestingly, around 14 genes from these proteins are clustered in humans (chromosome 1q21), and their dysfunction has been implicated in many diseases including neurodegenerative disorders and cancers.

A second calcium binding domain to be identified is the C2 domain. This protein structure contains approximately 120 amino acids folded into an

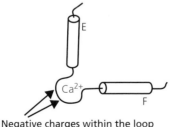

Negative charges within the loop
hold the calcium ion

Figure 9.2 The structure of an EF hand. Two α-helices are connected by a loop structure that provides a calcium ion binding site.

anti-parallel β structure. It is often involved in the association of a protein with a lipid bilayer. Calcium binding to the C2 structure increases a protein's affinity with the membrane, so allowing the protein to become membrane associated, whereas before it was not, and so effectively altering the cellular location of the protein. Membranes involved commonly include the plasma membrane, where proteins would become associated with the inner surface, and this is seen with protein kinase C (discussed in Chapter 6). However, proteins may become associated with other membranes such as the outer surface of mitochondria. It has been estimated that there are over 600 human proteins that may have C2 domains, emphasizing the importance of calcium binding to the control of the functioning of a host of proteins.

As discussed above, calcium binding proteins can potentially be filling two roles in the formation of a calcium signal. They might simply be acting as buffers, so when the calcium concentration alters, free calcium ions are not left in solution to have an effect. Alternatively, the proteins might be acting as the trigger, mediating the rise in free calcium and turning it into the signal, allowing the message to be transmitted down the transduction chain, or in many cases being the last part of the signal transduction chain and acting on the signal when perceived. On binding calcium ions, either at EF hands or elsewhere, these proteins usually undergo a major alteration in their conformation, and this conformational change will undoubtedly expose active sites on the protein and so allow further reactions and propagation of the signal.

One of the most well studied and best characterized examples of a Ca^{2+} binding protein with a clearly defined role is troponin. Troponin is, in fact, a complex of three polypeptides, TnC, TnI and TnT of 18 kDa, 24 kDa and 37 kDa respectively. Along with the protein tropomyosin, troponin lies along the thin filaments of muscle fibres, and it is the TnC subunit that binds Ca^{2+} ions. TnC has two domains that are homologous, and each has two EF hand binding sites for Ca^{2+}, those in the C-terminal domain being of high affinity whereas those in the N-terminal domain are of lower affinity. The domains are held together by an α-helix, the whole molecule showing high structural similarity to calmodulin (see below). On binding of Ca^{2+} to the low affinity sites, that is when Ca^{2+} has been released from the sarcoplasmic reticulum, a conformational change takes place in the TnC polypeptide, which is transmitted through the other polypeptides in the complex and to tropomyosin. A change in the orientation of tropomyosin relieves the steric hindrance it caused over the actin/myosin interaction, allowing muscle function.

Calcium control of muscle

It is interesting to note here that calcium is also involved in the control of glycogen breakdown, so calcium controls the movement of muscle, and also controls the availability of glucose for the formation of ATP to drive the muscle movement.

There are many other examples of calcium binding proteins, and a few will be highlighted here. One is a calcium binding protein called calbindin-D_{28k}. Another important, and rather interesting, example is calsequestrin. This 44 kDa protein is incredibly acidic, having over a third of its amino acids as either aspartic acid or glutamic acid, and, amazingly, each calsequestrin can bind to 43 Ca^{2+} ions. It is found in the sarcoplasmic reticulum, where it controls the concentration of free Ca^{2+} and so helps to keep down the calcium gradient across the membrane. Also present here is a polypeptide known as high-affinity Ca^{2+} binding protein.

But what is it that calcium is controlling in the cell? To answer that question would take a long time as calcium seems to have a plethora of actions, some of which will be discussed, almost in passing, in the other chapters of this book. Perhaps, before answering it briefly, a quick discussion of some of the experimental approaches used to elucidate the role of calcium in cells would be useful. Many of the actions of calcium ions have been unravelled by the use of calcium ionophores, such as A23187 or ionomycin. Such chemicals effectively punch holes in the plasma membrane and let calcium ions flood into the cell, so mimicking the sudden rise of Ca^{2+} concentration, which the cell could achieve by the release of intracellular stores or by allowing calcium ions in from outside by a signal mediated response. Therefore, a researcher would measure the baseline rate of activity, add A23817, and see what happens. If rates are modulated, it suggests that calcium is involved. If not, perhaps calcium has no bearing on the functioning being studied. Alternatively, calcium ions may be removed from the extracellular medium by the addition of chelating agents, such as EDTA, and therefore this will prevent the cell from using the plasma membrane channels as a means to increase cytoplasmic calcium. Radioisotopes can be used to measure calcium concentrations in cells, and with the use of fluorescent probes in conjunction with confocal microsopy (see later in this chapter), the concentrations over time of calcium, and the exact location in cells of the ions, can be determined. By the use of such methods, the diversity of the roles of calcium ions in the signalling of cells can be discerned.

Therefore, by studying calcium per se, and by the use of binding studies, and bioinformatic analysis, the answer to our question is revealed. Some, but by no means all, of the proteins that are known to have their activity modulated by the Ca^{2+} signalling systems are listed in **Table 9.1**.

Certainly, calcium is involved in the activation of Ca^{2+}/calmodulin-dependent protein kinase, protein kinase C, Ca^{2+}-dependent phosphatases (see **Chapter 6**), and in the activation of nitric oxide synthase (see **Chapter 10**) leading to the propagation of intracellular and intercellular signals. It is also involved in the direct activation of many enzymes, such as pyruvate dehydrogenase, as well as being involved in more complex patterns of activity such as the control of the cell cycle. Spikes of calcium have been shown to be essential for triggering the completion of meiosis and the initiation of mitosis in some cells, whereas depletion of calcium stores can arrest cells in the G_0-G_1 phases.

Calcium buffering proteins and disease

The lowering of the levels of some calcium binding proteins in the cytoplasm, such as calbindin, has been implicated in the onset of some diseases, for example Alzheimer's disease.

Table 9.1 Some of the proteins that are controlled by calcium ion concentrations.

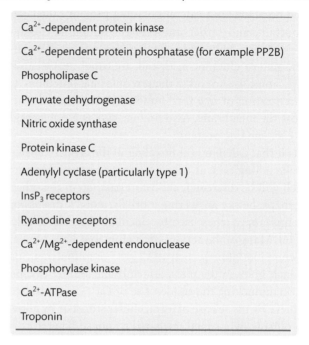

Ca^{2+}-dependent protein kinase
Ca^{2+}-dependent protein phosphatase (for example PP2B)
Phospholipase C
Pyruvate dehydrogenase
Nitric oxide synthase
Protein kinase C
Adenylyl cyclase (particularly type 1)
InsP$_3$ receptors
Ryanodine receptors
Ca^{2+}/Mg^{2+}-dependent endonuclease
Phosphorylase kinase
Ca^{2+}-ATPase
Troponin

Gene expression can be under the control of calcium concentrations, and here again calcium binding proteins such as calreticulin may mediate the changes in concentration into the final signal.

9.2 Calmodulin

Calmodulin is an extremely important protein, which is involved in the mediation of the Ca^{2+} signal in a vast number of regulatory pathways, including the regulation of metabolic activity, the regulation of other signal pathways, such as nitric oxide generation, and the regulation of gene expression.

It is sometimes found as an integral part of a enzyme complex, as in phosphorylase kinase, where it is known as the δ subunit. Often, it even appears to be involved in the contradictory regulation of an activity where it is responsible for an increase in the signal and a decrease at the same time. One example of this can be seen with its role in the control of cAMP levels, where it can increase the production of cAMP via the activation of adenylyl cyclase, while at the same time activating its breakdown by the stimulation of Ca^{2+}/calmodulin (CaM)-dependent phosphodiesterase. However, with tissue-specific expression of enzyme isoforms, such situations are almost certainly not as simple and clear as they might appear. Interestingly, calmodulin can also lead to the activation of the plasma membrane and endoplasmic reticular

Ca^{2+} pumps, which will cause the cessation of the calcium signal. Calcium, through calmodulin, therefore appears to control its own destiny.

In humans there are at least three calmodulin genes, found on chromosomes 2, 14 and 19. However, there are also a family of calmodulin-like proteins.

Calmodulin is a small, relatively acidic protein, which has the capacity to bind four molecules of Ca^{2+}. It has an affinity for Ca^{2+} of approximately 10^{-6} M, which lies between the resting Ca^{2+} concentration of calcium in the cell, usually less than 10^{-7} M, and the activated concentration, which might reach in extreme cases as high as 10^{-5} M. The shape of the protein molecule resembles a dumbbell, with two globular regions connected by a long, flexible and very mobile α-helix (**Figure 9.3A**). A representation of the structure of the N-terminal domain of calmodulin is shown in **Figure 9.3B**.

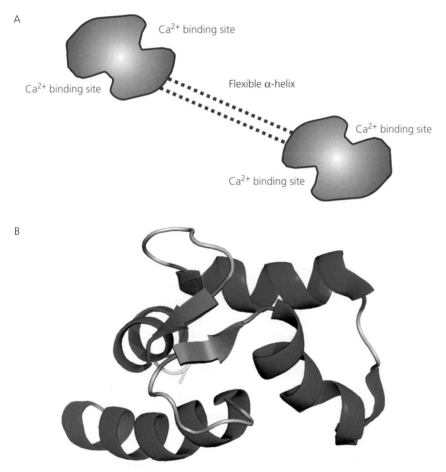

Figure 9.3 A. The domain structure of calmodulin showing the calcium binding sites. B. The proposed structure of the N-terminal domain of calmodulin obtained using NMR. Structure was obtained from the RCSB Protein Data Bank (www.rcsb.org/pdb/) PDB ID: 1f55 (Ishida, H., Takahashi, K., Nakashima, K., Kumaki, Y., Nakata, M., Hikichi, K. and Yazawa, M. (2000) Solution structures of the N-terminal domain of yeast calmodulin: Ca^{2+}- 39, dependent conformational change and its functional implication. *Biochemistry*, 13660–13668).

Each of the globular domains of the protein contains two EF hand regions, with their characteristic helix-loop-helix topology (**Figure 8.2**), each of which can bind to one molecule of Ca^{2+}. The two EF hands within each globular region are connected by a short antiparallel β-sheet region. Interestingly, despite the similarity in the two regions their affinity for Ca^{2+} differs. The C-terminal domain has the higher affinity and will bind to Ca^{2+} first, leading to major conformational changes within the molecule. These structure changes reveal two hydrophobic patches, one in each half of the protein. It is these patches that can interact with the molecule, another protein, which calmodulin will control.

Synthetic peptides

Small peptides (synthetic peptides) which have the same sequence as potential interaction regions in proteins, can be used to bind to potential binding sites and therefore stop native binding. By the use of a range of such peptides, usually based on areas of a protein of interest, the exact nature of protein/protein interactions can be determined. Their use as therapeutics has also been patented, although getting such peptides into cells is a challenge.

With the use of synthetic peptides to block binding, proteolysis experiments to remove potential binding sites from proteins and the expression of cDNAs, it has been attempted to map the areas on proteins that are controlled by calmodulin, which bind to the hydrophobic patches revealed in calmodulin after it has bound to Ca^{2+}. Most of these calmodulin binding sites involve an α-helix of 16–35 amino acids within the interacting protein. If such an α-helix is viewed from one end, it can been seen that it possesses a polarity, with the majority of the polar, mainly basic, amino acids lying down one side, with the majority of the hydrophobic amino acids lying on the opposite side (**Figure 9.4**) The binding regions also usually contains a potential phosphorylation site.

In the model organism *Dictyostelium discoideum*, over 60 calmodulin binding proteins have been identified. Such proteins span a wide range of functions and cellular locations, highlighting the importance of this calcium control protein. A similar situation is certain to exist in most eukaryotes.

Figure 9.4 Looking end-on down an α-helix that may be involved in the binding of a protein to activated calmodulin.

Besides the regulation of calmodulin by the binding of Ca^{2+} to the four binding sites, it has also been shown that calmodulin is itself phosphorylated *in vivo*. The central helix is phosphorylated at two sites, with a third site located in the third Ca^{2+} binding region. It is thought that the interaction of the calmodulin with its target protein is reduced on phosphorylation and this might be caused by an inhibition of the structural changes that occur in the calmodulin molecule on Ca^{2+} binding, with the hydrophobic patches not being revealed properly.

9.3 The plasma membrane and its role in calcium concentration maintenance

The plasma membrane is crucial to the maintenance of the intracellular calcium ion concentration, $[Ca^{2+}]_i$. Not only does it remove Ca^{2+} from the cell, but it also lets it back in again as required to create a signal. Therefore, these two activities need to be discussed.

Firstly, let us turn to its role in the removal of Ca^{2+} from the cell. The plasma membrane contains two types of pumps, which actively discharge Ca^{2+} from the cell, helping to maintain the low cytoplasmic Ca^{2+} concentration. One of these uses ATP as an energy source, whereas the other uses the electrochemical gradient that occurs across the membrane. The one that uses ATP is the plasma membrane Ca^{2+} ATPase (PMCA) pump (**Figure 9.5**). It is a P-type ATPase, using the energy of one or two molecules of ATP to drive the transport of one molecule of Ca^{2+} out of the cell. The pumping cycle uses a phosphorylated intermediate (hence its classification as a P-type pump), involving the transient transfer of a phosphoryl group to an aspartate residue on the polypeptide. This phosphorylation of the protein causes a conformational change needed to transport the Ca^{2+}

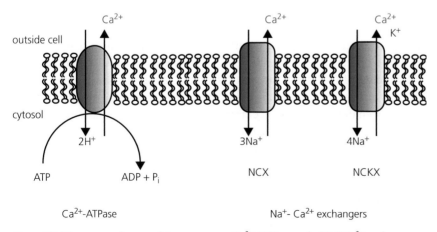

Figure 9.5 Plasma membrane calcium pumps: a Ca^{2+}-ATPase and a Na^+/Ca^{2+} exchanger.

ions across the membrane, and dephosphorylation would restore the protein back to its former state ready for another round of Ca^{2+} pumping. The net result is the breaking and loss of the high energy phosphate bond of ATP, the energy being used to drive the Ca^{2+} pumping activity of the ATPase.

The structure of the ATPase protein is thought to have 10 transmembrane spanning domains. Like many proteins involved in signalling, several tissue specific isoforms exist in humans, arising from four genes, as well as alternative splicing of the gene transcripts.

The pumps that use the electrochemical gradient occurring across the membrane are the Na^+/Ca^{2+} exchangers, and they are also active in the expulsion of Ca^{2+} ions from the cell (Figure 9.5). There are two main types. The first one is a simple exchanger, moving one Ca^{2+} out for every three Na^+ ions that move inwards. This is known as the NCX. The second type co-transports one Ca^{2+} along with one K^+ ion, with the subsequent movement of four Na^+ ions inwards. This is known as the NCKX.

The Ca^{2+}-ATPase has a higher affinity for Ca^{2+} ions than the Na^+/Ca^{2+} exchangers, but the latter can pump Ca^{2+} ions at a much higher rate. A typical Ca^{2+}-ATPase has been shown to pump Ca^{2+} ions at a rate of approximately 30 s^{-1} in an *in vitro* situation, whereas the Na^+/Ca^{2+} exchangers may pump Ca^{2+} ions at a rate as high as 2000 s^{-1}. Therefore, after an influx of calcium, which perhaps constituted a signal, large amounts of Ca^{2+} can be expelled from the cell quite quickly, and when the Ca^{2+} ion concentration is micro-molar the slower but higher affinity ATPase can continue to decrease the cytoplasmic Ca^{2+} ion concentration further. The ability to reduce the $[Ca^{2+}]_i$ very quickly may be important for the maintenance of calcium spikes (see below).

The plasma membrane is also a major site of re-entry of Ca^{2+} back into the cell. One of the most simple ways for allowing calcium to enter the cells is through gap junctions, as discussed in **Chapter 1**. These would allow small molecules such as Ca^{2+} ions to move between cells, and therefore from cells of relatively high Ca^{2+} to those with relatively low Ca^{2+}. Such a mechanism would allow the propagation of a calcium signal across a tissue. However, in most cells the re-entry of calcium through the plasma membrane will be from the extracellular fluid. Such re-entry may be voltage-gated, so that the re-entry of the calcium ions will be controlled by the polarization of the membrane. In some cases it may also be controlled by the presence of other second messenger molecules.

Movement of Ca^{2+} back into the cells often involves voltage-gated channels (CaVs). Re-entry of calcium ions through the channels is rapid, allowing very quick changes in Ca^{2+} concentration inside the cells in the vicinity of the channels. As the name suggests, the channels are sensitive to voltage changes across the membrane. The structure of the channels has been investigated, and it appears that there is a domain comprising a helix-turn-helix, which is acting as the sensor for the voltage change.

A second site of re-entry of Ca^{2+} is through the transient receptor potential (TRP) ion channels. The topology of the proteins involved is such that they have six transmembrane regions, but these are arranged as tetrameric structures, which create a relatively non-selective pore through the membrane. In mammals there are numerous isoforms of these channels, which can be separated into six classes: TRPA; TRPC; TRPM; TRPP; TRPV; TRPML. They can be activated in response to many environmental cues, including temperature and pH, as well as being responsive to the presence of some compounds. Interacting partners in mammals for the TRP proteins include calmodulin, phospholipase C, S100 proteins and annexins, with activation also being muted by reactive oxygen species, ADP ribose, DAG and even mechanical stress. Their dysfunction has been noted in some kidney disease, but they have also been found to be influential in the development of organisms.

The depletion of the intracellular stores can also serve as a signal for the re-entry of Ca^{2+} through the plasma membrane, through what is referred to as store-operated channels (SOCs). This process is referred to as capacitative Ca^{2+} entry (CCE). Work has intensified to find the channels that are involved and mechanisms proposed, referred to as the Ca^{2+}-release-activated-current (I_{CRAC}) or alternatively, depletion-activated-current (I_{DAC}). Recently, proteins involved in this system have been identified. One of these is STIM1, which contains an EF hand for calcium binding and has a single transmembrane region but resides primarily in the ER. However, it appears that it can modulate the activity of another protein found in the plasma membrane, and which can form a pore, a protein called Orai1. Hence, stimulation of STIM1 by calcium store depletion can alter the re-entry of Ca^{2+} through the plasma membrane via the Orai1 pore system. However, the exact details of the mechanisms of interactions of these proteins is still to be resolved.

Many second messenger molecules and mechanisms have also been suggested to be signals for opening of Ca^{2+} channels, including $InsP_3$, $InsP_4$, tyrosine phosphorylation, cGMP and G proteins, both the trimeric and small monomeric ones, but the exact mechanism has yet to be discovered. Messenger molecules, such as Ca^{2+} influx factor (CIF), have been suggested as a candidate for this role. CIF has been characterized as being a stable non-protein molecule, perhaps a negatively charged phosphorylated anion of less than 500 Da, but more studies will be required to fully elucidate its mode of action.

In plants, plasma membrane calcium channels have also been sought. Candidates here include products of the *AtTPC1*, *AtGORK* and *AtSKOR* genes in *Arabidopsis*, and members of the annexin family are likely to be involved.

■ The I_{CRAC} calcium ion re-entry mechanism was originally discovered in T-lymphocytes and mast cells.

9.4 **Intracellular stores**

Endoplasmic reticulum stores

A crucial aspect of calcium ion signalling is the maintenance of a very low cytoplasmic Ca^{2+} concentration, $[Ca^{2+}]_i$, the ability to allow it to suddenly rise, and then its subsequent restoration. All these aspects of the signalling process are facilitated by the presence of intracellular stores of calcium, in which Ca^{2+} can be sequestered, from which Ca^{2+} can be rapidly released, and then subsequently sequestered again, in a never ending cycle of potential signalling.

One of the major stores is the lumen of the endoplasmic reticulum, or in muscle tissues the sarcoplasmic reticulum (SR). Specialized areas of endoplasmic reticulum used for calcium storage have been observed in some cells and these have been called calciosomes.

As with the plasma membrane, Ca^{2+} needs to be pumped through the ER membrane, and then released back the other way at the appropriate time. In the membrane of the ER, calcium pumps pump calcium from the cytoplasm to the inside of the membranous network (**Figure 9.6**). These are commonly referred to as smooth endoplasmic reticulum calcium ATPase or SERCA pumps. In the specialized type of endoplasmic reticulum found in muscle, the sarcoplasmic reticulum, they are in fact a major protein component of the membrane, being approximately 80% of the integral proteins found in that membrane. The density of these pumps has been estimated to be as high as 25,000 per μm^2, and such data highlight the importance of this type of activity to the overall functioning of the cell. As the name suggests, this is an active process, pumping calcium from a region of very low concentration, the cytoplasm, to a region of relatively high concentration the inner part of the ER, the lumen. Therefore, an input of energy is required, and here the pumps use the breakdown of ATP. Again, like the PM equivalent, a phosphoryl intermediate is involved in the pumping steps (hence they are also P-type ATPases), that is, a phosphate group is added to an aspartate residue on the polypeptide causing a conformational change, resulting in the Ca^{2+} ions moving across the membrane where they can be released. Removal of the phosphate group

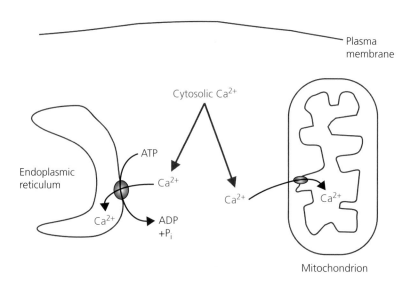

Figure 9.6 Intracellular calcium stores: the endoplasmic reticulum and mitochondria both sequester calcium ions.

again will restore the enzyme to its first conformation, allowing further rounds of ATP hydrolysis and Ca^{2+} pumping. It is thought that two Ca^{2+} ions are transported for every ATP hydrolyzed. Therefore, as would be expected, like the PMCA the SERCA is not only classified as a P-type ATPase, but also has ten transmembrane spanning regions defined in the polypeptide. However, despite this topological and functional similarity, very little sequence homology exists between the two types of Ca^{2+} pumps. The endoplasmic reticular protein appears to contain two main domains, with a smaller one on the lumen side of the membrane. One main domain is buried in the membrane and acts as a Ca^{2+} channel, the Ca^{2+} ions binding near the centre line of the membrane. The other major domain contains the ATPase activity on the cytoplasmic side of the membrane and the site of phosphorylation.

When a functional Ca^{2+}-ATPase was reconstituted into a phospholipid membrane, it was seen to pump Ca^{2+} ions at a rate of approximately $30\ s^{-1}$. Interestingly, if a calcium gradient was artificially set up across the membrane, but the wrong way around, the ATPase could be forced to run in reverse and ATP formation was seen if ADP was present. Such potential for reversibility is often seen with ATPases, but it is hard to imagine this happening *in vivo*, and probably remains a curiosity which can be used to unravel the molecular mechanisms involved.

The SERCA pumps, at least in some mammals, arise from three genes that appear to be expressed in a tissue-specific manner. The gene products SERCA1 and SERCA2 are expressed in different types of muscle, whereas SERCA3 is expressed in non-muscle tissue. Thapsigargin, a lactone that also contains tumour promoting activity, has been identified as a highly specific inhibitor of SERCA activity, trapping the pump in a Ca^{2+} free state, and has been useful as a biochemical tool for the study of these activities as it does not inhibit plasma membrane pumps. Alternatively, orthovanadate can be used as a more general inhibitor of P-type ATPases.

The activity of the Ca^{2+}-ATPases is probably not just continuous, but is controlled. Such regulation would allow the calcium to be removed quickly, or perhaps more slowly if the response was required to be elongated. An increase in cytoplasmic Ca^{2+} accelerates the activity of the ER pump, whereas an increase of free Ca^{2+} inside the ER inhibits the activity of the pump, with half the activity being lost if the Ca^{2+} reaches approximately 300 µM. Control of the pumps may also come from a receptor mediated mechanism.

Once inside the ER, the Ca^{2+} is sequestered and buffered, keeping the concentration of free ions down. As discussed above, one such sequestering protein is calsequestrin, which has an amazing affinity for calcium. No role for the sequestering of the calcium ions has been assigned other than as a store, which can be released back to the cytoplasm at the appropriate moment. However, as one of the major roles of the ER is in the sorting and trafficking of proteins, it has been suggested that the calcium ions might have a dual role here, although there is little evidence of this. Certainly, if free Ca^{2+} was allowed to be high, then the normal functioning of the ER would be disrupted, and that of course would be detrimental to the cell. Therefore, the ER has a crucial role in calcium signalling, but this role cannot be to the hindrance of its other functions.

Therefore, calcium uptake by the ER is an energy-dependent pumping mechanism, but there also has to be a mechanism to allow it quick release when required as a signal. As discussed in the last chapter, one of the breakdown products of phosphatidylinositol 4,5-bisphosphate (PtdInsP$_2$) breakdown is inositol trisphosphate (InsP$_3$), the major role of which is the release of calcium from the intracellular stores, in particular the ER. To be able to do this, the machinery of the ER has to recognize and respond to InsP$_3$. The ER, in fact, contains InsP$_3$ receptors (InsP$_3$R), which facilitate the release of the calcium. The receptors are tetrameric in conformation with a C-terminal domain, which spans the membrane of the ER. Each subunit has an approximate molecular weight of 310 kDa and contains a cationic pore, which shows relatively little selectivity, but allows the movement of Ca^{2+} through the membrane. Each of the four subunits also contains an arginine and lysine rich region at the N-terminal end, which can bind one InsP$_3$ molecule, acting as a ligand binding site for the protein. Therefore, analogies can be drawn here with the receptors of the plasma membrane (Chapter 5). The InsP$_3$ acts as the ligand, which binds to the receptor, and so opening a channel for the movement of Ca^{2+}. The N-terminal region, which is on the cytoplasmic side of the membrane has binding sites for two ATP molecules and at least one Ca^{2+} ion.

As is regularly seen with other signalling proteins, the InsP$_3$ receptor can exist in several isoforms. In mammals, at least four genes encode InsP$_3$ receptors with significant homology between them. Expression of the receptors, as might be expected, differs between different tissues.

However, InsP$_3$ receptor activity is not only controlled by the binding of InsP$_3$, but also very importantly the passage of Ca^{2+} through these proteins is modulated by Ca^{2+} itself, Ca^{2+} being referred to as a co-agonist. If InsP$_3$ concentrations are low, relatively low cytoplasmic Ca^{2+} concentrations accelerate Ca^{2+} release, but at higher cytoplasmic Ca^{2+} concentrations the release of Ca^{2+} through InsP$_3$R is inhibited. Therefore, if plotted, the release response to Ca^{2+} concentrations would give a bell-shaped curve, with a maximum around 0.2–0.3 µM (Figure 9.7). This is a very important phenomenon, because, as discussed below, the passage through InsP$_3$R is not the only way to increase cytoplasmic calcium ion concentrations. If Ca^{2+} is released through other channels, perhaps from non-ER-like stores, then the arrival of Ca^{2+} at the InsP$_3$R can influence whether more Ca^{2+} is released through them from the ER, or SR. Therefore, by relatively low calcium rises having a control on Ca^{2+} release, small changes can very rapidly be converted to large changes, in what could be referred to as a feed-forward type control. Alternately, inhibition at high Ca^{2+} might also be important, giving a level of moderation.

Control of InsP$_3$R is not only through rises in [Ca^{2+}]$_i$, but may also be via phosphorylation, for example, through cAMP-dependent protein kinase.

InsP$_3$R proteins

InsP$_3$R proteins were originally discovered as glyco-phosphoproteins involved in development in mutant mice. They were originally termed P400.

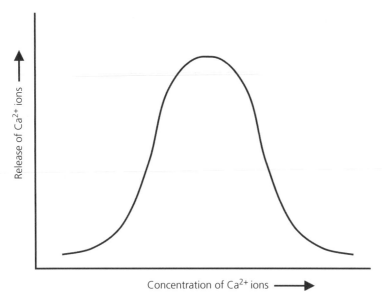

Figure 9.7 Ca^{2+} induced Ca^{2+} release from $InsP_3R$ in the presence of lower concentrations of $InsP_3$.

Its activity is also modulated by the presence of Mg^{2+}. For particular interest to those that drink lots of strong coffee, caffeine has been shown to inhibit Ca^{2+} release from $InsP_3R$ in response to $InsP_3$, whereas heparin, better known for its anti-coagulant properties, competitively inhibits $InsP_3R$ by binding to the $InsP_3$ binding sites. Interestingly, ethanol is also a potent inhibitor of $InsP_3R$. In fact, $InsP_3Rs$ have been reported to have a multitude of binding partners, suggesting that in fact they exist in large complexes, and therefore can have their activity modulated in a wide range of ways, no doubt being subtly different in various cells types and tissues. One of these proteins is called $InsP_3R$-binding protein released with $InsP_3$ (IRBIT), which is released when $InsP_3R$ binds to $InsP_3$. IRBIT itself seems to have downstream signalling with subsequent binding proteins identified, such as Na^+/HCO_3^- cotransporter 1 (NBC1).

It is also interesting to note that $InsP_3$-like receptors have been found in the plasma membrane of some cells, but their role there has not been fully elucidated. Clearly, as discussed above, re-entry of Ca^{2+} into the cell through the PM, having been expelled previously, is very important and this suggests one mechanism by which this might happen.

It is worth a quick note to point out that as well as $InsP_3$, other inositol phosphate metabolites might be involved in the release and control of Ca^{2+}. For example, inositol 1, 3, 4, 5- tetrakisphosphate ($InsP_4$) has been shown to affect the Ca^{2+} influx in cells and also to be an inhibitor of Ca^{2+} ATPase pumps. Such inositol compounds are further discussed in Chapter 8.

Leaving the inositol metabolites for a while, the release of Ca^{2+} from intracellular stores can also involve proteins other than $InsP_3R$. The other calcium ion release channel found in the ER is the ryanodine receptor (RyR). Like the InsP$_3$R, the RyR is also a tetramer, but this time the subunits are much bigger, having a molecular weight of up to 560 kDa, and the receptor has been purified as a 30S complex, which highlights its size. Three genes (*ryr-1*, *ryr-2* and *ryr-3*) code for separate isoforms of the RyR, with expression of the three being very tissue-specific. The gene *ryr-1* is expressed in skeletal muscle, whereas *ryr-2* is found expressed in cardiac muscle. The gene *ryr-3* represents a non-muscle form. Of the proteins encoded by these genes, RyR1 and RyR2 show a large amount of sequence homology, but the third form, RyR3 is much smaller, at less than 650 amino acids. The C-terminal ends of the subunits create the Ca^{2+} channel through the membrane, having between four and ten putative membrane spanning regions, whereas the N-terminal end protrudes into the cytosol. As the name suggests, these channels are sensitive to ryanodine, which is a plant alkaloid. Low concentrations of ryanodine cause opening of the RyR channels, but at higher, micromolar, concentrations of ryanodine the channels are once again closed. Opening of the RyR is also seen in the presence of caffeine, opposite to that seen with InsP$_3$R.

As ryanodine is a plant alkaloid, and unlikely to be an endogenous signalling molecule in mammals, what causes the RyR to open as a physiological response? In some tissues, for example striated muscle, the sarcoplasmic reticulum comes into close contact with the plasma membrane and protein complexes exist that connect the two membranes. These protein complexes are known as T-tubule feet, and contain RyRs. The other part of the foot structure is a plasma membrane associated voltage-sensitive receptor, such as the dihydropyridine receptor. This receptor will sense changes in the voltage across the membrane, and undergo a conformational change. This change in conformation of the plasma membrane associated receptor is sensed by RyR, as it is part of the same protein complex. A conformational change is induced in RyR, which results in channel opening and the release of calcium ions (**Figure 9.8**). This is a good example of conformational changes in one protein being transmitted across to another, having the desired effect.

Alternatively, like InsP$_3$R, RyRs are sensitive to low concentrations of Ca^{2+}, having a similar bell-shaped dependence on Ca^{2+} (as depicted in **Figure 9.7** for InsP$_3$R). Like InsP$_3$R, low concentrations of Ca^{2+} will induce the opening of RyR, whereas higher, mM, concentrations will inhibit it. Therefore, a voltage change across the plasma membrane may induce a small influx of Ca^{2+} from the outside, through a Ca^{2+} voltage-gated channel. This would be sufficient to induce the opening of RyR and the release of intracellular Ca^{2+} stores as illustrated in **Figure 9.8**.

Thirdly, are reports that RyRs are sensitive to, and open in, the presence of cyclic ADP-ribose (cADPr: discussed further below). RyR2 seems to be the target for cADPr as RyR1 is insensitive to this messenger molecule. It is possible that RyR3 is also controlled in this way. Control of RyR by calmodulin

A

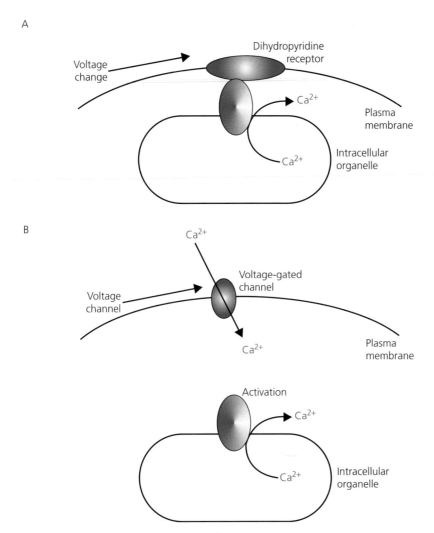

Figure 9.8 The mechanisms for release of Ca^{2+} through ryanodine receptors. A. Voltage changes are sensed by a plasma membrane associated receptor and protein conformational changes are relayed to the ryanodine receptor causing its opening. B. Voltage-gated Ca^{2+} channels allow small amounts of Ca^{2+} to enter the cell, which trigger the ryanodine receptors to open.

and phosphorylation has also been reported, although the exact influence of these has to be established.

Interestingly, all the Ca^{2+} channels so far studied contain an amino acid sequence -Thr-X-Cys-Phe-Ile-Cys-Gly-. This is found in the C-terminal ends of the polypeptides and may play a role in the opening of the channel. It also seems to bestow thiol sensitivity onto the channels, the SH groups of the cysteine residues being open to possible oxidation. It is conceivable, therefore, that such sequences are targets for reactive oxygen species (see discussion in **Chapter 10**).

Mitochondrial calcium metabolism

Calcium uptake by mitochondria

It should be noted that mitochondria have an outer membrane as well as a quite impermeable inner membrane. Although Ca^{2+} movements through the inner membrane need pump proteins, Ca^{2+} seems to readily diffuse through the outer membrane through large pore structures.

As well as sequestration of Ca^{2+} by the endoplasmic reticulum, other organelles also actively take up calcium ions. One of the most significant is the mitochondria. In some cases, as cytoplasmic calcium rises it triggers events that need more energy, so by mitochondria taking up calcium ions, calcium can be used to trigger an increase in mitochondrial activity, and a subsequent rise in ATP production. Mitochondria can acquire calcium through the use of a Ca^{2+} uniporter, driven by the proton gradient across the inner mitochondrial membrane, which is maintained by the respiratory electron transport chain (**Figure 9.6**). One such channel has been named the mitochondrial Ca^{2+} channel (MiCa). It has relatively high affinity for Ca^{2+}, and the mitochondrial Ca^{2+} can rise to approximately 0.5 mM. Once pumped in the calcium ions are not free, but appear to form a "gel" with phosphate, so allowing the calcium to be somewhat immobile, but in a state that can be rapidly released if required. Efflux from the mitochondria is via a calcium/sodium exchanger (Ca^{2+}/Na^+), although the exact mechanism is uncertain, perhaps being $2Na^+$ for each Ca^{2+} (with no net movement of charge), or $3Na^+$ for each Ca^{2+} (with net movement of charge).

The proton gradient

The mitochondrial electron transport chain uses energy from the transfer of electrons to pump protons across the inner mitochondrial membrane, and although extensively this H^+ gradient is there to drive ATP synthesis, it can also be utilized by many other proteins. For example, ADP and ATP are transported using its energy, as is phosphate, whereas it can also be utilized for the transport of calcium ions. In this latter case, the protein is referred to as a uniporter as it has movement of ions in one direction only, as opposed to an antiporter, where movement of one molecule in one direction is accompanied by the movement of another in the opposite direction.

As well as being a way of sequestering Ca^{2+} from the cytoplasm, and its release back, and therefore a useful organelle for use in the control of cytoplasmic calcium, like the ER, the Ca^{2+} in the mitochondria is also used to control the activity of enzymes, such as pyruvate dehydrogenase. This enzyme acts as a crucial link between glycolysis and the citric acid cycle

(Kreb's cycle), and hence is involved in the production of ATP. Calcium ion concentrations in the mitochondria have also been linked to the onset of degenerative diseases, where a lack of control can lead to a depletion of ATP in neuronal tissue leading to neuronal death.

9.5 Gradients, waves and oscillations

An interesting and challenging aspect of the calcium signalling story is that the concentration of calcium is not uniform throughout the cytoplasm, or indeed in the compartment in which it might be sequestered. Intuitively, it would be expected that if calcium ions rush in from the outside of the cell through the plasma membrane then the concentration would rapidly dissipate throughout the cytoplasm by diffusion. However, this is clearly not the case. It has been estimated that calcium ions only migrate for approximately 50 μs, a distance of only 0.1–0.5 μm. By that time, the ions would have been picked up by local binding proteins. However, such estimations will be altered by the saturation of the binding proteins and by the uneven distribution of such proteins. Some buffering proteins might be free in the cytoplasm, whereas others might be bound and immobile, again altering the equation. Therefore, a gradient of Ca^{2+} concentration may extend away from the site of Ca^{2+} release. Such gradients are called micro-gradients as they occur within a space of 1–10 μm. The highest concentrations of Ca^{2+} may reach up to 1 μM. As well as these micro-gradients, very high local concentrations will be found around an open channel. Mobilization through a channel may well be more efficient than the diffusion of the Ca^{2+} ions through the cytosol, resulting in a very steep gradient extending away from that channel. These gradients will decay within tens of nanometres of the channel opening, and are therefore referred to as nano-gradients. Typically, such nano-gradients will only last for less than a millisecond, but they could have Ca^{2+} concentrations as high as 100 μM. Gradients of Ca^{2+} have been proposed to be important in the control of cell migration, exocytosis, ion transport and gap junction regulation. Such uneven distribution of $[Ca^{2+}]_i$ is often visualized using fluorescent probes in conjunction with confocal microscopy (see discussion below). "Hot-spots" of Ca^{2+} are seen, often near the point of perception of a signal or stress. Therefore, these calcium signals are local, not pertinent to the whole cell, and global measures of both $[Ca^{2+}]_i$ and resulting enzyme activities are often erroneous—such data have to be interpreted with caution.

Other phenomena that remain puzzles are the observations that the calcium concentrations can be measured as oscillations or waves. That is, the calcium ion concentration in the cytoplasm, for example, appears to rise steeply, drop steeply, and then rise again, in a series of oscillations.

Therefore, it might not be the overall concentration of the calcium that actually constitutes the signal, but either the amplitude or frequency of the concentration oscillation. Such oscillations, if created locally in the cell, perhaps on a particular part of the ER, will not necessarily be static either, and may move through the cell, like a wave of Ca^{2+}. Again, this gives new parameters to what the signal might be: above a threshold of Ca^{2+} or the frequency of the wave.

So how could a wave of Ca^{2+} be created in the cell? The creation of a wave might use the following as a typical mechanism. It is probable that local concentrations of Ca^{2+} are created around the opening of a channel in the endoplasmic reticulum, which in turn may cause the activation of channels in the locality, so releasing Ca^{2+} further along the membrane. Simultaneously, such local concentrations will inhibit the first channel from releasing any more calcium, whereas the local Ca^{2+} ATPases will start to actively remove the ions, once again lowering the Ca^{2+} concentration in the location of the first channel to open. The new release further along the membrane can then initiate the same response with the channels and ATPases even further along the membrane, and so the wave propagates out from the site of initiation. Such a scheme could propagate a wave of calcium across the membrane as illustrated in **Figure 9.9**. Interestingly, it has been reported that over-expression of the Ca^{2+} ATPase pump of endoplasmic reticulum increases the frequency of the concentration waves recorded. Such waves may move through the cytosol at speeds of 5–100 µm/s, and, interestingly, wave speed may depend on the concentration of agonist applied to the cell.

Importantly, such waves have been reported to travel from cell to cell through gap junctions, which, as we saw in **Chapter 1**, allow the passage of small signalling molecules from cell to cell, and therefore such a mechanism may constitute a longer range signalling mechanism than first envisaged. Such signalling is important to consider when we discuss apoptosis. Here, if one cell is programmed to die, the organism may wish to prevent the spread of the "death" signal, and Ca^{2+} movement through gap junctions will be prevented.

Having considered how waves might be created and propagated, let us turn back to oscillations. These were first demonstrated in blow-fly salivary glands, where it was observed that the Ca^{2+} concentration could be measured as a series of spikes and that the frequency of the spikes increased as the agonist concentration increased. Subsequent research has revealed that the patterns of the oscillation depend on many factors, including the agonist concentration, the receptor type activated and the Ca^{2+} buffering capacity of the cell.

Two main questions remain; firstly; what characteristics of such spikes are important as signals, and, secondly, how do such oscillations occur? Spikes may vary in several ways. Their frequency may change as seen above, that is, they can be closer together, or even further apart. Alternatively, they may have a greater or smaller amplitude, or the calcium levels within the spikes may be elevated for longer periods of time, that is, the individual single spikes last longer.

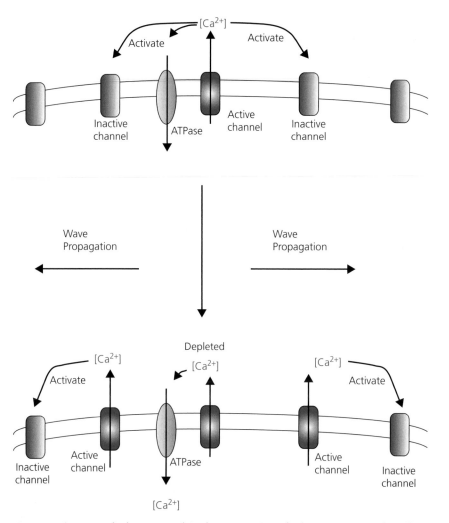

Figure 9.9 A proposed scheme to explain the propagation of calcium ion waves along the membranes of cells.

All of these scenarios have been found to be important in different cells and during different signalling events. To answer the second question, several models have been put forward to try to explain this phenomenon. One such model is known as the $InsP_3$-Ca^{2+} cross-coupling model (ICC model). Here, an extracellular signal would lead to the production of $InsP_3$ as outlined above (**Figure 9.1**), and this would lead to the subsequent rise in cytoplasmic Ca^{2+} concentration. This increase in $[Ca^{2+}]_i$ ions may then stimulate the activity of phospholipase C leading to more $InsP_3$, and hence a greater release of Ca^{2+} from intracellular stores. However, $InsP_3R$ is inhibited by high concentrations of Ca^{2+}, and therefore once the cytoplasmic Ca^{2+} reaches a critical level, the $InsP_3R$ channels are closed and no further Ca^{2+} release takes place. The cell now actively removes the calcium ions from the cytoplasm, as the pumps would be activated, and so restores the cell back to the unstimulated situation. The reduction in cytoplasmic Ca^{2+} would also lead to a reduction of phospholipase C activity and a reduction of $InsP_3$ in the cytoplasm. The presence of agonist can once again initiate a spike in calcium release. The release of Ca^{2+} from the intracellular stores in this model would be a consequence of the spiking in phospholipase C activity. However, research has shown that even if the cytoplasmic $InsP_3$ concentration is kept constant, the calcium ion concentration can still be seen to oscillate. Therefore, further models are needed.

Two further models are based on the proposed presence of a single pool of Ca^{2+} able to be released, the one-pool model, or the presence of two independent pools of Ca^{2+} that could be released, the two-pool model. One pool implies that all the calcium is sequestered in organelles which are equal, that is, the kinetics of activation, uptake and release are the same. Two pools implies that there is more than one type of calcium sequestering organelle, where the kinetics of activation, uptake and release are different. This allows for the release from one pool before the other, or the re-uptake into one before the other. For example, early release from one pool can cause rises in $[Ca^{2+}]_i$, which induces a later release in another. Of course, the true situation in some cells might be the presence of multiple pools, each of which has different kinetics, and which can be activated or inhibited by different arrays of molecules.

In the one-pool model (**Figure 9.10**), the initial release of Ca^{2+} is relatively slow. However, because opening of $InsP_3$ receptors by $InsP_3$ is enhanced by the presence of Ca^{2+}, as the cytoplasmic Ca^{2+} rises, so the release of more Ca^{2+} through $InsP_3R$ is accelerated. This gives the rapid rise of Ca^{2+} seen as the early part of the spike. However, higher concentrations of Ca^{2+} inhibit $InsP_3Rs$ and the channels close, allowing no further rise in cytoplasmic calcium. A drop is seen as the calcium is re-sequestered and removed from the cytoplasm by the various calcium pumps, perhaps activated by the high $[Ca^{2+}]_i$. It has also been suggested that the rise in Ca^{2+} causes the activation of phospholipase A_2, with the concomitant release of arachidonic acid. This could also potentially inhibit $InsP_3R$, and so play a role in the formation of the calcium spike.

As discussed above, the two-pool model suggests that Ca^{2+} may be sequestered into different intracellular stores that have receptors which differ in their sensitivity to $InsP_3$. In this model (**Figure 9.11**), only the most sensitive $InsP_3Rs$ will cause a release, just from those independent stores. Other stores that are not releasing Ca^{2+} can still be actively taking in calcium, and so their internal Ca^{2+} concentrations will rise, whereas cytoplasmic calcium buffers will also become saturated. Once the concentrations inside the second store reach a critical level, their membrane pumps are shut off and the cytoplasmic calcium level rises, as no more Ca^{2+} can be buffered and no more can be sequestered. Above a threshold cytoplasmic Ca^{2+} concentration, all the $InsP_3Rs$ open and all the stores release calcium. Hence, the cytoplasmic

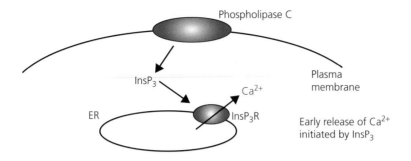

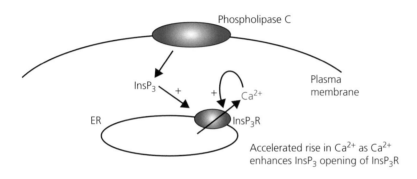

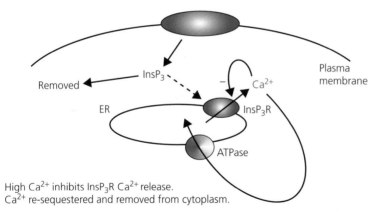

Figure 9.10 A simplified scheme showing the control and role of calcium ion concentrations within the cell.

calcium ion concentration rises sharply, seen as the early phase of the spike. Again, as in the one-pool model, a high cytoplasmic Ca^{2+} concentration will result in the inhibition of $InsP_3Rs$ and the cessation of Ca^{2+} release. Pumps in the membranes will now actively re-sequester and remove Ca^{2+} from the cytoplasm, giving the down part of the spike.

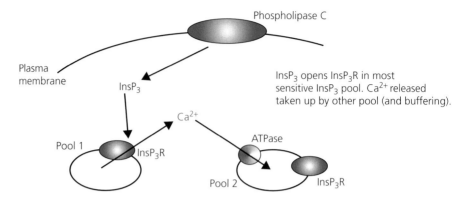

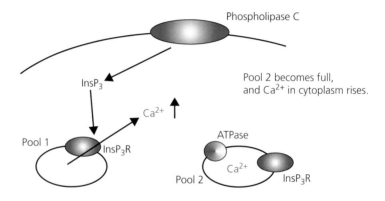

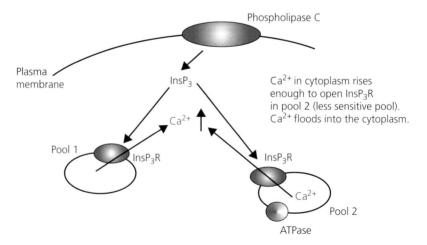

Figure 9.11 Two-pool model of calcium ion oscillations. The two pools here have InsP$_3$Rs of differing sensitivities, but other models using two pools of calcium have also been proposed.

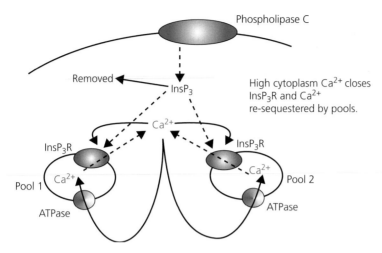

Figure 9.11 (Continued).

A modification of the two-pool model may be that one store contains $InsP_3$ receptors, whereas the other contains ryanodine receptors. It must also be remembered that it is not only the release and pumping of calcium at the intracellular store membranes which is important, but also the plasma membrane will play a vital role in these events.

Clearly, the control of $[Ca^{2+}]_i$ is complicated, and no doubt differs in subtle if not major ways in different cells. It is likely that a true consensus model will not be drawn up to encompass the full plethora of calcium signalling, but underlying principles will remain, and $[Ca^{2+}]_i$ will remain a central key to many signal transduction pathways.

Early calcium signalling research centred on $InsP_3$ as the main compound produced and able to release sequestered Ca^{2+}, but more recently several other molecules, from apparently disparate locations, have been found to be involved in calcium signalling. These will be briefly discussed below.

9.6 **Sphingosine-1-phosphate**

As well as the classical route for calcium release through the production of $InsP_3$, another lipid metabolic pathway can also lead to the release of Ca^{2+} from intracellular stores, this being the sphingosine pathway as discussed in Chapter 8. Here, one of the main active second messengers appears to be sphingosine-1-phosphate, which can mobilize Ca^{2+} from internal cell stores in an inositol-independent manner, and in a way which is independent of extracellular calcium. Sphingosine 1-phosphate can be produced by sphingosine kinase from sphingosine, a kinase which as expected is dependent on ATP as a source of phosphoryl group and can be competitively inhibited by DL-threo-dihydrosphingosine. It is likely that the kinase is located within the endoplasmic reticulum membrane, where sphingosine-1-phosphate will have its action. Removal of sphingosine-1-phosphate will be by lyases and phosphatases.

However, it should also be noted that sphingosine-1-phosphate may have many other signalling roles in an organism, many of which stem from its presence in body and tissue fluids, rather than its role as an intracellular calcium ion releasing compound.

Another metabolite that has been reported to be important in calcium ion oscillations in the cell is sphingosylphosphorylcholine (SPC), and although it seemed to be the same Ca^{2+} stores released by this metabolite as those released by $InsP_3$, the response was not mediated by $InsP_3$ receptors.

9.7 Cyclic ADP-ribose

As well as lipid derivatives such as $InsP_3$ and sphingosine-1-phosphate, which can be used by cells to mobilize calcium to increase $[Ca^{2+}]_i$, there are other small signalling molecules which also can be used. One of the most studied is cyclic ADP-ribose (cADPr). cADPr is synthesized by a family of enzymes, ADP-ribosyl cyclases, which use oxidized nicotinamide adenine dinucleotide (NAD^+) as a substrate. These enzymes have been found in both invertebrate and mammalian cells. cADP-ribose has been shown to release Ca^{2+} in sea urchin eggs and, as mentioned above, may be an agonist for type 2 ryanodine receptors (RyR2) and type 3 ryanodine receptors (RyR3). Therefore, it has been proposed that cADPr has an important second messenger role in many cells. The effect of cADPr on ryanodine receptors, and therefore Ca^{2+} release, can be inhibited by the presence of either ruthenium red or 8-amino-cADPR.

Levels of cADPr may well be modulated by cGMP in some instances, giving a control mechanism that will be physiologically important. cGMP is the product of guanylyl cyclase, which may be under the control of the presence of the gas nitric oxide (NO). NO may be produced by the same cell or may be produced by a neighbouring cell. The production and roles of NO are discussed further in **Chapter 10**. Therefore, a transduction pathway can be drawn, in which NO causes increases in cGMP, which lead to increases in cADPr, causing a rise in intracellular calcium, which brings about the physiological response. An example of this is seen in plants, where such a pathway has been suggested to control, for example, stomatal movements, which themselves control gas exchange and water loss.

9.8 Nicotinate adenine-dinucleotide phosphate

Recently, nicotinate adenine-dinucleotide phosphate (NAADP$^+$), a deaminated derivative of NADP$^+$, has been shown to have the capacity to release Ca^{2+} from intracellular stores. It is synthesized by the same family of enzymes that produce cADPr, that is, the ADP-ribosyl cyclases. Significantly, the ability of NAADP$^+$ to release calcium from stores is independent of the release triggered by InsP$_3$ and cADPr. For example, the release triggered by NAADP$^+$ was not inhibited by heparin, which would interfere with InsP$_3$ responses, and was not affected by the presence of ruthenium red, used to block the cADPr response. Such results suggest a separate pathway that a cell could use for signalling the release of calcium ions. The response was selectively inhibited by the presence of thionicotinamide-NADP$^+$, which will no doubt be of great use in the elucidation of the exact role of NAADP$^+$ in calcium signalling in many cells. Very recent work has shown that NAADP$^+$ concentrations are raised in cells following stimulation, and revealed that the NAADP$^+$ effects were centred on lysosomal-like vesicles, separate from the membranes targeted by InsP$_3$ and cADPr. Furthermore, it was shown that Ca^{2+} released from NAADP$^+$ stores could further influence the release of Ca^{2+} from InsP$_3$ and cADPr sensitive stores. However, the identity of the NAADP$^+$ receptor, and mechanisms used to control NAADP$^+$ levels in cells, at the moment remain elusive.

9.9 Fluorescence detection and confocal microscopy

Methods of detection of many molecules involved in biological functions have often relied on the presence of reporter molecules, and often these can be visualized because they are fluorescent. Sometimes the fluorescence is constitutive, and the research aims to find it or not. However, with many reporter probes for signalling molecules which are fluorescent, they have altered fluorescent characteristics depending on whether the reporter has bound or reacted with the signal or not. This might be seen as an increase in fluorescence, or a decrease, depending on the reporter used. For example, such fluorescent probes are widely used to detect reactive oxygen species and reactive nitrogen species, as discussed in **Chapter 10**, and are also widely used to detect the presence, or rather the amount, of Ca^{2+} that resides in a particular location in the cell.

There is a plethora of fluorescent probes now available to detect Ca^{2+}, such as *Fura-1* and *Fluo-3*, but it is certainly beyond the scope of the discussion here to cover them all. However, these chemicals all possess the ability to bind to free Ca^{2+} ions, and, on binding, their fluorescent spectral characteristics change. These changes can then be readily detected by several techniques. The light emission change may be simply detected on a gross scale by the use of a fluorimeter, which will monitor the change in emitted light if the sample is excited with light of a pre-set wavelength, either fixed or scanning. However, this technique is limited in that a population of cells is required to obtain a detectable signal. Fluorescence microscopes can overcome this problem, as single cells can be focussed on, but, interestingly, such studies often show that two cells from the same tissue treated with the same stimulant may not necessarily respond in the same way. An even more powerful technique is the use of a confocal scanning laser microscope. Here, the light source is from a laser as opposed to a normal white light source, and the laser is driven to scan in the X–Y plane. The laser is focussed onto the sample, enabling a single image to be obtained from a very small area of the sample, often a single cell. More excitingly, the fluorescence emission is also focussed back to the detector, usually a photomultiplier tube, which sends a signal to a computer, which itself creates the image on a VDU screen. The inherently grey image can then be enhanced by the use of pseudo-colour imaging, enabling the users to get an easy-to-see scale of fluorescence in the field of view. The power of this type of system comes from the fact that the laser is not only scanned onto the horizontal plane, but also, because the signal is re-focussed on its way back from the sample, the image is extremely finely focussed in the vertical plane. Therefore, the microscope can be used to scan in the Z plane (vertical) too. Hence, single cells, or indeed whole tissues, can be optically sectioned allowing a three-dimensional image of the sample to be created. These techniques have led to great advances in the study of the role of intracellular calcium as a signal. The concentration of Ca^{2+} ions inside organelles can now be visualized, giving an insight into the changes in these concentrations in, for example, the nucleus. New roles of calcium signalling, such as in the control of gene expression, will undoubtedly come to light as the use of such equipment expands. However, one of the most puzzling findings, discussed above in more detail, is that the concentration of calcium in the cytoplasm is uneven. Ca^{2+} is not uniformly distributed throughout the cytoplasm, rather it is seen as hot-spots. Such imaging has revealed that other signalling molecules also reside temporarily in hot-spots, and therefore our view of the cytoplasm as being the "watery bit of the cell" where all things are even and able to diffuse, needs to be revised.

Another useful Ca^{2+} probe is the photoprotein aequorin. This protein emits light on the binding of Ca^{2+}, and can be purified from the luminous jellyfish *Aequorea forskalea*. It can be used to sense calcium concentrations in the nanomolar to micromolar range.

■ Summary

- Calcium ions are unusual signalling molecules in that they are neither created nor destroyed, but the signal comes from a rapid change in their distribution (**Figure 9.12**).

- Cells usually expend large amounts of energy actively removing Ca^{2+} ions, using pumps in their plasma membrane, which maintains a gradient where the intracellular Ca^{2+} concentration is maintained at less than 10^{-7} M, with the extracellular Ca^{2+} concentration in the region of 2×10^{-3} M.

- Pumps in the endoplasmic reticulum, sarcoplasmic reticulum and mitochondria are also used to sequester Ca^{2+} into intracellular stores.

- A signal arises when the cytoplasmic Ca^{2+} concentration is allowed to suddenly rise, often by the release of Ca^{2+} from the intracellular stores, commonly mediated by the presence of $InsP_3$ derived from the breakdown of lipids on the plasma membrane, an event which itself is tightly controlled.

- Although $InsP_3$ has been shown to have a major role in calcium release, it is not the only means by which this can result. As well as $InsP_3$ receptors, ryanodine receptors are involved in calcium release from intracellular stores, and other signalling molecules such as $NAADP^+$, cADPr and sphingosine-1-phosphate have been found to be important in many systems.

- Along with the presence of Ca^{2+} ions, the signal commonly relies on the presence of calcium ion binding proteins.

- Calcium ion binding proteins quite commonly contain polypeptide domains known as EF hands.

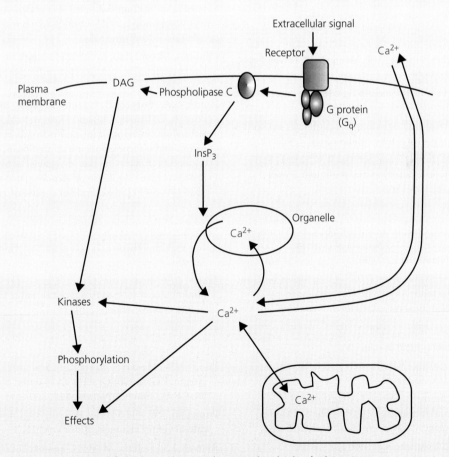

Figure 9.12 A simplified scheme showing the control and role of calcium ion concentrations within the cell.

- The most common and certainly ubiquitous calcium ion binding protein is calmodulin. This protein has two Ca^{2+} binding domains (each capable of binding to two Ca^{2+} ions) connected by a flexible α-helix.

- Research into the movements of Ca^{2+} in cells has revealed the presence of hot-spots of high Ca^{2+}, oscillations in Ca^{2+} concentrations, and waves of high Ca^{2+} moving through cells.

- Calcium is known to control many cell functions, including regulation of protein kinase C, Ca^{2+}/calmodulin-dependent protein kinase, cell cycles, metabolic enzymes and even other signalling routes such as nitric oxide synthase.

- The ubiquity of calcium signalling and its central importance in transduction cascades suggests that much more work will be carried out before the role of Ca^{2+} in cells is fully understood.

→ Further reading

Berridge, M.J., Lipp, P. and Bootman, M.D. (2000) The versatility and universality of calcium signalling. *Nature Reviews Molecular Cell Biology*, 1, 11–21.

Clapham, D.E. (1995) Calcium signaling. *Cell*, 80, 259–268.

Clapham, D.E. (2007) Calcium signaling. *Cell*, 131, 1047–1058.

Clapham, D.E. and Sneyd, J. (1995) Intracellular calcium waves. In *Advances in second messengers and phosphoprotein research*, A.R. Means (ed.), Raven Press, New York, pp. 1–24.

Dominguez, D.C. (2004) Calcium signalling in bacteria. *Molecular Microbiology*, 54, 291–297.

Feske, S. (2007) Calcium signalling in lymphocyte activation and disease. *Nature Reviews Immunology*, 7, 690–702.

Heizmann, C.W., Fritz, G. and Schäfer, B.W. (2002) S100 proteins: structure, functions and pathology. *Front Bioscience* 7, d1356–1368.

Lemmon, M.A. (2008) Membrane recognition by phospholipid-binding domains. *Nature Reviews Molecular Cell Biology* 9, 99–111.

Nicholls, D.G. and Ferguson, S.J. (2002) *Bioenergetics3*. Academic Press, London, UK. ISBN 0125181213 [in particular Sections 8.2.3, 8.2.4 on calcium accumulation by mitochondria, and Section 9.3.2 on measuring calcium concentrations].

Oh-hora, M. and Rao, A. (2008) Calcium signaling in lymphocytes. *Current Opinion in Immunology*, 20, 250–258.

Rutter, G.A., Burnett, P., Rizzuto, R., Brini, M., Murgia, M., Pozzan, T., Tavaré, J.M. and Denton, R.M. (1996) Subcellular imaging of intramitochondrial Ca^{2+} with recombinant targeted aequorin: significance for the regulation of pyruvate dehydrogenase activity. *Proceedings of the National Academy of Sciences USA*, 93, 5489–5494 [an example of the use of aequorin].

Calmodulin

Babu, Y., Sack, J.S., Greenhough, T.J., Bugg, C.E., Means, A.R. and Cook, W.J. (1985) 3-dimensional structure of calmodulin. *Nature*, 315, 37–40.

Catalano, A. and O'Day, D.H. (2008) Calmodulin-binding proteins in the model organism *Dictyostelium*: A complete & critical review. *Cellular Signalling*, 20, 277–291.

Finn, B.E. and Forsen, S. (1995) The evolving model of calmodulin structure, function and activation. *Structure*, 3, 7–11.

James, P., Vorherr, T. and Carafoli, E. (1995) Calmodulin-binding domains: just two faced or multi-faceted. *Trends in Biochemical Sciences*, 20, 38–42.

The plasma membrane and its role

Clapham, D.E. (2003) TRP channels as cellular sensors. *Nature*, 426, 517–524.

Frischauf, I., Schindl, R., Derler, I., Bergsmann, J., Fahrner, M. and Romanin, C. (2008) The STIM/Orai coupling machinery. *Channels (Austin)*, 2, 261–268.

Kiselyov, K. and Muallem, S. (1999) Fatty acids, diacylglycerol, Ins(1,4,5)P$_3$ and Ca^{2+} influx. *Trends in Neurosciences*, 22, 334–337.

Prakriya, M., Feske, S., Gwack, Y., Srikanth, S., Rao, A. and Hogan, P.G. (2006) Orai1 is an essential pore subunit of the CRAC channel. *Nature*, 443, 230–233.

Putney Jr., J.W. and McKay, R.R. (1999) Capacitative calcium entry channels. *BioEssays*, 21, 38–46.

Ramsey, I.S., Delling, M. and Clapham, D.E. (2006) An introduction to TRP channels. *Annual Review of Physiology*, 68, 619–647.

White, P.J., Bowen, H.C., Demidchik, V., Nichols, C. and Davies, J.M. (2002) *Biochemica et Biophysica Acta*, 1564, 299–309.

Yeromin, A.V., Zhang, S.L., Jiang, W., Yu, Y., Safrina, O. and Cahalan, M.D. (2006) Molecular identification of the CRAC channel by altered ion selectivity in a mutant of Orai. *Nature*, 443, 226–229.

Intracellular stores

Choe, C.-u. and Ehrlich, B.E. (2006) The inositol 1,4,5-trisphosphate receptor (IP$_3$R) and its regulators: sometimes good and sometimes bad teamwork. *Sci STKE*, 363, re15.

Finch, E.A. and Goldin, S.M. (1994) Calcium and inositol 1,4,5-trisphosphate-induced Ca^{2+} release. *Science*, 265, 813–815.

Meissner, G. (1994) Ryanodine receptor/Ca^{2+} release channels and their regulation by endogenous effectors. *Annual Review Physiology*, 56, 485–508.

Mikoshiba, K. (2007) IP$_3$ receptor/Ca^{2+} channel: from discovery to new signalling concepts. *Journal of Neurochemistry*, 102, 1426–1446.

Shirakabe, K., Priori, G., Yamada, H., Ando, H., Horita, S., Fujita, T., Fujimoto, I., Mizutani, A., Seki, G. and Mikoshiba, K. (2006) IRBIT, an inositol 1,4,5-trisphosphate receptor-binding protein, specifically binds to and activates pancreas-type Na$^+$/HCO$_3^-$ cotransporter 1 (pNBC1). *Proceedings of the National Academy of Sciences USA*, 103, 9542–9547.

Toyoshima, C., Sassabe, H. and Stokes, D.L. (1993) Three-dimensional cryo-electron microscopy of the calcium ion pump in the sarcoplasmic reticulum membrane. *Nature*, 362, 469–471.

Gradients, waves and oscillations

Allbritton, N.L. and Meyer, T. (1993) Localised calcium spikes and propagating calcium waves. *Cell calcium*, 14, 691–697.

Tsunoda, Y. (1991) Oscillatory Ca^{2+} signalling and its cellular function. *New Biology*, 3, 3–17.

Sphingosine-1-phosphate

Alvarez, S.E., Milstien, S. and Spiegel, S. (2007) Autocrine and paracrine roles of sphingosine-1-phosphate. *Trends in Endocrinology & Metabolism*, 18, 300–307.

Brinkmann, V. (2007) Sphingosine 1-phosphate receptors in health and disease: Mechanistic insights from gene deletion studies and reverse pharmacology. *Pharmacology & Therapeutics*, 115, 84–105.

cADP-ribose

Fliegert, R., Gasser, A., and Guse, A.H. (2007) Regulation of calcium signalling by adenine-based second messengers. *Biochemical Society Transactions*, 35, 109–114 [also includes NAADP signalling].

Higashida, H., Salmina, A.B., Olovyannikova, R.Y., Hashii, M., Yokoyama, S., Koizumi, K., Jin, D., Liu, H.-X., Lopatina, O., Sarwat Amina, S., Islam, M.S., Huang, J.-J. and Noda, M. (2007) Cyclic ADP-ribose as a universal calcium signal molecule in the nervous system *Neurochemistry International*, 51, 192–199.

Nicotinate adenine-dinucleotide phosphate

Lee, H.C. (2005) NAADP-mediated calcium signalling. *Journal of Biological Chemistry*, 280, 33693–33696.

Rutter, G.A. (2003) Calcium signalling: NAADP comes out of the shadows. *Biochemical Journal*, 373, e3-e4 [and papers cited within].

Confocal microscopy

Rizzuto, R., Brini, M., Murgia, M. and Pozzan, T. (1993) Microdomains with high Ca^{2+} close to $InsP_3$-sensitive channels that are sensed by neighbouring mitochondria. *Science*, 262, 744–747.

Williams, D.A. (1993) Mechanisms of calcium-release and propagation in cardiac-cells: Do studies with confocal microscopy add to our understanding? *Cell Calcium*, 14, 724–735.

Web held images

Many companies, for example those that sell the machinery, and those that sell the probes, have confocal/Ca^{2+} images on their web pages. Many research teams also show their images on university held web pages, so the use of a search engine with an author's name could easily reveal images from confocal microscopy work. A good place to start would be the site of *Molecular Probes Inc.*

Reactive oxygen species, reactive nitrogen species and redox signalling

10

Compounds such as reactive oxygen species (ROS) and reactive nitrogen species (RNS) are now known to be important as signals in many systems. Early research in this field concentrated on the role of these compounds as destructive agents, targeted against either invading pathogens, or the cellular components themselves. However, more recent research has revealed a role of ROS and RNS in the control of a host of cellular functions. Although the action of many of these molecules is not fully understood, this chapter discusses the current state of our understanding of how they are generated and perceived, and how they propagate their messages. The main components involved here include those used in phosphorylation (see Chapter 6) and cyclic nucleotides as discussed in Chapter 7.

10.1 Introduction

The cellular production of reactive chemical species, especially those based on the reduced states of molecular oxygen, had been known for several years, but it was probably the realization that endothelium-derived relaxing factor (EDRF) was in fact nitric oxide, and that it had profound physiological effects, that started a new area of research. That is, the role of such chemicals in cell signalling.

There are basically two groups of such chemicals:

- Reactive oxygen species, referred to as ROS or AOS (active oxygen species), which includes the superoxide anion ($O_2^{\bullet-}$) and hydrogen peroxide (H_2O_2).
- Reactive nitrogen species (RNS), which is mainly thought of as the nitric oxide radical $NO^{\bullet}$. However, this can gain and lose electrons to give the NO^- and NO^+ species too.

At first glance these are rather surprising groups of molecules to be involved in cell signalling, and, on the face of it, these compounds would not appear to have the right criteria to be good signalling molecules. Moreover, they have the potential capacity to be detrimental to the cells in which they are formed, as well as to the cells around them. However, research papers appear regularly citing ever increasing amounts of experimental evidence for roles for these molecules in the control of cellular functions, including apparent direct effects on gene transcription. Nitric oxide, for example, has been shown to regulate guanylyl cyclase activity, control neurotransmitter release and to act as a neurotransmitter, as well as to have a role in a bacterial and tumouricidal capacity.

If they are signalling molecules they should be able to exhibit some of the characteristics of signals that were outlined in Chapter 1. That is, they should be relatively small, move easily from their site of production to their site of action, and have a unique and defined effect. They should also be able to be removed relatively easily and quickly. In fact, these molecules, are all characterized by being small inorganic molecules, and as such can diffuse easily to their site of action. Furthermore, neither nitric oxide nor hydrogen peroxide are charged so they can diffuse through membranes, with the potential to move from one cell to another. However, they generally are extremely reactive and have known reactivity towards biological materials, often causing the latter's alteration and loss of function. Molecules such as superoxide and hydrogen peroxide have well-established classical functions in the body, in these cases, in the destruction of invading organisms as a major part of host defence. In fact, this is not unique to the animal kingdom, with more and more literature now showing a similar role for superoxide and hydrogen peroxide in plants, possibly leading to local cell death and areas of necrosis. However, this characteristic means these molecules are relatively short lived, and almost by default removed once their signal is perceived.

So it can be seen that even if on face value they seem unlikely to be signalling molecules, if their production is controlled and their presence perceived, then they can add to the cells' repertoire of signals. The following discussion looks at their production, what they might be controlling and how they might be doing it.

10.2 **Nitric oxide**

The small gaseous molecule nitric oxide (NO$^\bullet$) has been found to be a significant signalling molecule, where its use is not restricted to one particular tissue but has functions in various and diverse locations. Originally a factor that caused the relaxation of cells was described, known as endothelium-derived relaxing factor (EDRF), but in 1987 Moncada and colleagues realized that this activity was mediated by the molecule nitric oxide. Furthermore, it was found that some treatments for angina, for example, the use of nitroglycerine, are mediated through nitric oxide. The nitroglycerine or other organic nitrates are converted to NO$^\bullet$, which causes relaxation of the blood vessels, hence increasing the heart's blood supply. More recently, some functions of brain and other nervous tissue have been shown to be inhibited by compounds that reduce the production of NO$^\bullet$. Although not really signalling, macrophages appear to involve the use of NO$^\bullet$ in part of their host defence mechanism. Therefore, the production and roles of nitric oxide have become popular research projects.

Nitric oxide is in fact a free radical, commonly written as NO$^\bullet$ (the superscript dot denoting its radical status here). That is, it contains an unpaired electron in its outer electronic orbital. This leads to its increased reactivity as this is an unfavoured electronic state, and one from which a molecule will be "keen" to escape. Here, the NO$^\bullet$ can gain or lose an electron to form NO$^+$ or NO$^-$ (as discussed above) or can be converted to nitrates or nitrites. Alternatively, it can react with other potential signals such as superoxide to form the compound peroxynitrite, or glutathione to produce S-nitrosoglutathione (GSNO). Peroxynitrite is very reactive, and may be involved in either signalling events or cell death, whereas GSNO has been suggested to be a form of NO$^\bullet$ that may be transported around an organism, perhaps in its vascular system.

As NO$^\bullet$ is so reactive, with a half-life in the order of only 5–10 seconds, its effects are usually local. However, it can readily diffuse across membranes, which means that its signalling action is not restricted to the cell of origin, but neighbouring cells can also feel the effect. The signal is usually turned off by its deactivation, where NO$^\bullet$ is converted to nitrates and nitrites by oxygen and water.

There are several enzymes that can potentially produce NO$^\bullet$, but in animals the most likely is the enzyme nitric oxide synthase (NOS). Here, NO$^\bullet$ is formed by the oxidation of L-arginine (**Figure 10.1**). The guanidine group of arginine is oxidized in a process, which uses five electrons, resulting in the formation of L-citrulline and nitric oxide through an intermediate step in which hydroxyarginine is formed. This intermediate remains tightly bound to the enzyme and is not released. Usefully for experimental design, the enzyme activity can be inhibited by the addition of L-N$^\omega$-substituted arginines such as L-N$^\omega$

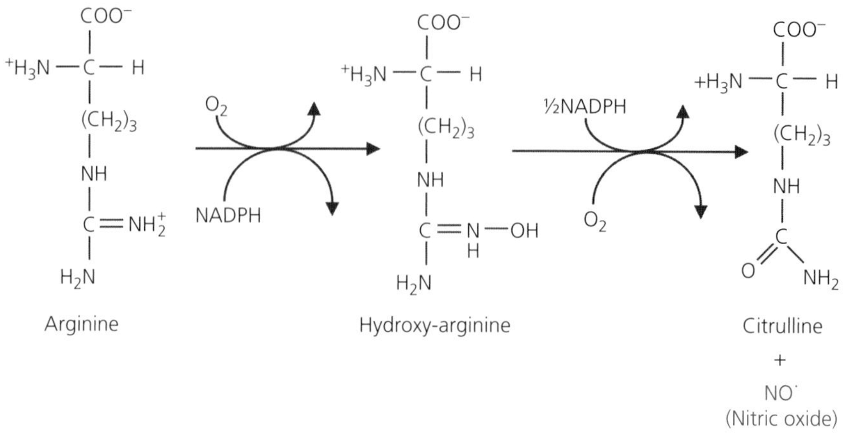

Figure 10.1 The production of nitric oxide from arginine as catalyzed by nitric oxide synthase.

aminoarginine (L-NAA) or L-N$^\omega$ methylarginine (L-NMA). These substituted analogues act competitively with the binding of arginine, although long exposure to some of these compounds leads to irreversible inhibition of the enzyme. Therefore, such compounds can be used to assess the role of NOS in NO$^\bullet$ production in new species or under new conditions.

However, nitric oxide synthase is not a single enzyme entity, but rather a family of isoforms. Using the observation that calmodulin was required for activation of the enzyme, Bredt and Snyder first purified the enzyme from brain tissue in 1990. This was denoted bNOS: a protein of approximately 160 kDa. This led to the purification by others of isoforms from macrophages (macNOS) and endothelial cells (eNOS: approximately 133 kDa). Subsequent Southern blot analysis has suggested that mammalian genomes contain three genes that encode for NOS isoforms. Two of these are expressed constitutively, an endothelial type enzyme (eNOS) and a neuronal type (nNOS), the latter of which has an expression pattern that includes non-neuronal tissues. The third gene encodes an inducible form (iNOS), which includes the enzyme seen in macrophages.

Cloning of the nitric oxide isoenzymes has revealed that they all share a close homology to the enzyme cytochrome P450 reductase (**Figure 10.2**): a NADPH binding site has been identified, as well as areas for FAD and FMN binding, and comparison of the sequences reveals consensus sequences for the binding of these co-factors. Purification of NOS has shown that each monomer has one FAD and one FMN, although, like other enzymes, the FAD can slowly dissociate and has to be exogenously added to obtain full activity *in vitro*. The similarity to the cytochrome P450 system continues in that NOS also contains a haem prosthetic group, that is, iron protoporphyrin IX, as we saw with guanylyl cyclase in **Chapter 7**. Like many haem containing enzymes, NOS can react with and be inhibited by carbon monoxide, giving a characteristic CO binding

A.

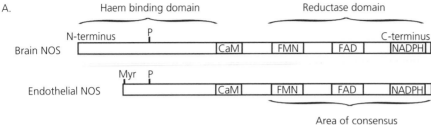

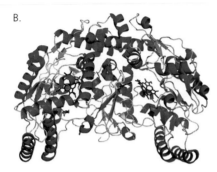

Figure 10.2 A. The domain structure of isoforms of nitric oxide synthase with the area of homology to cytochrome P450 reductase highlighted. CaM; calmodulin binding site, FMN, FAD, NADPH; binding sites for these prosthetic groups/co-factors. B. The proposed structure of the oxygenase domain of nitric oxide synthase, obtained using X-ray diffraction. Often large proteins such as NOS can be cleaved into smaller sections for structural studies. Therefore, the reductase domain can be studied separately. Structure was obtained from the RCSB Protein Data Bank (www.rcsb.org/pdb/) PDB ID: 1zvl (Matter, H., Kumar, H.S., Fedorov, R., Frey, A., Kotsonis, P., Hartmann, E., Frohlich, L.G., Reif, A., Pfleiderer, W., Scheurer, P., Ghosh, D.K., Schlichting, I. and Schmidt, H.H. (2005) Structural analysis of isoform-specific inhibitors targeting the tetrahydrobiopterin binding site of human nitric oxide synthases. *Journal of Medicinal Chemistry*, 48, 4783–4792).

absorbance spectrum suggesting that the haem is attached via a cysteine residue. Such binding of carbon monoxide is characteristic of enzymes that have the capacity to bind to oxygen, as seen with cytochrome oxidase and haemoglobin. It appears that the first step of the catalysis is the binding of the arginine to the haem group, with subsequent oxidation reactions.

In the NOS structure, between the binding regions for FMN and FAD and that for haem binding, is a region used for calmodulin binding, which in some cases confers calcium ion control on the enzyme activity. At the N-terminal end of this particular part of the polypeptide is a trypsin sensitive region. Digestion with trypsin yields two domains, one from the N-terminal end of NO˙ synthase, which can bind to arginine and contains haem, and a second containing the NADPH and flavin binding regions. It has been suggested therefore that the synthase is a bi-domain enzyme, with one domain containing the oxygenase activity and the other containing the reductase activity, where the two domains may even be able to function independently of each other. A representation of the structure of the oxygenase domain is shown in **Figure 10.2A**. The enzyme probably exists as dimers with the oxygenase domain of one subunit interacting with the reductase domain of the other. However, the exact stoichiometry of the electron transfer has yet to be fully determined, with electron leakage possible to molecular oxygen which would yield the superoxide ion: also an unstable free radical (see below).

The purification of the brain enzyme showed that tetrahydrobiopterin was tightly bound to NOS and it was thought that it must function in the catalysis as it could take part in an electron transfer role.

Not only was the binding of calmodulin to the enzyme crucial to its original purification, but it has become apparent that Ca^{2+} is important in the regulation of many isoforms of NOS. The concentrations of calcium involved are in the region to be expected for a calmodulin activated system, with an EC_{50} of $2–4 \times 10^{-7}$ M. Neither calcium nor calmodulin appear to affect arginine binding, but calmodulin binding seems to regulate the electron transfer activities of the enzyme.

■ The EC_{50} value is the concentration of an agonist that elicits a response which is 50% of the maximum response for that agonist.

As mentioned above, whereas some NOSs are produced by the cells constitutively, other forms are inducible, including NOS of macrophages, designated mNOS or iNOS. It is interesting to note that these isoforms contain calmodulin binding sites, but are in fact unaffected by calmodulin antagonists or Ca^{2+}, and indeed calmodulin is tightly bound to these enzymes even in the absence of calcium. Production of new NOS protein molecules of these isoforms can be stimulated by the presence of, amongst many other things, interferon-γ, interleukin-1 and lipopolysaccharide (LPS). Cloning of the transcription start site of this NOS gene has shown that two distinct regions contain LPS and interferon responsive elements. The LPS region lies up-stream about 50–200 basepairs of the transcription start site while region 2, responsive to interferon-γ, is about 900–1000 base pairs up-stream from the start site. Other response elements have also been reported, with such data indicating that levels of iNOS protein in cells are carefully regulated. Furthermore, recently cells other than macrophages have also been shown to contain the inducible form of the enzyme.

As well as control by calcium ions, other control mechanisms for NOS also exist. Consensus sequences for phosphorylation by cAMP-dependent protein kinases (PKAs) have been found in the NOS of brain and endothelial cells, although the macrophage form of the enzyme seems to lack them. cAMP-dependent protein kinase, protein kinase C, cGMP-dependent protein kinase and Ca^{2+}/calmodulin-dependent protein kinase can all phosphorylate the neuronal form of the enzyme, and such phosphorylation results in a decrease in NOS activity.

It appears that it is not only the enzymatic activity that is regulated by phosphorylation, but also the subcellular distribution of the enzyme in endothelial cells. The NOS of these cells is primarily located in the plasma membrane. On addition of bradykinin, a signalling cascade leads to the enzyme becoming phosphorylated and it is then translocated to the soluble fraction of the cells, albeit in an inactive state. This mechanism would ensure that the enzyme is only active when it is located in a place where the nitric oxide is released from the cell, that is when it is associated with the plasma membrane. Enzyme location may also be influenced by other factors. For example, the eNOS enzyme is myristoylated at the N-terminal end, and if this site is removed by site-directed mutagenesis the enzyme alters its location from the membrane to the cytosol. There are also reports that the NOS polypeptide may be palmitoylated.

But what does nitric oxide actually do once it is made and released? One of the main cellular targets of NO˙ is the enzyme guanylyl cyclase (Figure 10.3), the enzyme responsible for the production of cGMP, itself an important signalling molecule, as discussed in Chapter 7. It is usually the soluble form of the cyclase that is the target of NO˙, but some data have been reported that suggest the membrane form may also be NO˙ regulated to some extent. The activation of guanylyl cyclase is caused by binding of NO˙ to the haem group of the enzyme, the presence of iron within the haem group being the key here, leading to the production of cGMP and, of course, an increase in intracellular cGMP concentration, assuming it is not removed as quickly by the appropriate phosphodiesterase. cGMP might be responsible for the regulation of serine/threonine kinases, cGMP-dependent protein kinase in particular, or it may regulate the activity of some phosphodiesterases with a resultant modulation of any cAMP response, as described in Chapter 7. In vascular tissues, the activation of guanylyl cyclase by NO˙ causes muscle relaxation, probably by the activation of a cGMP-dependent protein kinase acting on myosin light chains. One of the well-studied physiological effects in which NO˙ plays a major role is the control of smooth muscle contraction, and how this controls the flow of blood through the vessels, with probably the most cited example

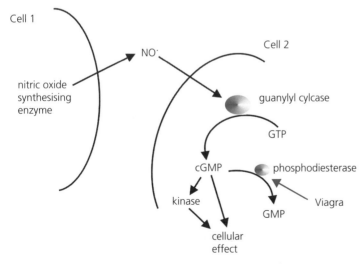

Figure 10.3 A scheme showing how NO might fit into a signalling pathway.

Haem groups (alternatively spelt as heme)

The prosthetic group haem is used commonly in biological systems, in a variety of manners. In haemoglobin, the haem is responsible for oxygen chelation, whereas in cytochromes it is involved in electron transfer. NO can bind to many haem-containing proteins, and so alter their function. This might lead to changes in electron transport chain activity in mitochondria for example. NO binding to haemoglobin is used as an assay for the presence of NO.

being that of penile erection. For men with dysfunctional penile responses, the drug Viagra™ is available. It is, in fact, a misconception that Viagra™ either releases NO˙, or causes generation of NO˙. What it does is inhibit phosphodiesterase activity, in particular that of phosphodiesterase V. Therefore, the signalling pathway leads to NO˙ production and NO˙ will activate guanylyl cyclase, resulting in elevated cGMP levels, but these are soon removed by phosphodiesterase activity. By inhibiting this phosphodiesterase, the cGMP persists for longer in the cell, and the "NO˙ signal" is enhanced. The place where Viagra™ acts is schematically shown in Figure 10.3. Women too are now thought to be able to be helped by Viagra™, with the drug altering the blood supply to the uterus. Interestingly, some people who take Viagra™ have found that they have affected eyesight. This might sound bizarre, until the signalling pathway that relays the message from the light receptors to the brain is studied (see Chapter 12, and previous discussion in Chapter 7). Here, the signalling pathway also involves cGMP, but this time a different isoform of phosphodiesterase, type VI, which is obviously effected by Viagra in some individuals.

In some enzymes NO˙ is also able to bind iron that is not associated with haem groups, for example, in enzymes containing iron-sulphur complexes, such as NADH-ubiquinone oxidoreductase—otherwise known as complex I of the mitochondrial electron transport chain. Its interaction with such non-haem iron has been implicated in regulation of the translation of some proteins, and even in the alteration of rates of DNA synthesis. Furthermore, the binding of NO˙ to ferritin can cause the liberation of free iron, which in the presence of oxygen free radicals may lead to lipid peroxidation of membranes through a mechanism involving the production of hydroxyl radicals (see below). This would potentially be very detrimental to many of the activities within a cell.

In neuronal tissues NO˙ is involved in many facets of the neuronal physiology. It has been implicated in the regulation of neurotransmitter release for example, and, interestingly, both the differentiation and regeneration of neurones also might be affected by NO˙. It has been reported that brain NOS is transiently expressed after neuronal injury. NO˙ can itself also act as a specific neurotransmitter. However, if the NO˙ concentration rises above its normal very low levels, neurotoxicity can result.

Besides its role as a cell signalling molecule, like superoxide and its by-products, NO˙ can also act in tumouricidal and bactericidal capacities and it has been reported that viral replication is also inhibited by NO˙.

Other enzymatic sources of NO˙

Nitric oxide synthase is not the only source of NO˙ in biological systems. One enzyme that for many years has been studied for its role in hydrogen peroxide production is xanthine oxidoreductase (XOR: otherwise referred to as xanthine oxidase (XO) or xanthine dehydrogenase). This enzyme can also generate NO˙, particularly under anaerobic conditions. Therefore, here we have an enzyme that can generate ROS if oxygen is plentiful, or RNS if there is no oxygen. Both sets of products are powerful signalling molecules, and therefore this enzyme appears to have the ability to switch between signalling pathways, dependent on the oxygen levels. Moreover, if oxygen is low, both ROS and RNS are produced simultaneously, with the subsequent generation of peroxynitrite, a mode of action thought to be involved in anti-microbial activity.

Anti-microbial activity of xanthine oxidoreductase

As XOR can simultaneously produce NO˙ and ROS, then peroxynitrite can be generated. XOR is found in cells, but also in extracellular fluids, particularly milk and it has been shown by at least two research groups that milk has anti-microbial activity for which XOR is thought to be the main protagonist.

NO˙ generation by plants has been known for a long time, but it was only in 1998, when NO˙ release was shown to be involved in pathogen defence responses, that the research area opened up. However, the source of NO˙ in plants has been difficult to determine. A look at the *Arabidopsis* genome database will not reveal the presence of an obvious NOS enzyme gene. Although NOS inhibitors in some cases do inhibit both NO˙ generation and NO˙-induced effects, it is clear that there is more than one major source of NO˙ in plants. The main candidates are nitrate reductase and a NOS-like enzyme. Using mutant plants where the nitrate reductase activity is very low, it has been shown for example that this enzyme is key to NO˙ generation, which controls water loss in *Arabidopsis*. Like XOR, this is a molybdenum and flavin containing enzyme. The second potential NO˙-generating enzymes in plants, which are NOS-like, have still not been identified. One of the likely candidates is now thought to be a G protein, perhaps controlling NO˙ generation in some way but not being directly involved. Therefore, the exact nature of a NOS-like enzyme has still to come to light for plants, if indeed such an enzyme truly exists.

10.3 Reactive oxygen species: superoxide and hydrogen peroxide

The early discoveries in 1933 by Baldridge and Gerard that phagocytic cells had an increase in oxygen consumption when stimulated, have led to a great interest in the molecular events that take place on the activation of these cells. Sbarra and Karnovsky in 1959 overturned the earlier view that the oxygen was used for increased respiration, and, by 1961, Iyer and his group noted that the oxygen consumption was accompanied by an increase in the hexose monophosphate shunt leading to the production of NADPH. It is now known that the oxygen taken up by phagocytic cells, such as neutrophils, is, in fact, used in the production of superoxide ions, and, as discussed below, many more reactive oxygen species (ROS) as a result of further reactions. The production of superoxide, however, involves the direct enzymatic reduction of molecular oxygen by a complex situated in the plasma membrane of phagocytic cells: this enzyme complex is known as the NADPH oxidase. This enzyme has been most well characterized in neutrophils, but is now studied in many cell types.

The importance of this biological activity in neutrophils is clearly demonstrated by the disease chronic granulomatous disease (CGD). This is a genetic disease that manifests itself as a defect of the NADPH oxidase complex with the complete abolition of any superoxide production. People suffering from this disorder suffer from recurrent bacterial and fungal infections, and, until relatively recently, died at an early age.

The primary product from the NADPH oxidase enzyme is the superoxide ion where the electrons are supplied by intracellular NADPH:

$$2O_2 + NADPH \rightarrow 2O_2^{\bullet -} + NADP^+ + H^+$$
superoxide ion

The extra electron supplied to molecular oxygen is in an unpaired state, and hence like nitric oxide this ion is classified as a free radical. Again, this electronic state is relatively unstable and the new ion is consequently reasonably reactive. For example, it readily undergoes dismutation with the formation of hydrogen peroxide, not itself a free radical but often grouped in with the oxygen free radicals in discussions:

$$2O_2^{\bullet -} + 2H^+ \rightarrow H_2O_2 + O_2$$
hydrogen peroxide

Although this reaction can occur spontaneously, especially at low pH, it is also catalyzed by an enzyme called superoxide dismutase (SOD). This enzyme has two main forms, a copper/zinc containing form that resides in the cytosol of cells and a manganese containing form that is located in mitochondria.

Both superoxide ions and hydrogen peroxide are reactive towards biological materials, although it is probably the results of a further cascade of reactions

■ Hydrogen peroxide is not itself a free radical but is often grouped in with the oxygen free radicals in discussions.

that cause the most damage. Superoxide has been shown to cause biological oxidation, especially in hydrophobic environments, whereas hydrogen peroxide, a relatively weak oxidizing agent, has been found to inactivate some enzymes, usually by the oxidation of thiol groups (see discussion below). In a similar manner to the removal of superoxide by SOD, hydrogen peroxide is catalytically destroyed by either the enzyme catalase or by the glutathione cycle, and hence both superoxide and hydrogen peroxide can be removed quickly once their potential signals are no longer needed – as discussed above, this is a good characteristic of any potential signal.

The importance of the abolition of superoxide is demonstrated by the fact that all aerobic organisms appear to contain at least one isoform of superoxide dismutase, life in the presence of oxygen being dependent on the destruction of harmful oxygen free radicals. However, the real danger comes when superoxide and hydrogen peroxide react together with formation of hydroxyl radicals:

$$O_2{}^{\bullet-} + H_2O_2 \rightarrow OH^{\bullet} + OH^- + O_2$$
hydroxyl radical

This reaction is catalyzed by the presence of iron ions, Fe^{2+} or copper ions, Cu^{2+}, proceeding via the Haber–Weiss reaction or Fenton reaction. Hydroxyl radicals are extremely reactive, the free unpaired electron making it extremely unstable. In fact, it has been mooted that hydroxyl radicals are the most reactive chemicals found in biological systems. Damage can be caused to a cell in the form of oxidation of proteins, oxidation of bases that can lead to DNA strand breakage, direct attack of deoxyribose moieties and lipid peroxidation as mentioned above. Usually such reactions proceed via removal of hydrogen atoms from organic molecules, often leading to the formation of new radicals and possibly a cascade of further free radical reactions. In the case of lipid peroxidation this can lead to membrane dysfunction.

Other oxygen free radicals can also be formed as a consequence of the production of superoxide. These include singlet oxygen (O_2^1), which can lead to the destruction of carotenes, haem proteins and membrane lipids, and also in phagocytic cells where the haem containing enzyme myeloperoxidase is present, the production of hypochlorite is seen.

As above when the production of NO was discussed, it should be emphasized that superoxide can also react with nitric oxide to produce a very reactive compound, peroxynitrite.

$$NO^{\bullet} + O2{}^{\bullet-} \rightarrow OONO^-$$
peroxynitrite

Therefore, it is possible that the production of superoxide ions modulates the role of nitric oxide, albeit with the production of a very reactive compound.

Despite this apparent cascade of dangerous chemicals produced as a consequence of the release of superoxide, both superoxide ions and hydrogen peroxide appear to be released in non-phagocytic cells. Here, the function is almost certainly not in host defence. The levels of superoxide are in the order of only 1% of that of neutrophils, whereas the production is sustained for very long periods of time, possibly constitutively by some cells. Different cell types might use these free radicals in different ways, but several lines of evidence point to a cell signalling role in many instances.

It is not only the species in the animal kingdom that show this activity, but many more species besides. For example, there is now great interest in the release of ROS from plants, where they seem to be involved in a range of physiological responses, from the defence against pathogen attack, the control of stomatal apertures and hence water loss, and the growth and gravitropic responses of roots.

Evidence for superoxide and hydrogen peroxide acting as a signal

How, if at all, are these reactive, destructive molecules being used as a way of signalling within the cells? Studies are often done by adding exogenous ROS or by the addition of free radical scavengers and such conditions can promote or reduce the proliferation of cells in culture respectively. Early work by Crawford and colleagues showed that the addition of H_2O_2 or xanthine oxidase/xanthine stimulated the family of genes, c-*fos* and c-*myc* and switched on DNA synthesis. Other groups reported that H_2O_2 increased the expression of the genes c-*fos*, c-*jun*, *egr*-1 and JE in other cell lines. There was also an increase in the level of AP-1 DNA binding activity. AP-1 is a transcription factor, a complex composed of *jun* and *fos* gene products. Of significance, it was found that the transcription factor NF-κB was activated by H_2O_2 and it was suggested that ROS were serving as second messengers, which directly mediated the release of the IκB, inhibitory subunit, from NF-κB. However, there is still some doubt as to whether this a direct interaction, or mediated by a signal transduction pathway.

However, there is little doubt now that ROS are acting as signals. Pathways that can lead to altered gene expression, such as those involving mitogen-activated protein kinases (MAP kinases) and JAK/STATs signalling have been shown to be activated by ROS, whereas phosphatase activity has been shown to be inhibited. MAP kinases are rapidly activated in response to several external factors, promoting growth and differentiation. In neutrophils, H_2O_2 caused an increase in tyrosine phosphorylation and it was concluded that activation of the MAP kinase was due to stimulation of tyrosine and perhaps threonine phosphorylation of the kinase mediated by a MAP kinase kinase, MEK, which was also concomitant with the inhibition by H_2O_2 of a MAP kinase phosphatase, shutting down the dephosphorylation of MAP kinase.

With the advent of the genome sequencing projects and the ready availability of gene sequences, global gene expression studies have been undertaken, showing that ROS both induce and reduce the expression of a plethora of genes. Approximately 3% of the genes expressed have their rates of transcription altered, the 97% not being altered being good evidence that the cells are not just being killed.

ROS are also implicated in the onset and maintenance of apoptosis programmes, and so there is no doubt that in animals the production and perception of ROS is critical to the functioning of many cells. Reports in the literature also suggest that H_2O_2 can cause the activation of both guanylyl cyclase and phospholipase D. An increase in the activity of guanylyl cyclase results in the formation of cGMP as was seen above with NO, whereas increased phospholipase D activity would instigate the production of lipid derived signals. Certainly, the production of prostaglandins and thromboxanes have been seen to be stimulated by the presence of hydrogen peroxide in some cell types.

Similar signalling by reactive oxygen species has been noted in plants too. H_2O_2 has been shown to induce cellular protection genes and to act as a diffusible element that switches on gene expression in adjacent cells, in this case the genes for glutathione S-transferase and glutathione peroxidase, both of which code for products used in the protection of cells. Recently, as with animal cells, global gene expression studies have been undertaken. In *Arabidopsis* just under 200 genes had their expression increased, whereas the expression of around 65 was depressed, out of the approximately 8000 genes studied.

As well as the activation of the expression of genes, H_2O_2 has been shown to activate MAP kinases here as well. Defined isoforms of MAP kinases have been identified in plants that are activated by hydrogen peroxide and a recent report also shows that a kinase OXI1 is instrumental in H_2O_2 signalling here.

Recent work in both yeast and plants has shown that histidine kinases (see **Chapter 6**) might be involved in hydrogen peroxide signalling into the cell, and it is possible that such a mechanism is more widespread than presently appreciated.

The NADPH oxidase complex

One of the most well-characterized sources of superoxide, and hence other reactive oxygen species (ROS) is from the enzyme NADPH oxidase. It is, in fact, an enzyme complex catalyzing the one electron reduction of molecular oxygen to superoxide. The preferred electron donor is NADPH (K_m approx. 50 mM), not NADH (K_m approx. 500 mM), hence the name of the complex.

NADPH oxidase is a membrane bound complex consisting of a short electron transport chain, containing only two redox active groups, a FAD moiety and a cytochrome (**Figure 10.4**), although as we shall see, both the haem and FAD group reside on the same protein, and the cytochrome is referred to as a flavo-cytochrome.

The cytochrome involved is a *b*-type cytochrome, located in the plasma membrane and specific granules of neutrophils, and it has been found primarily in the plasma membrane of many other cells, such as fibroblasts and mesangial cells from the kidney. The cytochrome is unique in that it has a very low midpoint potential, originally measured as $Em_{7.0} = -245$ mV, considerably lower

Oxidase enzymes

In humans there are at least six versions of the NADPH oxidase known – NOX enzymes. Originally only thought to be in phagocytic cells, sequence analysis has revealed other oxidase enzymes, some of which have a bigger large subunit, containing other domains and functions—the so called DUOX proteins.

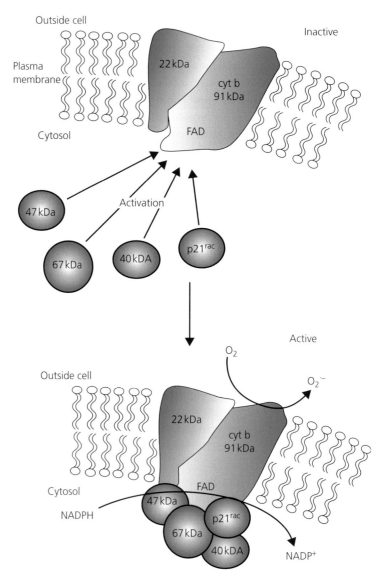

Figure 10.4 A schematic representation of the NADPH oxidase complex and its activation by translocation of several cytosolic components to the plasma membrane.

than that of other eukaryotic cytochromes. This extremely low mid-point potential is, however, sufficiently low to facilitate the reduction of molecular oxygen to superoxide and it has also been shown that the cytochrome is kinetically competent to act as the direct electron donor to oxygen. Because of this unique characteristic, the unusually low mid-point potential, the cytochrome is often referred to as cytochrome b_{-245} but it is also referred to by the wavelength of its α-band absorbance in the visible spectrum, that is cytochrome b_{558}. In fact, more recent research has shown that the cytochrome has two haem groups, of different mid-point potentials, and hence the name cytochrome b_{-245} might be misleading.

The cytochrome is also unusual in being a heterodimer, containing a small α-subunit of approximate molecular weight 22 kDa. The protein has been named p22-*phox*, the *phox* referring to the fact that it was characterized from a phagocytic cell oxidase. The larger β-subunit of the cytochrome has an approximate molecular weight of 76–92 kDa, and is highly glycosylated. It is referred to as gp91-*phox* and has the haem associated with it, with each haem being held between two α-helices that span the membrane, two histidines being used to hold each haem. Therefore, to support this topology, gp91-*phox* is believed to contain several membrane spanning domains.

For a long time the site of flavin binding was also contested, but it is now known that the FAD group also binds to the gp91-*phox* subunit, albeit in a domain that resides on the cytosolic side of the membrane. This domain also contains the NADPH binding site. Therefore, the large gp91-*phox* contains all that there is needed to allow electron transfer from the NADPH in the cytoplasm, through the FAD, subsequently onto the haem and finally to the oxygen on the other surface of the membrane, where superoxide can be released.

Besides the membrane bound flavo-cytochrome, the oxidase also requires the presence of several cytosolic proteins for full activity (**Figure 10.4**). By studying patients with CGD, two polypeptides have been identified as being integral with NADPH oxidase activity. These are of 47 kDa, p47-*phox* and 67 kDa, p67-*phox*. Both these proteins are primarily found in the cytosol of resting neutrophils, but on activation are translocated to the plasma membrane. Both have been cloned and sequenced and both contain SH3 domains (see **Chapter 1**), which are probably used in their interaction with the other polypeptides in the complex. It is thought that the binding of the cytosolic proteins to the membrane complex in some way controls or facilitates electron flow though the flavo-cytochrome. Recently another cytosolic NADPH oxidase component containing a SH3 domain has been recognised. This is a 40 kDa polypeptide, p40-*phox*, which appears to form an activation complex with p47-*phox* and p67-*phox* which as a unit translocates to the plasma membrane to associate with the flavo-cytochrome. Sequence data reveal that this polypeptide shares a large region of homology with the N-terminus of p47-*phox*, but interestingly no CGD patients appear to have been identified that lack this protein, suggesting that it might have other roles, not just being associated with the NADPH oxidase system.

Other cytosolic proteins required for full activity of the NADPH oxidase were found by the reconstitution of activity from mixtures of cytosol and membranes. Besides the p47-*phox* and p67-*phox* it was found that a heterodimeric complex was required. This complex contained a G protein related to Ras, p21^{rac} and a GDP-dissociation inhibition factor, Rho-GDI. These two proteins undergo a cycle of dissociation and association with the subsequent turnover of GTP as discussed more fully in **Chapter 5**.

Other sources of superoxide

The NADPH oxidase is not the only cellular source of superoxide, and in fact the electron leakage from most electron transport chains can give up electrons to oxygen with the formation of oxygen radicals. The mitochondrial electron transport chain, for example, can lose electrons from complex I (NADH-ubiquinone oxidoreductase) or complex IV (cytochrome oxidase), as can the photosynthetic pathways of plants. It has been estimated that the average human doing average exercise will generate 2–3 g of superoxide ions per day! However, there are no reported cases of the production of oxygen radicals from this source being used in a constructive way, unless their use during apoptosis is considered. Furthermore, this electron leakage has been implicated in the reduction of the activity of these complexes, and it is thought that the damage caused may lead to a further increase in the production of oxygen radicals. Such a compounding effect is thought to lead to cellular death in some degenerative diseases such as Parkinson's disease, Alzheimer's disease, and, in fact, in the general ageing process.

A more likely source of oxygen free radicals is from the enzyme xanthine oxidoreductase. This is a molybdenum and iron containing flavoprotein, which catalyzes the oxidation of hypoxanthine to xanthine and then to uric acid. Molecular oxygen is used as the oxidant, and what can be almost considered as a by-product is hydrogen peroxide. In fact, the addition of hypoxanthine and xanthine oxidase to cultured cells is a good experimental way of introducing hydrogen peroxide to the proximity of cells. However, under low oxygen tension this enzyme can also produce NO$^{\bullet}$ (see above).

10.4 Redox signalling and molecular mechanisms of hydrogen peroxide signalling

It is clear, therefore, that both physiological effects of ROS can be seen, and that several proteins have been identified which have their activity altered by ROS. However, it is still not clear how ROS are perceived by cells. Although some effects on guanylyl cyclase are seen, this is not the primary way in which ROS are acting. Similarly, some researchers question whether ROS do have a direct effect on some proteins such as NF-κB. Therefore, how is ROS perceived, and how does it affect protein function?

One of the ways in which ROS might be acting is on the redox state of the cell. Cells contain a large concentration of glutathione and other redox active compounds in their cytoplasms and organelles, and the majority of this glutathione is maintained in a reduced state. Therefore, if the Nernst equation is used the redox poise of the cytoplasm of a cell can be calculated - it is found, in fact, to be very negative, around −250 mV. ROS will react with the glutathione, oxidizing it. This has two consequences. Firstly, the amount of available glutathione is lowered, perhaps so that it can no longer partake in certain reactions, and, secondly, the redox state of the cytoplasm becomes oxidizing. It has been suggested that certain cysteine thiol groups on proteins are maintained in a reduced state, until the redox state of the solution around them alters, allowing them to become oxidized and therefore alter the conformation of the protein in which they reside (see **Figure 10.5**). This would certainly be an effective mechanism to alter the activity of certain proteins. During apoptosis of animal cells, the redox state of the cytoplasm is said to become approximately 70 mV more oxidizing, certainly enough to affect a thiol group that has an appropriate mid-point potential.

Alternatively, ROS may act on the proteins directly. Some of the chemistry in which thiol groups can partake is shown in **Figure 10.5**. The −SH group can be oxidized to the sulphenic acid group, and further oxidized to sulphinic acid and then to sulphonic acid. Each form of oxidation may have its own effects on the proteins function. If there are two thiol groups being oxidized, and they are in close proximity, once the sulphenic acid groups have been formed this allows a further reaction to create a disulphide bridge, again with ramifications for the structure and function of the polypeptide, or polypeptides if the disulphide can form across subunits. In most cases the chemistry is reversible, and, as such, the oxidation of the thiol is akin to phosphorylation, in that it can be toggled between the two states, which is ideal for cell signalling where a protein's activity may need to be rapidly turned on and off.

Exciting work on the mechanism by which hydrogen peroxide might inhibit the activity of phosphatases has also been published. While studying the oxidized forms of the enzyme, where it was proposed that the cysteine residue in the active site is oxidized, an interesting cysteine derivative was found, the sulphenyl-amide intermediate (**Figure 10.5**). This was formed by a reaction of the sulphur of the cysteine linking to the nitrogen of the serine, which was next in the amino acid chain (**Figure 10.6**). This species lacked the oxygen and its formation was fully reversible, whereas it has been suggested that the irreversible forms, sulphinic acid and sulphonic acid, were only formed at higher hydrogen peroxide concentrations, as observed by peptide analysis using MALDI-TOF mass spectroscopy. Also importantly, the sulphenyl-amide intermediate could react with glutathione to produce glutathionylated derivatives, and hence, potentially, further signals (see **Figure 10.5**).

> **Nernst equation** This equation can be used to calculate either the proportions of a compound that exists in its oxidized state and reduced state, or alternatively can be used to calculate the redox state of a solution.

MALDI-TOF mass
spectroscopy

Matrix assisted laser desorption
ionization-time of flight mass
spectroscopy, along with elec-
trospray ionization mass spectroscopy (ESI-MS) are extremely
powerful ways to analyze proteins, either to identify them,
or study changes that might
have taken place.

Irreversibility of higher oxidation states of thiols

Although the formation of the sulphinic acid and sulphonic acid groups was thought
to be irreversible, recent work has revealed an enzyme that can catalyze the oxidation
of sulphinic acid to sulphenic acid, showing that reversibility is possible, and therefore
the formation of such higher oxidation states on thiol groups could be involved in cell
signalling processes.

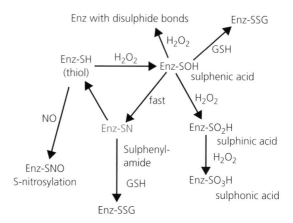

Figure 10.5 Some of the proposed reactions of protein cysteine residues. Much of the
chemistry appears to proceed through the –SOH (sulphenic acid) intermediate. Note that
the reverse reactions are not indicated here, but many, although not all, of these steps are
reversible.

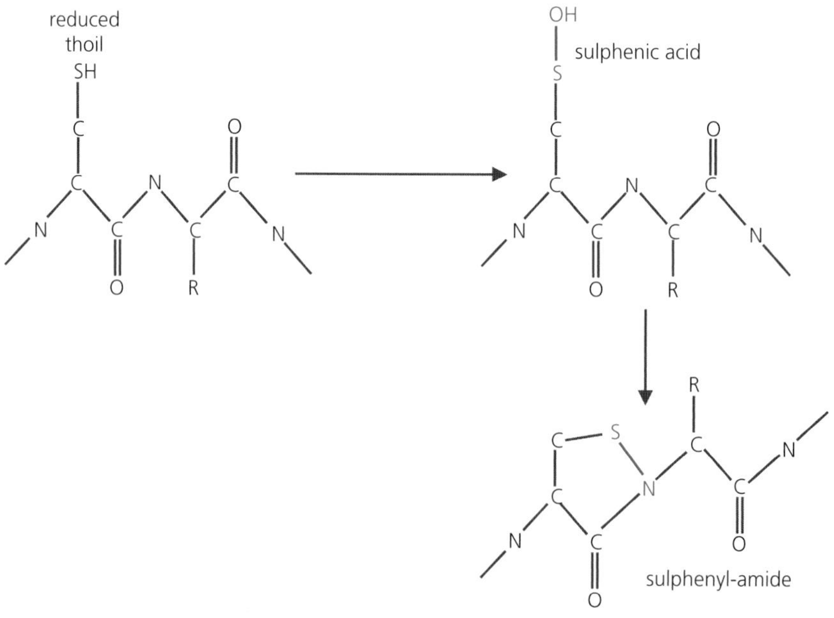

Figure 10.6 The formation of the sulphenyl-amide derivative.

A mechanism such as the formation of the sulphenyl-amide intermediate highlights the fact that hydrogen peroxide and proteins can directly interact, with the alteration of protein conformation and therefore activity of the protein. However, it is unlikely that phosphatases are unique in this, and that they perceive all the hydrogen peroxide signalling needed, and so "hydrogen peroxide perception proteins" need to be identified.

10.5 **Measuring ROS and RNS**

In Chapter 9, the use of fluorescent probes was discussed as a way of finding the location of Ca^{2+} ions, but the same principles can be used here to study the generation of ROS and RNS. Specific probes such as diaminofluorescein diacetate (DAF-2 DA) have been instrumental in measuring the release of NO from many cell types, and allow terrific images to be obtained. For example, DAF-2 DA has shown that only certain cells release NO in response to hormones, such as guard cells in response to ABA in leaves. Often these compounds are supplied to the cell in an inactive form, and are converted by the cell to an active form. DAF-2 DA is not sensitive to NO, but becomes sensitive once de-esterified to DAF. However, DAF will not be taken in by the cell. So, to study NO generation in cells, DAF-2 DA is added, and it is taken up by the cells, which then metabolize it to DAF. This not only releases the active compound, but also traps it inside the cell where it is required. Similar compounds are available to study ROS too, although great care has to be taken as not all compounds used for ROS measurements are very specific.

Delivery of compounds as esters

Often an ester form of a compound is used to deliver compounds or probes to cells. Ester forms are often more membrane-permeable, and they are taken up by the cell, de-esterified and trapped in the cell, unable to re-cross the membrane and escape. Once in the cell, the de-esterified probe can monitor the signalling molecule to which it is sensitive. The esters themselves are often not sensitive to the signalling molecule, and any ester not taken up can be simply washed off, so that it is unable to interfere with the data being sought.

To elucidate redox changes in cells, green fluorescent protein has been engineered, and so it should be possible to get such a protein expressed in the cell, and the redox environment monitored.

Of course fluorescence is not the only way of measuring ROS, RNS and redox, and fluorescence is notoriously difficult to quantify. Many other methods also exist for studying the release of these reactive compounds, including the haemoglobin assay for NO, and luminol luminescence for ROS.

10.6 Carbon monoxide and other compounds

Recent evidence shows that guanylyl cyclase is not only under control by the gas nitric oxide, and possibly hydrogen peroxide, but also that its activation might be modulated by another gas, carbon monoxide (CO). For example, evidence has been presented to show that this system may be involved in regulation of insulin secretion, as well as in control of corticotropin-releasing hormone release in the hypothalamus. Using an isolated form of soluble guanylyl cyclase from bovine lung, it was shown that carbon monoxide could bind to the haem group of the enzyme to form a 6 co-ordinate complex. CO has also been implicated in the control of blood vessel relaxation, which is how NO was originally identified as a signalling molecule, in platelet function and in the control of gene expression, which could lead to a plethora of down-stream events. CO can be generated by haem oxygenase.

In plants, too, CO has been seen to have effects. It was shown to be beneficial for seed germination under salt stress conditions, and in this system CO appeared to be involved in the regulation of genes encoding proteins involved in oxidative stress.

The last compound to be considered here is hydrogen sulphide (H_2S). It can be produced either chemically or enzymatically, often with the starting substrate for its formations emanating from the diet, perhaps from fruits, beans, and in particular onions and garlic. Enzymes such as cystathionine gamma-lyase in smooth muscle and cystathionine beta-synthase in the brain can generate H_2S, and, although physiological effects have been reported, the exact mechanisms in which this compound partakes have yet to be fully unravelled.

■ Summary

- A group of small inorganic molecules has been discovered that are produced by cells and diffuse to neighbouring cells, or act on the cell that produced them, where they have a role in the control of many cellular functions.

- Here, signalling molecules include nitric oxide, superoxide ions, hydrogen peroxide and carbon dioxide.

- Nitric oxide is produced in animals primarily by the enzyme nitric oxide synthase, with arginine as the substrate.

- Cloning of the enzyme has shown that it contains areas of homology with the enzyme cytochrome P450 reductase. Both contain domains for binding NADPH, FMN and FAD, whereas, like cytochrome P450, nitric oxide synthase also contains haem.

- The original role of nitric oxide was reported as a cellular relaxation factor, and it is now known that one of its functions is control of cGMP production by guanylyl cyclase.

- Other enzymes that can generate NO˙ include xanthine oxidoreductase and nitrate reductase.

- Oxygen free radicals such as superoxide can be produced by the electron leakage of many electron transport chains, including the mitochondrial complexes.

- Large amounts of superoxide are produced by the enzyme NADPH oxidase, and its role in the killing of invading pathogens is highlighted by its absence in the disease, chronic granulomatous disease.

- Reactive oxygen species, including the non-radical hydrogen peroxide, have been shown to control cellular proliferation and the rates of transcription of cells, perhaps either through a direct action on transcription factors or by the stimulation of MAP kinase-type pathways.

- The molecular mechanisms by which hydrogen peroxide and nitric oxide relay their messages are now being unravelled, but include nitrosylation and oxidation of proteins.

- Other gaseous compounds such as CO and H_2S are now coming to light as important signals in many organisms, including humans.

→ Further reading

Hancock, J.T. (ed) (2008) *Redox-Mediated Signal Transduction: Methods and Protocols*. Methods in Molecular Biology series: Vol. 476. Humana Press, Totowa, New Jersey. ISBN 978-1-58829-842-3 [a collection of discussions and methods].

Lambeth, J.D., Krause, K.H. and Clark, R,A. (2008) NOX enzymes as novel targets for drug development. *Semin Immunopathology*, 30, 339–363.

Nicholls, D.G. and Ferguson, S.J. (2002) *Bioenergetics3*. Academic Press, London, UK. ISBN 0125181213 [an excellent book on redox].

Nitric oxide references

Bredt D.S. and Snyder S.H. (1990) Isolation of nitric oxide synthase, a calmodulin-requiring enzyme. *Proceedings of the National Academy of Science, USA*, 87, 682–685.

Burnett, A.L., Lowenstein, C.J., Bredt, D.S., Chang, T.S.K. and Snyder, S.H. (1992) Nitric oxide: A physiologic mediator of penile erection. *Science*, 257, 401–403.

Desikan, R., Griffiths, R., Hancock, J.T. and Neill, S.J. (2002) A new role for an old enzyme: Nitrate reductase-mediated nitric oxide generation is required for abscisic acid-induced stomatal closure in *Arabidopsis thaliana*. *Proceedings of the National Academy of Science, USA*, 99, 16319–16324.

Furchgatt, R.F. (1995) Special topic: nitric oxide. *Annual Review Physiology*, 57, 659 [a collection of several excellent articles on the production and role of nitric oxide].

Godber, B.L.J., Doel, J.J., Durgan, J., Eisenthal, R. and Harrison, R. (2000) A new route to peroxynitrite: a role for xanthine oxidoreductase. *FEBS Letters*, 475, 93–96.

Hancock, J.T., Salisbury, V., Ovejero-Boglione, M.C., Cherry, R., Hoare, C., Eisenthal, R. and Harrison, R. (2002) Antimicrobial properties of milk: dependence on the presence of xanthine oxidase and nitrite. *Antimicrobial Agents and Chemotherapy*, 46, 3308–3310.

Neill, S.J., Desikan, R. & Hancock, J.T. (2003) Nitric oxide signalling in plants. *New Phytologist*, 159, 11–35.

Palmer, R.M.J., Ferrige, A.G. and Moncada, S. (1987) Nitric oxide release accounts for the biological activity of endothelium derived relaxing factor. *Nature*, 327, 524–526.

Stuehr D.J., Cho H.J., Kwon, N.S., Weise, M.F. and Nathan C.F. (1991) Purification and characterisation of the cytokine induced macrophage nitric oxide synthase, an FAD containing and FMN containing flavoprotein. *Proceedings of the National Academy of Science, USA*, 88, 7773–7777.

Zemojtel, T., Fröhlich, A., Palmieri, M.C., Kolanczyk, M., Mikula, I., Wyrwicz, L.S., Wanker, E.E., Mundlos, S., Vingron, M., Martasek, P. and Durner, J. (2006) Plant nitric oxide synthase: a never-ending story? *Trends in Plant Science*, 11, 524–525 [discussion on the NOS from plants].

Superoxide and hydrogen peroxide

Baldridge, C.W. and Gerard, R.W. (1933) The extra respiration of phagocytosis. *American Journal Physiology*,103, 235–236.

Cooper, C., Patel, R. P., Brookes, P. S., Darley-Usmar and V. M. (2002) Nanotransducers in cellular redox signaling: Modification of thiols by reactive oxygen and nitrogen species *Trends in Biochemical Sciences*, 27, 489–492.

Groemping, Y. and Rittinger, K. (2005) Activation and assembly of the NADPH oxidase: a structural perspective. *Biochemical Journal*, 386, 401–416.

Harman, D. (1972) The biologic clock: the mitochondria? *Journal Geriatric Society*, 20, 145–147.

Iyer, G., Islam, M.F. and Quastel, J.H. (1961) Biochemical aspects of phagocytosis. *Nature*, 192, 535–542.

Jacob, C., Holme, A.L. and Fry, F.H. (2004) The sulfinic acid switch in proteins. *Organic Biomolecular Chemistry*, 2, 1953–1956.

Levine A., Tenhaken, R., Dixon, R and Lamb, C. (1994) H_2O_2 from the oxidative burst orchestrates the plant hypersensitive disease resistance response. *Cell*, 79, 583–593.

Kiley, P.J. and Storz, G. (2004) Exploiting thiol modifications. *PLoS Biology*, 2, e400.

Neill, S.J., Desikan, R. & Hancock, J.T. (2002) Hydrogen peroxide signalling. *Current Opinion in Plant Biology*, 5, 388–395.

Rhee S.G., Bae, Y.S. Lee S. -R., Kwon, J. (2000) Hydrogen peroxide: a key messenger that modulates protein phosphorylation through cysteine oxidation. *Science's Signal Transduction Knowledge Environment*, http://stkesciencemag.org/cgi/content/full/OC_sigtrans;2000/53/pe1

Salmeen, A., Anderson, J.N., Myers, M.P., Meng, T.-C., Hinks, J.A., Tonks, N.K. and Barford, D. (2003) Redox regulation of protein tyrosine phosphatase 1B involves a sulphenyl-amide intermediate. *Nature*, 423, 769–773.

Sbarra, A.J. and Karnovsky, M.L. (1959) The molecular basis of phagocytosis. *Journal of Biological Chemistry*, 234, 1355–1362.

Schafer, F.Q. and Buettner G.R. (2001) Redox environment of the cell as viewed through the redox state of the glutathione disulphide/ glutathione couple. *Free Radical Biology Medicine*, 30, 1191–1212.

Schreck, R., Rieber, P. and Baeuerle, P.A. (1991) Reactive oxygen intermediates as apparently widely used messengers in the activation of the NF-κB transcription factor and HIV-1. *EMBO Journal*, 10, 2247–2258.

Heyworth, P.G., Cross, A.R. and Curnutte, J.T. (2003) Chronic granulomatous disease. *Current Opinion in Immunology*, 15, 578–584.

Van Montfort, R.L.M., Congreve, M., Tisi, D., Carr, R. and Jhoti, H. (2003) Oxidation state of the active-site cysteine in protein tyrosine phosphatase 1B. *Nature*, 423, 773–777.

Woo, H.A., Chae, H.Z., Hwang, S.C., Yang, K.-S., Kang, S.W., Kim, K. and Rhee, S.G. (2003) Reversing the inactivation of peroxiredoxins caused by cysteine sulfinic acid formation. *Science*, 300, 653–656.

Measuring ROS and RNS

Hancock, J.T. and Jones, O.T.G. (1994) Assays of plasma membrane NADPH oxidase. *Methods in Enzymology*, 233, 222–229.

Østergaard, H., Henriksen, A., Hansen, F.G. and Winther, J.R. (2001) Shedding light on disulphide bond formation: engineering a redox switch in green fluorescent protein. *EMBO Journal*, 20, 5853–5862.

Carbon monoxide and other compounds

Dulak, J. and Józkowicz, A. (2003) Carbon monoxide - a "new" gaseous modulator of gene expression. *Acta Biochimica Polomica*, 50, 31–47.

Jacob, J., Anwar, A., and Burkholz, T. (2008) Perspective on recent developments on sulfur-containing agents and hydrogen sulfide signaling. *Planta Medica*, 74, 1580–1592.

Liu, K., Xu., S., Xuan, W., Ling, T., Cao, Z., Huang, B., Sun, Y., Fang, L., Liu, Z. Zhao, N. and Shen, W. (2007) Carbon monoxide counteracts the inhibition of seed germination and alleviates oxidative damage caused by salt stress in *Oryza sativa*. *Plant Science*, 172, 544–555.

Piantadosi, C.A. (2008) Carbon monoxide, reactive oxygen signaling, and oxidative stress. *Free Radical Biology and Medicine*, 45, 562–569.

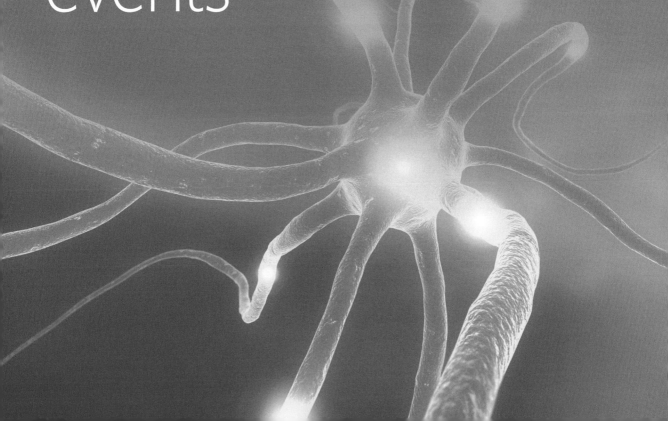

Part 3

Selected examples of signalling pathways and events

Insulin and the signal transduction cascades it invokes

11

The preceding chapters in this book describe the components that may be recruited to transmit signals in biological systems. However, most commonly a variety of such components must come together to form a signalling pathway. In this chapter, insulin signalling is used as an example of such a system. As will be discussed, once insulin has been produced, transported, and perceived, a surprisingly diverse signalling transduction mechanism may be used to bring about the final alteration of cellular function. Dysfunction here may cause diabetes, and therefore a discussion of insulin highlights the importance of signalling to normal cellular function.

11.1 The insulin signalling system

In the preceding chapters the separate components of possible signal transduction pathways have been discussed. Often, although a component is found in one pathway, it has profound effects, or sometimes even more important effects, in another transduction pathway. However, none of these signalling components works in isolation, and therefore it is important to illustrate how they might come together in a pathway. Here, insulin is used as an example of an extremely important hormone that results from, and leads to, complex signalling. It is made in one cell, stored until needed, released when the secretory mechanism is suitably stimulated, travels to its target cell, is recognized by the extracellular surface of that cell, and through a cascade of various messengers, results in many cellular effects. Such a scheme, resulting in the movement of glucose transporters in cells is shown in **Figure 11.1**. Some of the effects resulting in insulin binding are listed in **Table 11.1**.

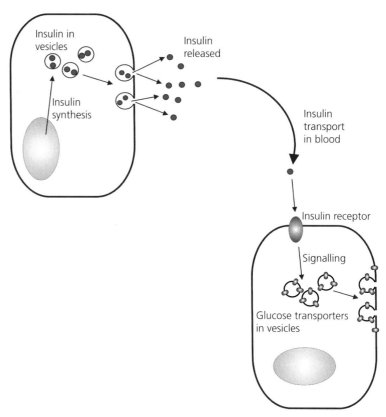

Figure 11.1 A simplified overview of insulin signalling. Insulin is made, and stored in one cell until needed. Under control of a signal transduction pathway it is released, travels to its site of perception on another cell, and then more signalling controls the movement of glucose transport proteins to the plasma membrane to alter glucose uptake rates.

Table 11.1 A list of some of the cellular effects induced by insulin.

Phosphorylation of IRS1
Activation of PtdIns 3-kinase
Activation of Ras
Phosphorylation of kinases: e.g. MAP kinases; ribosomal S6 kinase
Phosphorylation of phosphatases: e.g. PP1
Phosphorylation of metabolic enzymes: e.g. glycogen
Dephosphorylation of metabolic enzymes: e.g. glycogen synthase/phosphorylase kinase
Translocation of proteins: GLUT-4; insulin receptors
Regulation of gene expression
Modulation of protein synthesis

Some of the insulin-induced effects are seen through modulation of cytosolic enzymes, whereas other effects include control of gene expression inside the nucleus. In general, insulin will promote anabolic processes of a cell while causing a reduction in catabolic processes. For example, insulin will promote the synthesis of glycogen and fatty acids while simultaneously inhibiting their breakdown. An alteration of insulin signalling, either its production, its detection, or subsequent signalling, can lead to the disease diabetes mellitus.

Insulin is produced by the cells of the islets of Langerhans. These are cell clusters found in the pancreas. As well as insulin, these cells are responsible for the production of another important hormone, glucagon. Glucagon, a single polypeptide chain hormone, along with insulin, is responsible for the regulation of concentration of glucose in the blood stream of mammals.

The route for the production of insulin has been studied in a human tumour of the islets cells, where insulin was produced in large amounts. Using tritiated leucine incorporation into the insulin polypeptide allowed analysis of the various steps, and therefore this process is well characterized. Production starts with the synthesis of a single polypeptide chain known as preproinsulin. At the N-terminal end of this molecule is a 19 amino acid signal sequence, which is relatively hydrophobic and directs the polypeptide to the endoplasmic reticulum. In the lumen of the endoplasmic reticulum the 19 amino acids are proteolytically removed to form proinsulin. Proinsulin is subsequently passed through the Golgi apparatus of the cell and into secretory granules, where a 33 amino acid stretch of polypeptide is removed from the middle of the chain. When analyzed, even from different species, the sequence removed was found to contain a -Lys-Arg- sequence at the C-terminal end with an -Arg-Arg- sequence at the N-terminal end – such conserved amino acids serve as recognition sequences for the protease responsible for the cleavage. Once removed, this leaves two polypeptide chains, an A chain of 21 amino acids and a B chain of 30 amino acids, which are connected by two disulphide bridges. A third disulphide bridge also spans across two cysteine residues in the smaller chain of amino acids, the A chain (**Figure 11.2**). The amino acid sequence of insulin was determined in 1953 by Frederick Sanger, while its structure was studied by Dorothy Hodgkin, both of whom won the Nobel prize.

The three-dimensional structure of insulin, when determined to a resolution of 1.9Å, showed that as well as the three disulphide bonds, the structure is held together by salt links and hydrogen bonds. The molecule assumes a basically globular structure, with only the N-terminal and C-terminal ends of the B chain reaching out into solution.

It is worth noting that insulin is not totally unique, and that a second molecule shares great similarity to insulin, that is insulin-like growth factor (IGF-1). This is a hormone that is similar in both sequence and structure to insulin produced in the liver. Its release is controlled by pituitary hormones and its role seems to be in the control of an organism's growth.

Anabolic and catabolic
Anabolic refers to reactions that require energy, such as the synthesis of fats, whereas catabolic refers to reactions that transform cellular fuels, such as fats, to usable energy. The sum of anabolism and catabolism may be referred to as metabolism.

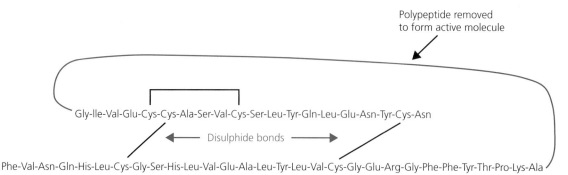

Figure 11.2 The amino acid sequence of insulin. The disulphide bonds are indicated by the solid lines, and the connecting peptide, which is removed, is shown by the blue line.

Once made, the insulin is contained in secretory granules in the cell, until such time as it is needed. Translocation of these granules, and their fusion with the plasma membrane, allowing the release of the insulin into the extracellular fluid is both under the control of other hormones and in response to neuronal signals. Hormones arriving at the insulin containing cells will bind to a receptor, and lead to a signal transduction pathway, resulting in the insulin release, in the manner shown in **Figure 1.2** and **Figure 11.1**. Kinases such as 5'-AMP-activated protein kinase (AMPK) are thought to be involved here, where inhibition of AMPK in the presence of glucose activates insulin secretion. The resultant level of insulin measurable in the blood is usually very low, only in the order of 10^{-10} M, and it is this concentration that must be perceived by the insulin-responsive cells.

As stated earlier the effects of insulin are fairly numerous. Some of the effects are seen relatively quickly, whereas others can take a matter of hours. This suggests a complicated signalling pathway, or set of pathways, that is used by insulin, not a straight path leading to a single effect. The immediate effects of insulin include modulation of activities of various enzymes in different metabolic pathways, some of which are listed in **Table 11.2**, and also include an increase in the rate of glucose uptake.

Usually such effects are stimulated by blood insulin levels in the order of 10^{-9} to 10^{-10} M. Longer exposure of cells to higher concentrations of insulin (approximately 10^{-8} M) will induce protein synthesis of enzymes involved in glycogen synthesis in the liver, enzymes needed in triglycerol synthesis in adipose tissue, and in some cells such as fibroblasts will stimulate proliferation. However, such high concentrations are generally not physiological, although the measurement of local concentrations of signalling molecules is often not carried out.

Clearly, insulin is released by one cell type, and has profound effects elsewhere, in other tissues and cells. Therefore, one of the key events is the perception of insulin by cells. Insulin is detected by the target cell mainly by the insulin receptor. This receptor was first purified by Pedro Cuatrecasa.

Table 11.2 A list of some enzymes that have their activity modulated through the action of insulin.

Enzyme	Metabolic pathway involved	Effect of insulin on activity
Phosphorylase kinase	Glycogen metabolism	Decrease
Pyruvate kinase	Glycolysis	Decrease
Lipase	Lipid breakdown	Decrease
Glycogen synthase	Glycogen synthesis	Increase
Pyruvate dehydrogenase	Citric acid cycle	Increase
Acetyl-CoA carboxylase	Fatty acid synthesis	Increase
Hydroxymethylglutaryl-CoA reductase	Cholesterol synthesis	Increase

Local concentrations

Often experiments are carried out to measure the concentration of hormones or other signals using a group of cells, perhaps from tissue culture or a whole tissue, so an average concentration is recorded, but the actual concentrations present around the surface of a cell, in the immediate locale of a receptor, are not measured. Usually these are extremely hard to measure, so it is possible that the concentrations of signals may be much higher, or lower, than thought.

The receptor has a binding affinity for insulin of approximately 10^{-10} M, the same order of magnitude as the concentration of insulin in the blood that needs to be detected. It is a glycoprotein that contains two large, α subunits of 135 kDa and two smaller β subunits of 95 kDa, and therefore would be described as having an $\alpha_2\beta_2$ subunit structure. The β subunits are integral to plasma membrane, whereas the α subunits reside on the outside of the cell membrane and are held to the β subunits by disulphide bonds as depicted in Figure 11.3. As was discussed in **Chapter 5**, receptors are often found in dimers, especially following ligand binding. Even though the insulin receptor is a tetramer, perhaps it would be interesting to view it as a dimer, where each part of the dimer has two subunits, an α and a β, and therefore the insulin receptor can be thought of as, in fact, a dimer of dimers. In this way, perhaps it is not too unusual, and ligand binding is similar to many other receptors. In fact, even though there are two different subunits in the insulin receptor, the α and β subunits of the receptor are synthesized as a single polypeptide chain of 1382 amino acids. The precursor peptide has a signal peptide at the N-terminal end, which is cleaved off, and this leaves the α subunit, followed by a tetrapeptide with the sequence -Arg-Lys-Arg-Arg-, followed by the sequence of the β subunit. The tetrapeptide is highly basic and serves as a recognition signal for the protease that processes the polypeptide.

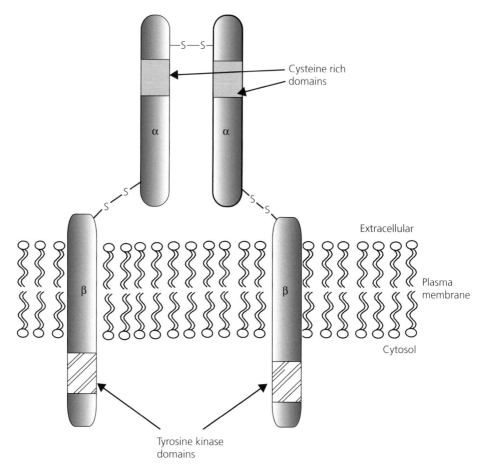

Figure 11.3 A model of the structure of the insulin receptor.

Insulin recognition by cells is not limited solely to the insulin receptor, but insulin can also bind to the IGF-1 receptor. This receptor is very similar to the insulin receptor, but insulin binds to the IGF-1 receptor approximately 100 times less well. Conversely, IGF-1 can bind to the insulin receptor, but again the binding is about 100 times less.

So what does the binding of insulin to its receptor invoke inside the cell? One of the effects of insulin is to increase glycogen synthesis by cells. Here, the synthesis of glycogen is catalyzed by an enzyme called glycogen synthase, which has its activity controlled by phosphorylation (see **Chapter 6** and **Figure 6.2** in particular). In the phosphorylated state the synthase is inactive, and dephosphorylation activates it. Therefore, to get activation requires the activity of a phosphatase, in this instance protein phosphatase 1 (PP1). The phosphatase itself is controlled by phosphorylation, but in this case phosphorylation increases the kinase's activity. So how does insulin fit here? Insulin binding to its receptor activates tyrosine kinase activity (discussed in more detail below), which leads to the activation of what has been referred to as insulin-sensitive protein kinase or insulin-stimulated protein kinase (ISPK). This kinase phosphorylates PP1, activating it, and therefore the active PP1 can dephosphorylate glycogen synthase, activating it, and the synthesis of glycogen is increased. As discussed in earlier chapters, futile cycles in metabolism are to be avoided, and PP1 also has action on phosphorylase to prevent the instant breakdown of the glycogen formed.

As well as the rapid effects on glycogen metabolism, one of the major effects of insulin on cells is the resultant increased capacity of the cell to be able to take up glucose, and so leading to lowering of blood glucose concentrations, a so-called hypoglycaemic effect. Within approximately 15 minutes of the application of insulin to adipocytes, their rate of uptake of glucose will have increased between 10 and 20-fold. This short time span would not be sufficient to allow for the synthesis of new protein and hence new glucose transporters. However, analysis has shown that the number of glucose transporters found in the plasma membrane will have greatly increased during this time. Glucose transporters, otherwise known as the GLUT-4 protein, are synthesized and targeted to endosome-like vesicles. They are made as integral membrane proteins, as their role is to transport glucose across the plasma membrane, and as they travel through the cell they remain as integral membrane proteins of vesicles. However, these vesicles remain in the cytoplasm, where the GLUT-4 proteins have no chance to function, until the cell is stimulated by insulin. Once stimulated, the vesicles will be transported to the plasma membrane, where the membranes fuse and the GLUT-4 transporters become part of the plasma membrane where they can function in the rapid uptake of glucose. On the removal of the insulin signal, the GLUT-4 proteins can be endocytosed back into the interior of the cell, and so the uptake of glucose can once again be reduced. Recent work where the transporters have been tagged with a fluorescent marker, has allowed this movement of the GLUT-4 polypeptides to be visualized using a confocal microscope.

So what is the link between the insulin receptor and the vesicles containing the GLUT-4, and to investigate the pathway in a logical order, perhaps, how does the insulin receptor relay the message, from the recognition and binding of insulin to the control of the intracellular mechanisms it affects? The α chains of the receptor contain cysteine rich domains and it is on these subunits that insulin binding takes place. The β subunits, on the other hand, contain domains with tyrosine kinase activity, and hence the receptors are classed as receptor tyrosine kinases (RTKs: see Chapter 6). On binding of insulin to the receptor, there will be a defined change in the conformation of the subunits of the receptor, relayed from the binding site on the α subunits to the β subunits, and subsequently through the membrane to have an influence on events on the intracellular side of the plasma membrane. The first response is that the receptor phosphorylates itself on at least three tyrosine residues, that is, it autophosphorylates (Figure 11.3). This has two results. Firstly, it activates the receptor to phosphorylate other cellular proteins.

Secondly, even if the insulin is subsequently removed from the receptor, the receptor remains active unless the receptor itself is dephosphorylated by a phosphatase.

However, one of the key events in signal transduction from the insulin receptor is multi-phosphorylation of a protein known as insulin receptor substrate 1 (IRS1), as illustrated in Figure 11.4. The IRS proteins are actually a family of at least four proteins (IRS1–IRS4), all having PH domains and phosphotyrosine binding domains. IRS-1 is a protein of 130 kDa, and can be phosphorylated by either the insulin receptor or the IGF-1 receptor. IRS1 is highly serine phosphorylated and also can be highly tyrosine phosphorylated by these receptors. Six of the sites of tyrosine phosphorylation on this protein have the neighbouring sequence of -Tyr-(Pro)-Met-X-Met- (where X could be a variety of amino acids).

IRS1 acts as a relay protein, and, once phosphorylated, and therefore activated, IRS1 can interact and modulate the activity of different pathways. One of the things that IRS1 does is to bind to SH2 domains. As IRS1 is phosphorylated on tyrosine residues, this creates binding sites for the SH2 domains of other proteins (see Chapter 1). One such SH2 domain is that of the regulatory

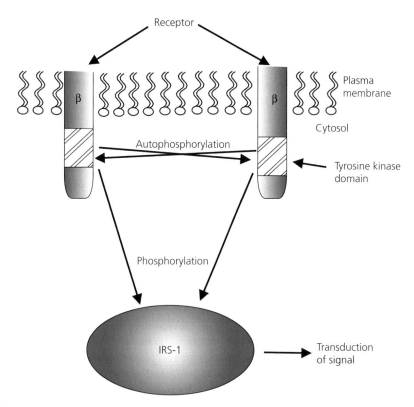

Figure 11.4 A schematic representation of the phosphorylation events that result from the activation of the insulin receptor.

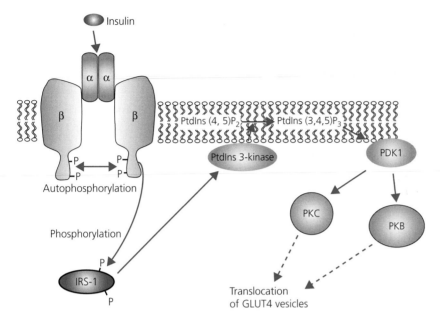

Figure 11.5 One of the pathways in which insulin signals to control the movement of glucose transporters to the plasma membrane. Here, PtdIns 3-kinase control is the key.

subunit of PtdIns 3-kinase, causing its activation. PtdIns 3-kinase is also phosphorylated on a tyrosine by the insulin receptor. As discussed in **Chapter 8**, activation of PtdIns 3-kinase will lead to the production of phosphorylated forms of membrane phosphatidylinositol lipids: on the 3 position of the inositol ring. Particularly, PtdIns(4,5)P$_2$ will be converted to PtdIns(3,4,5)P$_3$, which in turn will activate 3-phosphoinositide-dependent kinase (PDK1). This will then be able to phosphorylate and activate protein kinase B (PKB) and perhaps also PKC (see **Figure 11.5**). Downstream signalling from these two kinases will lead to translocation of the GLUT4 transporter proteins to the plasma membrane through movement of the GLUT4 containing vesicles, as shown in **Figure 11.1**.

Another substrate of the insulin receptor is the Cbl/CAP complex. Cbl is a proto-oncogene product, and it becomes associated with the insulin receptor aided by the protein CAP. Cbl is phosphorylated by the activated insulin receptor, and dissociates from the receptor, upon which the Cbl/CAP complex moves to become associated with lipid rafts. Flotillin in the lipid raft associates with the Cbl/CAP proteins, and further signalling is invoked that leads to the translocation of the GLUT4 containing vesicles to the plasma membrane (**Figure 11.6**). The exact nature of the signalling down-stream from the lipid raft has yet to be elucidated.

As discussed above, insulin signalling uses a wide range of components, including kinases, phosphatases, adaptor proteins, lipid rafts, and as yet to be discovered components, to control the rates of metabolism and the intake of glucose. However, there is another pathway that has been investigated, one

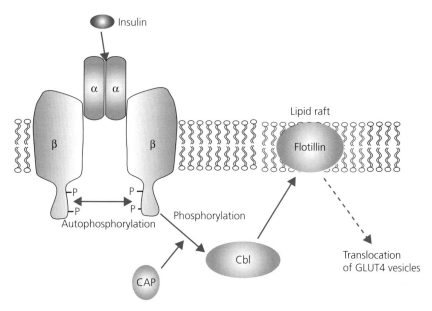

Figure 11.6 An alternate signalling pathway used by insulin, the CAP/Cbl pathway. CAP, Cbl activating protein.

Lipid rafts

Lipid rafts are sub-areas of phospholipid membranes, which have been shown to have a structure different from the rest of the membrane. They are like islands of a particular structure, floating in the sea of the membrane. Particular proteins, and hence functions have been associated with these rafts, and much interest is now focussed on what they actually do, and also how their function might be modulated, perhaps in the form of new drug therapies.

■ Similar pathways, different details: it is perhaps worth noting that the scheme illustrating the insulin-induced pathway leading to gene expression, **Figure 11.7** is similar to that shown in previous chapters, e.g. **Figure 6.8**.

very similar to that used by other growth factor-like molecules, leading to events in the nucleus. This pathway, of course, also starts with the perception of insulin by its receptor, and also involves the phosphorylation of IRS-1. Phosphorylation of tyrosine residues in proteins such as IRS-1 creates binding sites for SH2 domains, and this is used here. Once phosphorylated, IRS-1 interacts with an adaptor protein called GRB2, through the latter's SH2 domains. The GRB2 protein also contains two SH3 domains, which can further bind to a guanine nucleotide releasing protein, in this case Sos (see **Chapter 7**). Sos can, once activated, catalyze the release of GDP from the monomeric G protein Ras, and therefore allow the G protein to take up its active state in which GTP is bound. Ras can then activate the kinase Raf, which leads to the phosphorylation and activation of MAP kinase cascades. Such a pathway is illustrated schematically in **Figure 11.7**.

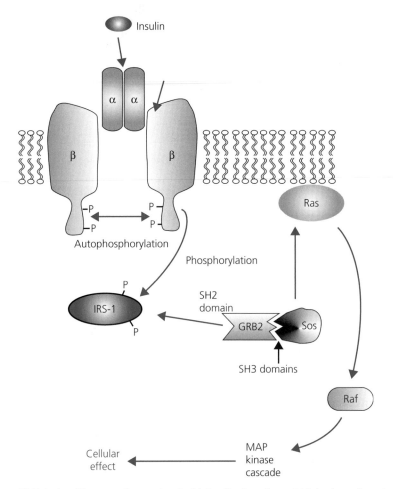

Figure 11.7 A signalling cascade associated with insulin detection, which leads to alterations of gene expression. Binding of insulin to its receptor leads to autophosphorylation of the receptor and phosphorylation of a relay protein, IRS1. The phosphotyrosines formed on IRS1 interact with the SH2 domains of the adaptor GRB2 and in a similar fashion to the EGF receptor pathway lead to sequential involvement of Sos, Ras and then a MAP cascade, which transmits the signal into the cell, often resulting in enhanced transcription.

One of the end results in the activation of this cascade is the alteration of gene expression. MAPKs will, once activated in cytoplasm, translocate to the nucleus, where they have their action, that of phosphorylating the next protein in the pathway, perhaps a transcription factor. In fact, the promoter regions of the genes that have their expression altered by the presence of insulin at the cell surface have also come under close scrutiny. Termed insulin-responsive elements (IREs), such regions have been determined in several genes including those for phosphoenolpyruvate carboxylase (PEPCK), amylase, liver pyruvate kinase and glyceraldehyde-3-phosphate dehydrogenase (GAPDH). In general, they seem to contain two eight-basepair elements, which are AT rich, and both of these need to be present; although some genes, such as that for liver pyruvate kinase, do not follow this pattern. By the use of fusion of the luciferase gene to the relevant promoters, and looking for light emission from cells where the gene product is formed, and then by transforming cells with mutated forms of the proteins involved in the MAP kinase pathway, Tavaré and colleagues have illustrated the importance of some of the above signal transduction components in insulin signalling.

Once activated by the insulin receptor, IRS1 can also interact with other proteins, such as a protein called Syp. This is a tyrosine phosphatase and it has been proposed that it is involved in the dephosphorylation of the IRS1 protein itself, and therefore involved in turning off the insulin-induced signal.

Like many receptors, once it has done its job, the insulin receptor itself is internalized and down regulated (see Chapter 5). Radiolabelled (^{125}I) insulin was found to be taken into the cell quite rapidly and after 5 minutes up to 30% of the insulin that had bound to the cell was associated with internal vesicles. Once internalized, the environment of the receptor protein will change, there will be an alteration in the conformation of the protein, and the internalized insulin will dissociate from the receptor, then generally will be destroyed by the cell. While the receptor is still in the phosphorylated state it can continue to signal to the cell, even though it has been internalized, but normally it will become dephosphorylated and can then be translocated back to the plasma membrane for another round of signalling. Several phosphatases have been identified that may be involved in the dephosphorylation of the insulin receptor. Of course, recycling the receptors is a sensible strategy for the cell, as discussed further in Chapter 5.

■ Summary

- Insulin has varied and profound effects on many aspects of a cell's functioning, ranging from the control of metabolic pathways, the stimulation of protein translocation and the control of gene expression.

- Insulin is composed of two peptide chains, held together by disulphide bonds, hydrogen bonds and salt bridges.

- Insulin is made by the cells of the islets of Langerhans of the pancreas as a single gene product, which is modified by cleavage.

- Once made, insulin is stored in vesicles until needed. On stimulation of the cells, insulin is released and washed around the body by the blood, until it reaches its target.

- Insulin is detected by cells mainly by the insulin receptor, but can also bind to the receptor for insulin-like growth factor-1.

- The insulin receptor is a tetramer with an $\alpha_2\beta_2$ subunit structure. The α subunits are on the outside of the cell and are responsible for insulin binding, whereas the β subunits contain transmembrane domains along with tyrosine kinase domains.

- Binding of insulin to its receptor results in autophosphorylation of the receptor on tyrosine residues, and also results in the phosphorylation of other cellular proteins, such as IRS1.

- Activation of IRS1 leads to activation of other proteins through the interaction of its phospho-tyrosyl residues with the SH2 domains of other polypeptides.

- The PtdIns 3-kinase pathway is recruited by insulin, leading to the formation of PtdIns(3,4,5)P$_3$, activation of PDK1, activation of PKB/C and translocation of GLUT4 to the plasma membrane.

- The translocation of GLUT4 has also been found to involve the Cbl/CAP complex, and lipid rafts.

- IRS1 also interacts with an adaptor protein such as GRB2, which can lead to activation of MAP kinase cascades and the modulation of gene expression.

- Insulin signalling can be turned off by the internalization of the receptor, which may be re-cycled back to the plasma membrane. Integral in this is the dephosphorylation of various proteins involved in insulin signalling.

→ Further reading

Baumann, C.A., Ribon, V., Kanzaki, M., Thurmond, D.C., Mora, S., Shigematsu, S., Bickel, P.E., Pessin, J.E. and Saltiel, A.R. (2000) CAP defines a second signalling pathway required for insulin stimulated glucose transport. *Nature*, 407, 202–207 [also see comment on pp. 147–148].

Da Silva Xavier, G., Leclerc, I., Varadi, A., Tsuboi, T., Moule, S.K. and Rutter, G.A. (2003) Role for AMP-activated protein kinase in glucose-stimulated insulin secretion and preproinsulin gene expression. *Biochemical Journal*, 371, 761–774.

Kimball, S.R., Vary, T.C. and Jefferson, L.S. (1994) Regulation of protein synthesis by insulin. *Annual Review Physiology*, 56, 321–348.

Litherland, G.J., Hajduch, E. and Hundal, H.S. (2001) Intracellular signalling mechanisms regulating glucose transport in insulin-sensitive tissues (Review). *Molecular Membrane Biology*, 18, 195–204.

Muretta, J.M. and Mastick, C.C. (2009) How insulin regulates glucose transport in adipocytes. *Vitamins and Hormones*, 80, 245–286 [and reviews cited within].

Myers, M.G. and White, M.F. (1996) Insulin signal transduction and the IRS proteins. *Annual Review Pharmacology Toxicology*, 36, 615–658.

Pearl, L.H. and Barford, D. (2002) Regulation of protein kinases in insulin, growth factor and Wnt signalling. *Current Opinion of Structural Biology*, 12, 761–767.

Pessin, J.E. and Saltiel, A.R. (2000) Signalling pathways in insulin action: molecular targets of insulin resistance. *Journal of Clinical Investigation*, 106, 165–169.

Roth, R.A. (1990) Insulin receptor structure. *Handbook Experimental Pharmacology*, 92, 169–181 [and other articles in that journal volume].

Whitehead, J.P., Clark, S.F., Ursø, B. and James, D. (2000) Signalling through the insulin receptor. *Current Opinion in Cell Biology*, 12, 222–228.

✺ Literature link

Molecular events of insulin action

Background information

Insulin was discovered in 1921 by F.G Banting, C.H. Best, J.J.R. Macleod and J.B. Collip, and since the early 1920s the injection of insulin has been used as a successful treatment for diabetes. Insulin is a peptide hormone and was soon crystallized by J.J. Abel and then its composition determined by F. Sanger. Its structure was solved by Dorothy Hodgkin several years later. Since then there has been an immense amount of work on understanding the processes of insulin secretion and perception, and studies to unravel the intracellular pathways that insulin initiates. It is now clear that there is more than one signal transduction pathway that leads to a cellular response which has been initiated by insulin.

The main focus of insulin research remains on diabetes, and the hunt for long-term treatments. As the signalling involving insulin requires production and release of the hormone from one cell type, and its perception at a different cell then the scope for defects is large, making diabetes a complex disease to understand.

Essay topic

Name the types of diabetes and briefly describe the molecular defects which underlie the different types of the disease. Discuss how a dysfunction in the mechanism involved in the production of a hormone can lead to a disease that appears to be similar to that caused by a defect in the hormone's perception. What are some of the long-term complications of diabetes and how does the use of animal models help in the research for finding the causes of such diseases?

Starter references

Alberti, K.G.M.M. and Zimmet, P.Z., WHO Consultation. (1998) Definition, diagnosis and classification of diabetes mellitus and its complications. Part 1: diagnosis and classification of diabetes mellitus. Provisional report of a WHO Consultation. *Diabetic Medicine.* 15, 539–553.

Chatzigeorgiou, A., Halapas, A., Kalafatakis, K. and Kamper, E. (2009) The use of animal models in the study of diabetes mellitus. *In Vivo,* 23, 245–258.

Morino, K., Petersen, K.F. and Shulman, G.I. (2006) Molecular mechanisms of insulin resistance in humans and their potential links with mitochondrial dysfunction. *Diabetes,* 55, S9–S15.

Rask-Madsen C and King GL. (2007) Mechanisms of disease: endothelial dysfunction in insulin resistance and diabetes. *Nature Clinical Practice Endocrinology and Metabolism* 2007, 3, 46–56.

Schalkwijk, C.G. and Stehouwer, C.D. (2005) Vascular complications in diabetes mellitus: the role of endothelial dysfunction. *Clinical Science (London),* 109, 143–159.

Perception of the environment

12

All cells must be able to perceive elements of their environment, and the same is true for whole organisms. For example, humans need to perceive light to be able to see, and this involves a complex signalling mechanism. Aspects of this signalling are used to illustrate how components described in earlier chapters may come together to form coherent signalling pathways. Other systems used by organisms to perceive their environment are also discussed briefly to highlight the importance and diversity of such systems and to show that most draw on the repertoire of components discussed previously in this book.

12.1 Introduction

One of the essential roles of cell signalling events is to enable the survival of the cell, and there is a need for all cells, whether they compose individual organisms or are part of a complex organism, to stay alive and prosper. Many cells are in a constant struggle to live, as they have a constant requirement for nutrients, and a need to avoid dangers and toxins. Integral to this, is the ability of the cell to respond to environmental changes imposed upon it. Such changes might be shifting of temperatures, for example, a simple organism, or plant, might have to survive the cold of winter, or the heat of the mid-day desert sun. In a similar manner, a cell might need to respond to an environmental change such as the arrival of chemicals that indicate the proximity of a source of food, or alternatively the proximity of danger, and as such smell becomes a crucial sense.

Therefore, whether an organism consists of a single cell or is a multicellular structure, it is imperative that cells, and organisms as a whole, can sense their environment. Many organisms have adapted their awareness of their environment to suit their niche in nature, for example, some animals such as moles and bats have very poor eyesight, not easily detecting and perceiving light, as they live either underground or hunt at night where little light is available. Others might have an acute sense of smell or very efficient hearing, such as animals that graze in the open and need to sense the arrival of a predator before it has a chance to get too close.

Some of the signals are released by individual organisms and sensed by others, such as the pheromones discussed in Chapter 4, whereas the perception of environmental signals is often akin to those used by the individual cells within the body, with the use of specific receptors, tuned to the particular environmental factor for detection, and the transmission of a change in the receptor into the interior of the cell and a subsequent cellular response, as discussed in Chapter 5 and subsequent chapters.

Here, a focus on the molecular signalling of the mammalian eye allows a discussion of how many of the components discussed in previous chapters might come together to allow an organism to sense one aspect of its environment, that is, light. Other organisms also have to sense light, of course, and most, if not all, need to have some sense of the array of chemicals around them, sensed by humans as smell and taste, and many need to have auditory senses too.

12.2 Photodetection in the eye

A sense that many of us take for granted is the ability to see. Photodetection requires that the light be perceived, and that a signal is created which the organism can translate, either into a response, in lower organisms, or an image in higher organisms. In humans, the eye is the specialist organ that undertakes this role, where the perception of the light is translated into electrical signals, which are unscrambled by the brain to create the image. Often, the response is a conscious act resulting from us seeing the image, a conscious act which, of course, is controlled by further cell signalling events. However, it is imperative that the image is continuous. Therefore, it is no good for the cell signalling mechanism involved to "see" the scene, and send that to the brain, and then stop. It needs to perceive the light, send the signal, and then reverse quickly to the ground state ready for the next photon of light to be perceived. Hence, the signalling needs to be rapidly activated, and rapidly reversed, and any proposed mechanism must take this into account. Clearly, such activation and deactivation of signalling components cannot be spontaneous, and there is a time lag in the ability to perceive photons. If a person watches a train pull into the station, the wheels often appear to reverse momentarily as the train slows. This is a consequence of the time lag that the cell signalling events have imposed on the photoreception of the eye. However, our brain has the ability to translate the signals into a continuous image, which appears to us to be seamless.

At the cellular level, photoreception in humans is the responsibility of two types of specialized cells, the rods and cones, so named because of their shape. These cells form a layer called the retina at the back of the eye, onto which the light is focussed by the lens. The rods are used in the perception of a wide range of wavelengths of light at low light levels, whereas the cones function in brighter light and fall into three classes, each with their own wavelength sensitivity. One senses blue light, one green light and a third red light. In humans, the wavelength maxima for the three classes of cones are 426 nm, 530 nm and 560 nm, although other species such as fish have slightly different wavelength maxima: 455 nm; 530 nm; 625 nm. If an animal lacks cones, then their perception of colours is compromised.

The rods and cones are single cells, but are bipolar, having an outer segment and an inner segment (**Figure 12.1**). The outer segment is basically an elongated bag containing flattened membranous sacks called discs, which lie one on top of the other perpendicular to the plane of the incident light. A rod may contain up to 1000 of these discs, which measure only about 16 nm thick. The inner segment contains the main body of the cell including the endoplasmic reticular machinery, mitochondria and a nucleus, but, as importantly, a synaptic body used for the transmission of any created signal to the nervous system (**Figure 12.1**).

The photoreceptive molecule, or chromophore, in rods is rhodopsin. This protein, called opsin, contains 11-cis-retinal as a prosthetic group (**Figure 12.2**). 11-Cis-retinal is derived from vitamin A (all-trans-retinol), which is obtained from the diet. In the protein it is attached via a protonated Schiff base linkage to a lysine residue, and it is this interaction, along with others, which leads the 11-cis-retinal to have an absorbance maximum of approximately 500 nm. Interestingly, rhodopsin also has an extremely large extinction coefficient, which means that it is very efficient at absorbing light, giving the eyes a fantastic light sensitivity.

The opsin protein is like many other membrane proteins in having seven membrane spanning α-helices (**Figure 12.3**). The retinal lies in the centre of the α-helical section and in the plane of the membrane. The N-terminal end of the polypeptide, which contains two N-linked oligosaccharides, lies in the intradiscal space, whereas the C-terminal end lies on the cytosolic side of the membrane. It is this region that is important in interactions with other proteins which transmit the signal to the down-stream part of the signal transduction cascade, and ultimately to the nervous system. For example, it is here that a G protein binds and is also phosphorylated on serine and threonine residues, leading to a deactivation of the molecule.

On receiving a photon of light, the 11-cis-retinal photoisomerizes to all-trans-retinal inducing a conformational change in the molecule. This means

Rhodopsin A protein comprised of a polypeptide called opsin, and a non-protein part (a prosthetic group) called 11-cis-retinal. The whole protein, with the polypeptide and prosthetic group together, is called rhodopsin.

Carrots and the eyes

During the Second World War it was said that eating carrots improved your eyesight in the dark, because the carrots were a good source of retinol. However, this was propaganda, as little is to be gained if a reasonably healthy diet is maintained.

Bacteriorhodopsin The rhodopsin related protein bacteriorhodopsin was one of the first membrane proteins to have its structure solved, and has become the model for seven-spanning proteins, such as the G protein-linked receptor family. Note similarities in the structures shown in **Figures 12.3 and 5.2**.

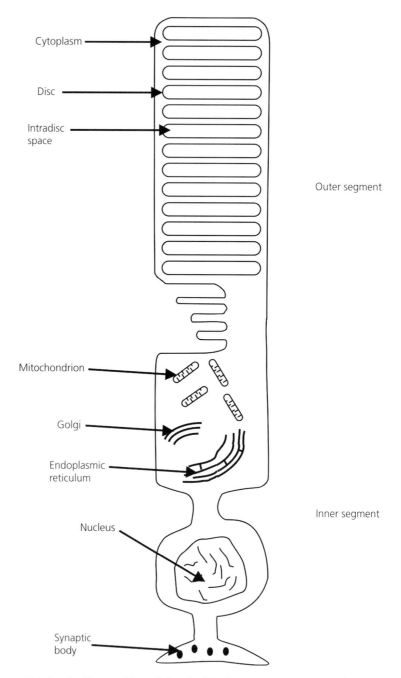

Cytoplasm

Disc

Intradisc space

Outer segment

Mitochondrion

Golgi

Endoplasmic reticulum

Inner segment

Nucleus

Synaptic body

Figure 12.1 A rod cell: one of the cells involved in photoreception in mammals.

that the Schiff base linkage to the protein moves by approximately 5 Å. This new molecule, called bathorhodopsin, undergoes several more conformational changes, each of which can be monitored by the changes in the maximal absorbance wavelength, until metarhodopsin II, otherwise known as photoexcited rhodopsin, is formed in which the Schiff base linkage has

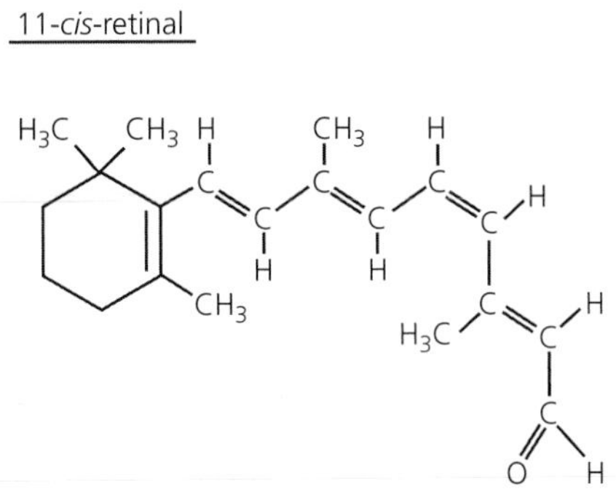

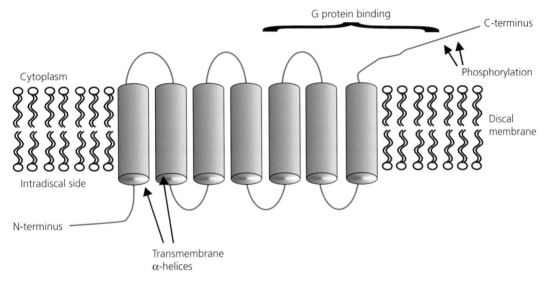

Figure 12.2 The molecular structure of 11-cis-retinal.

Figure 12.3 The domain structure of rhodopsin showing the similarity to seven span receptors.

become unprotonated. It is the formation of this form that triggers the resultant signalling cascade, resulting from the forced change in conformation of the opsin protein, a change perceived by the G protein to which it is bound.

Rhodopsin is re-formed by the hydrolysis of the all-trans-retinal from the opsin, its re-conversion back to 11-cis-retinal involves all-trans-retinol as an intermediate and then re-linkage to the opsin.

The same light sensitive system is also used in the cones, except here the opsin-like molecule has three hydroxyl group containing residues, which surround the 11-cis-retinal, and the alteration of its local environment is enough to alter its light absorbing characteristics and its wavelength maximum (as discussed above).

The C-terminal end of the rhodopsin molecule in the unexcited state is in association with a member of the trimeric G protein family. Here, the G protein is transducin, or G_t. This was, in fact, only the second G protein to be discovered, by both Mark Bitensky and Lubert Stryer. It is composed of an α subunit of 39 kDa, a β subunit of 37 kDa and a γ subunit of 8.5 kDa. The excited rhodopsin acts effectively like a GTP exchange factor, exchanging the GDP bound to the G protein for a GTP, so causing its activation. The G protein dissociates into the $G_t\alpha$ and $G_t\beta\gamma$ subunits. The $G_t\alpha$ subunit interacts with the inhibitory peptide of phosphodiesterase and so reduces the inhibited state of this enzyme. This enzyme normally resides as a complex of two catalytic subunits (α and β), which are inhibited by the presence of two inhibitory peptides, the γ subunits. This inhibitory restraint is removed by the interaction with $G_t\alpha$. The activated phosphodiesterase very rapidly and efficiently hydrolyzes cGMP in the cytosol of the cell's outer segment (Figure 12.4; also see section on phosphodiesterases in Chapter 7).

The subsequent drop in the cytosolic concentration of cGMP leads to a closure of cation-specific channels in the plasma membrane of the cells. In the dark, these channels are kept open by the binding of cGMP. The channel is a multipolypeptide complex, each subunit being of approximately 80 kDa. The cGMP binding is extremely cooperative, opening of the channel requiring interaction with at least three cGMP molecules. The open channels serve as a route for the return of Na^+ ions back into the outer segment of the cells down a large electrochemical gradient. Na^+/K^+ ATPases in the plasma membrane of the inner segments of the cells are used to maintain this gradient. Closure of the channels through the drop in cGMP levels means that the Na^+ ions are no longer able to return into the outer segment, and this results in a hyperpolarization of the plasma membrane. This hyperpolarization can be as great as 1 mV and is sensed by the synaptic body at the base of the cell and transmitted to the neurons, and so to the brain.

Although the $G_t\alpha$ subunit is often seen to be the active part of the G protein, as discussed in Chapter 6, the Gβγ is not without function. Here, the $G_t\beta\gamma$ subunit interacts with another protein known as the 33K protein or phosducin. Despite its name, this protein actually has a molecular weight of approximately 28 kDa. It is phosphorylated by cAMP-dependent protein kinase on a serine residue and dephosphorylated by phosphatase 2A, but interestingly this phosphorylation is light-dependent, being most highly phosphorylated in the dark. Binding to Gβγ phosducin regulates the recycling of the $G_t\alpha$ subunit, and so modulates the amount of G protein available for interaction with the rhodopsin, this function itself being regulated by phosphorylation.

As discussed in Chapter 1, one of the main uses of a signalling cascade, rather than a signal simply turning the final event on or off, is the potential for amplification of the signal as it is transduced along the pathway. The signalling events in the rods of the eye are a very good example of this, as a massive magnification of the signal is seen in this system, each step allows another degree of amplification. One rhodopsin molecule, once activated, can activate up to 500 transducin G proteins, whereas a massive amplification is seen in the cleavage of cGMP by the extremely active phosphodiesterase: many thousands of cGMP molecules can be removed by the perception of a single photon.

Stryer

Lubert Stryer, as well as carrying out research, also wrote text books. *Biochemistry* is now in its 5th edition, and, unsurprisingly, continues to have an excellent section on the signalling of the eye and the role of transducin, the protein Stryer discovered. N.B. *Biochemistry* by Stryer now has Jeremy M. Berg as the first author.

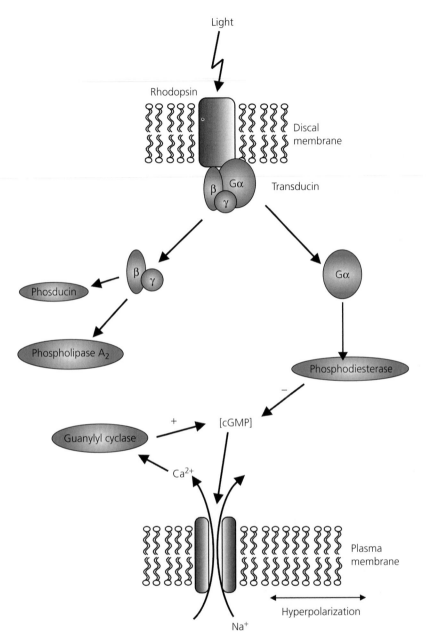

Figure 12.4 The signal transduction pathway in rod cells showing how light leads to a hyperpolarization of the plasma membrane.

It was stressed above that the system needs to be rapidly reversed and turned off, and this is achieved by several simultaneous events. Ca^{2+} is normally pumped out of the cell via an exchanger, but its re-entry is normally allowed into the outer segments via the cation-specific channels. On closure of these channels following the drop of cGMP, the intracellular Ca^{2+} levels drop and this in turn stimulates the synthesis of cGMP by the activation of guanylyl cyclase. Resurrection of the cGMP levels will once again open the cation channels and end the hyperpolarization of the membranes, with consequent cessation of the neuronal signal to the brain.

However, the rest of the signalling system needs to turn off too. Transducin has, like all heterotrimeric G proteins, an endogenous GTPase activity, and thus any bound GTP will be hydrolyzed allowing the reformation of the trimeric inactive G protein, assuming phosducin has not had an influence on the $G\beta\gamma$ levels. The transducin could then be re-activated by rhodopsin, allowing another round of signalling. However, rhodopsin can be turned off. It has been found that rhodopsin may be phosphorylated by rhodopsin kinase, on several serine and threonine residues at the C-terminal end of the polypeptide. Once phosphorylated, the rhodopsin can interact with a polypeptide called arrestin. Interaction with arrestin prevents the rhodopsin from interacting with the transducin and so prevents a further activation of the cascade. It is interesting to note that a similar mechanism is seen with the β-adrenergic receptors, and the β-adrenergic receptor kinase can phosphorylate the rhopdopsin molecule in a light-dependent manner, highlighting the structural similarity between these receptor molecules.

Although the signalling mechanisms discussed have been mainly elucidated by studies on the rods from the eye, the cones also have an analogous system. However, subtle differences are apparent, such as in the G proteins used. Rods express the G protein subunits α_{t-r}, β_1 and γ_1, whereas cones express a G protein consisting of a cone-specific α_t, β_3, and an γ subunit.

To complicate matters even more, phospholipase A_2 activity in rod outer segments is stimulated by light. It has been found that transducin $\beta\gamma$ subunits increase phospholipase A_2 activity, which was inhibited by transducin α subunits.

It can be seen, therefore, that light perception from the eye serves as a good example of how several signalling components work together to result in a coordinated response (Figure 12.4). Here, a light receptor (with analogies to a hormone receptor), a G protein, an enzyme, a small messenger molecule and an ion channel are all used to transmit the signal, with the cytosolic calcium concentration and another enzyme being involved in the return to the ground state. Other proteins also interact, and other enzymes are involved, giving overall a complex set of signalling events. It is not simply a linear cascade, but rather a complicated inter-play involving many actors.

It is not just the absolute intensity of the light to which we are sensitive, but rather changes in intensity, and therefore a biochemical mechanism is needed to explain how our eyes can so readily adapt to wide ranges of light intensity, maybe as much as a 10,000-fold difference. Work studying the photoreceptors of amphibians has shown that the activity of nitric oxide synthase and production of nitric oxide has an influence on the metabolism of cGMP, as discussed in Chapter 10 with other signalling pathways, and therefore such signalling molecules may well be involved in the adaptive nature of the eye. It is, and indeed needs to be, a complex signalling mechanism.

The signalling pathways involved in the light sensing organs of lower animals differs somewhat from that of vertebrates. In the fruit fly *Drosophila melanogaster*, the rhodopsin activates a trimeric G protein of the G_q type, which in turn activates phospholipase C. The phospholipase C catalyzes the breakdown of PtdInsP$_2$ in the membrane, and therefore produces DAG and InsP$_3$. The InsP$_3$ opens Ca^{2+} channels and allows release of Ca^{2+} from internal stores, which results in the propagation of the signal. Photo-adaptation and recovery also need the participation of an eye-specific protein kinase C.

12.3   Other environment perception systems

Plants also need to perceive light and respond to it in order to survive. Leaves on a tree, for example, are optimally aligned to maximize the amount of light captured for photosynthesis. If there is no light, plants usually fail to produce chlorophyll, and remain yellow and etiolated, but often grow tall in the search for light. With the restoration of light, the plant produces chlorophyll, a process which itself appears to be light catalyzed, and the plant can once again start photosynthesizing. If a leaf is in the shade, the plant might need to respond to increase the leaf area that is exposed to direct sunlight. Similarly, plants are able to detect the time of day, which can be seen, for example, with flowers opening and closing, and also the time of year. Plant cells have four main ways of monitoring light: phytochromes; UV-B receptors, UV-A receptors and blue light receptors. Phytochromes of higher plants, which are responsible for detecting red and far-red light, are coded for by at least five genes, assigned the names *PHYA-PHYE*. Down-stream signalling from such light perception will involve many of the signalling components discussed in the previous chapters.

As well as light, an organism must be aware of the presence of chemicals in the environment. Perhaps the presence of a chemical suggests the presence of a source of food, the proximity of a partner to mate with, a danger, or perhaps it can be ignored. Whichever response is required, the perception of the chemical is required, and organisms such as higher animals have specialist tissues and cells dedicated to the job.

Prokaryotes, such as *Escherichia coli,* have a need to detect molecules in the extracellular media and respond to their presence. Such molecules include attractants such as the sugars galactose and ribose, the amino acids serine, alanine and glycine, as well as dipeptides. Repellent molecules include acetate, leucine, indole and metals such as nickel and cobalt. The presence of these molecules is detected by receptors known as methyl-accepting chemotaxis proteins. These proteins form four homologous groups, but they all span the inner bacterial membrane, and are known as

Tsr, Tar, Trg and Tap. Sequence homology between the members of these receptor groups shows that the N-terminal end contains the receptor domain, which interacts with the ligand, with a membrane spanning domain on each side. The C-terminal end contains the domain that propagates the signal into the cell. The C-terminal end is also the site of methylation, which seems to be responsible for the sensitivity of the receptor, enabling the receptors to be receptive to a concentration gradient of the ligand rather than simply an on/off signal. Methylation is catalyzed by a methyltransferase known as CheR, whereas demethylation is catalyzed by CheB-phosphate, or CheB-P. The methyl group is supplied from *S*-adenosylmethionine, whereas the targets for transfer are glutamate residues. Different receptor classes are methylated differently, but, in general, between four and six methylation sites are seen. CheB-P, a methylesterase, removes the methyl group forming methanol. The CheB protein is phosphorylated in its most active state, that is as CheB-P, the non-phosphorylated polypeptide being a target of a kinase CheA.

Many higher animals have specialist chemical sensing mechanisms located in their noses. They are often extremely sensitive to the presence of a vast range of chemicals. It is not always clear, however, exactly what it is about the particular chemical that stimulates a certain response. Molecules that, on paper, appear to be very closely related structural analogues, often give very similar smell sensations, whereas chemicals that smell the same may have very differing structures. Despite this apparent paradox, in 1991 genes encoding odorant receptors were identified by Buck and Axel, and it is the proteins they encode that are responsible for the detection of volatile compounds in the air. These receptors are found in the neuroepithelial cells of the nose, with the nose containing as many as 10,000,000 neurons. As with many other receptors, signal transduction here is associated with a class of heterotrimeric G proteins, G_{olf}, which have their influence on adenylyl cyclase, and hence cAMP signalling.

In taste buds too, heterotrimeric G proteins are important. A G protein analogous to transducin of the rods and cones of the eye has been found here: the taste-specific G protein has been named gustducin. However, not all chemical detection by the taste buds uses the same mechanisms. Four basic tastes are recognizable, each leading to a different cellular response: salty tastes result from an increase in Na^+ movement through Na^+ channels; sour tastes are the result of the blockage of K^+ or Na^+ channels by H^+; sweet and bitter tastes on the other hand are mediated by the presence of G proteins.

Many organisms are responsive to sound, often over a large range of frequencies. Reports that the activity of certain enzymes are correlated to the intensity of sound are fascinating. Studies in the rat have shown that the activity of guanylyl cyclase in the inner ear is inversely related to the volume of sound to which the animals are exposed, suggesting an exciting signal transduction field.

Electronic noses

As vapour and volatile compounds are given off by organisms that cause disease, bio-sensors are being developed to sense these compounds, in machines referred to as "electronic noses". They may, for example, sense the presence of bacteria in humans, or fungi in plants, and so guide on the treatment of disease, or indeed storage of food-stuffs.

■ Summary

- The perception of the environment is as vital to whole organisms as it is for individual cells.

- The response to many environmental factors involves the use of signalling pathways and can be used to illustrate the coordination of signal transduction.

- Many organisms respond to light, either tailoring their growth to optimize their light capturing, as in plants, or to enable them to move around and avoid danger, as with animals.

- Plants monitor light by the use of phytochromes, UV-B, UV-A and blue light receptors.

- One of the most specialist, and well characterized, systems for detecting light is the mammalian eye.

- In the eye, photons of light cause the photoisomerism of retinal, resulting in activation of the opsin protein, which in turn leads to the activation of a trimeric G protein, G_t. The released $G_t\alpha$ causes the activat ion of a phosphodiesterase, which catalyzes the rapid breakdown of cGMP. The subsequent drop in cGMP in the cytoplasm leads to the closure of cation-specific channels in the plasma membrane and the formation of an electrochemical potential, propagated to the synaptic region of the cell. From here, electrical signals pass to the brain where the picture is deciphered.

- The signalling system in the eye has to be reversed to allow the perception of more light.

- The control of light perception in the rods and cones of the eyes is a complex inter-play of many signalling components, allowing a coordinated response across a wide range of light intensities.

- The presence of chemicals is an environmental factor that needs to be monitored.

- For prokaryotes, essential nutrients such as amino acids and sugars need to be sensed.

- Larger animals, such as mammals can detect air borne chemicals as smells as well as soluble chemicals as taste.

- The sense of smell appears to use receptors, linked to heterotrimeric G proteins, having their effect through adenylyl cyclase.

- Taste uses specialized receptor cells, but different types of taste appear to use different signal transduction pathways.

→ Further reading

Photodetection in the eye

Benovic, J.L., Mayor, F., Somers, R.L., Caron, M.G. and Lefkowitz, R.J. (1986) Light dependent phosphorylation of rhodopsin by β-adrenergic receptor kinase. *Nature*, 321, 869–872.

Berg, J.M, Tymoczko, J.L. and Stryer, L. (2003) *Biochemistry*, 5th edn, W.H. Freeman & Co, New York [Chap 32: a particularly good summary of the signalling of the eye].

Quail, P.H. (1994) Photosensory perception and signal transduction in plants. *Current Opinion Genetic Development*, 4, 652–661.

∞ Literature link

Perception of the environment

Background information

Organisms have a vital need to perceive their environment, and over many years there has been a research focus on unravelling the signal transduction processes in the mammalian eye. This has led to the realization that many receptors have a common structure, even if they are involved in completely different transduction pathways, and this work also aided in the discovery of G proteins. Furthermore, detailed study of a single system such as this can reveal how signalling components interact and control each other, leading to a coordinated response. Such research also shows that work on one system can really inform and aid in discoveries in related systems. However, the eye is not the only organ used by mammals to perceive their environment, whereas other organisms need to respond to environmental cues too.

Essay topic

Discuss how animals perceive chemicals in their surroundings, and where possible compare the signal transduction mechanisms used to those that have been identified for light perception by the human eye.

Starter references

Buck, L.B. (2004) Olfactory receptors and odor coding in mammals. *Nutrition Review*, 62, S184–188.

Eisthen, H.L. (1997) Evolution of vertebrate olfactory systems. *Brain Behaviour Evolution*, 50, 222–233.

Mure, L.S., Cornut, P.L., Rieux, C., Drouyer, E., Denis, P., Gronfier, C. and Cooper, H.M. (2009). Melanopsin bistability: a fly's eye technology in the human retina. *PLoS One*, 4, e5991.

Pellegrino, M. and Nakagawa, T. (2009) Smelling the difference: controversial ideas in insect olfaction. *Journal Experimental Biology* 212, 1973–1979.

Rouquier, S. and Giorgi, D. (2007) Olfactory receptor gene repertoires in mammals. *Mutat Research*, 616, 95–102.

Other environment perception systems

Amsler, C.D. and Matsumma, P. (1995) Chemotaxis signal transduction in *Escherichia coli* and *Salmonella typhimurium*. In *Two- Component Signal Transduction*, Hoch, J.A. and Silhavy, T.J. eds, American Society for Microbiology, Washington.

Buck, L. and Axel, R. (1991) A novel multigene family may encode odorant receptors: a molecular basis for odor recognition. *Cell*, 65, 175–187.

Gillespie, P.G. (1995) Molecular machinery of auditory and vestibular transduction. *Current Opinion Neurobiology*, 5, 449–455.

Kinnamon, S.C. (1988) Taste transducin: a diversity of mechanisms. *Trends in Neurosciences*, 11, 491–496.

McLaughlin, S.K., McKinnon, P.J. and Margolskee, R.F. (1992) Gustducin is a taste-cell-specific G protein closely related to the transducins. *Nature*, 357, 563–569.

Reed, R.R. (2004) After the Holy Grail: Establishing a molecular basis for mammalian olfaction. *Cell*, 116, 329–336.

Signalling in development and for the regulation of gene expression

13

Development of an organism is an extremely complex system, which has to be very carefully controlled. A multicellular organism has to grow from a single cell, but it would be no use to grow in a uniform manner from that single cell type, and end up as a ball of cells which are all identical. Cells change and differentiate into all the cells of the body, but, furthermore, all those cells have to be in the correct place. It may be interesting to note that in most cases the two feet of a human being are roughly the same size, their arms are the same lengths, and their bodies are in proportion. How is this achieved? How does your right arm know how long your left one is, and therefore to grow the same?

Clearly, to build a complex body from a single cell requires a vast amount of coordination, and that coordination comes about from the cell signalling that is taking place as the organism grows. The cell signalling mechanisms not only have to make sure that the body grows properly of course, but also have to manage the day-to-day running of the organism's activities. Cells next to each other will need to be aware of each other's existence, and also how the cells next to them may be changing. Cells need to communicate to cells further afield, and there may be polarizing signalling, to make sure growth has a particular pattern.

Regardless of the signalling that is taking place, the expression of the genes encoding the proteins that make up the cell will be key to not only what the cell becomes and what proteins are present within it, but also what the cell does, how it may communicate and how it might influence its environment and the cells around it. Therefore, control of gene expression is a key factor in the control of organism development. This chapter will give a glimpse into the complex world of development, starting with some mechanisms of controlling transcription factor activity and therefore the levels of gene expression, and then will discuss the some examples of cell signalling systems that are studied in a context of developmental biology. It is far from a definitive and comprehensive treatment of the topic, and therefore the reader is directed to the further reading at the end of the chapter for reviews and books on the topic, written by those directly in the field.

13.1 Introduction

Signalling is used in organisms from the moment that they are created by cell division or from the moment fertilization commences, and signalling will determine the development of the organism, its survival and possibly death. In higher organisms it is known that recognition is needed for the sperm to have a fruitful interaction with the egg, and, likewise in plants, pollen arriving at the stigma has to be recognized and an exchange of communication is key to successful pollen tube growth and subsequent fertilization.

Embyro development, the formation of tissues, organs, and all parts of the organism will be under the control of signalling events. It might seem a miracle that from a single cell a complex organism such as a mammal is created, but, unfortunately, although in the majority of cases it appears that the development runs smoothly, in some cases it does not, leading to disability or early death. Even once created, dysfunction of signalling pathways may lead to developmental problems. For example, many cell signalling pathways have been found to be involved in the formation of tumours and the overgrowth of cancer cells. The genes for many of these signalling proteins have been characterized, and, as when carrying out their normal function they are not cancer forming, but have the potential to become tumour forming, they are referred to as proto-oncogenes.

To develop an organism properly is an extremely complex job. From a single cell there needs to be the creation of different tissues and organs, and even within those there are complex arrangements of cells with specialist functions, for example the kidney is comprised of an outer layer, or cortex, with an inner section or medulla. The medulla has structures called nephrons, and each of these contains glomerula and Bowman's capsules, which themselves can be seen to have a finer structure. Once examined in detail it is hard to see how two cells in an organ could actually be identical, as each has a unique environment, is in contact with a unique set of other cells and has a defined function. Many cells are themselves not uniform in the shape or function. An excellent example would be the cells lining the gut of humans (Figure 13.1). The "outer" surface needs to be in contact with the gut contents, and have the functions of taking in nutrients and protecting the organism from the harsh environment of the gut contents. On the other hand the "inner" face of those same cells needs to be in contact with other cells, and be able to release those nutrients into the organism. When such cells align to form a barrier between the gut and the rest of the body, the interfaces between the cells have to be "sealed", with structures called tight junctions.

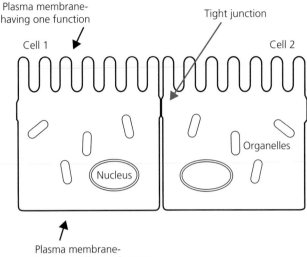

Plasma membrane-
having one function

Tight junction

Cell 1

Cell 2

Organelles

Nucleus

Plasma membrane-
having a different function

Figure 13.1 A schematic diagram of two cells that have a tight junction between them. Such a tight interface between cells aligned to form a "wall" can be found lining organs such as the gut.

Therefore, it can be seen that each individual cell of an organism has to be created to do its specific job, and to have correct shape and structure to enable it to fulfil that purpose. Furthermore, the complement of proteins inside each cell has to be such that it can partake in the functioning it is designed to do. Therefore, the expression of genes in the cells is paramount to creating the cell, and also in enabling it to survive and thrive.

In this chapter, some of the signalling involved in the control of gene expression and the developmental stages of organisms will be explained. Often this work has been carried out using simple organisms, in what are referred to as "model" organisms, as listed in Table 13.1. Clearly, once such work has been carried out the resulting data inform similar studies in higher and more complex organisms.

13.2 Transcription factors

The definition of a transcription factor would be a protein that controls the switching on and off of the expression of a gene. This will, therefore, dictate whether there is a transcript from that gene, when the transcript is produced and how much is produced, which will lead to the control in many cases of the amount of a certain protein present in a cell at any moment in time. Transcription factors are, therefore, able to bind to the relevant regions of the genome, that is the promoter regions of the genes, and regulate the level of expression.

Microarrays and proteomics

Once groups of cells have been identified and isolated, it can be useful to know the full complement of genes being expressed in those cells, and the full complement of proteins present, which can be determined by the use of microarrays and proteomics respectively. See **Chapter 3** for more details.

Table 13.1 Some of the common 'model' organisms and why they are attractive to use.

Model organism	Common name	Type of organism	Comment
Saccharomyces	Yeast	Fungus	- Large amount of genetic data, and expression profiles
Arabidopsis thaliana	Thale cress	Plant	- Complete genome sequenced - Large mutant collection
Medicago truncatula		Plant: Legume	- Used for studying root nodule development
Caenorhabditis elegans	Nematode	Invertebrate	- Cell lineage determined
Drosophila melanogaster	Fruit fly	Invertebrate	- Historically used for biological research
Xenopus tropicalis	African clawed frog	Vetebrate: amphibian	- Only diploid species of Xenopus - Small genome size - Short generation time
Danio rerio	Zebra fish	Vertebrate: fish	- Body is almost transparent
Mus musculus	Mouse	Vertebrate: mammal	- Easy to keep and use mammal

Further details can be found at: www.sanger.ac.uk/modelorgs/
N.B. There are a plethora of web pages on genetic resources based on model organisms, including databases of their complete genomes (when known).

Transcription factor expression

It should be noted that the complement of transcription factors in the cell is not fixed either, and microarray studies have shown that extracellular factors can alter the expression of the genes encoding the transcription factors themselves. This could be crucial for longer term adaptation.

As transcription factors are involved in the control of expression of genes, those genes may encode proteins that are involved in cell signalling events. Furthermore, the control of the transcription factors themselves is usually under the regulation of signal transduction pathways. Therefore, cell signalling pathways can be controlled by, for example, extracellular influences, which will lead to the control of transcription factors, which then influence the expression of cell signalling components, often components that were not present or involved in earlier signal transduction events. By this means of cell signalling mechanisms controlling the production of molecules involved in cell signalling, cells are able to change their complement of signalling components, either creating new components or modulating the levels of already existing ones. As such, cells are able to adapt, change to face new challenges and survive better than they might have been able to before.

However, as transcription factors have such an instrumental job of controlling which genes are expressed or not, coupled with the fact that organisms may have thousands of genes to be controlled all at the same time, and also coupled with the vast array of factors that may influence gene expression, it is no surprise to learn that the control of transcription factors is both wide ranging and complex, and so only a few examples will be discussed here.

Some of the main ways of activating transcription factors have been covered in preceding chapters, but a quick summary would have to include phosphorylation (Chapter 6), oxidation (Chapter 10) and steroid hormone binding (Chapter 5), with other mechanisms embedded in the examples below.

Activation by phosphorylation

cAMP has many of its actions through the activation of PKA, and one of the targets of PKA is the transcription factor CREB (cAMP-response element binding protein). Therefore, in response to hormone perception, which modulates cAMP levels in cells (see Chapter 7), transcriptional activity can also be regulated. CREB binds to DNA as a dimer, with each peptide contributing an α-helix to DNA binding. Once phosphorylated, CREB can bind to another protein called CRB-binding protein (CBP). This protein has several domains, including ones that can bind other proteins, so once CREB has been phosphorylated a complex can form, resulting in increased transcriptional activity.

MAP kinases were also seen in Chapter 6 to move to the nucleus, and targets of MAPKs include transcription factors such as Ets. The interaction between the kinase and the transcription factor may depend on the functioning of other proteins too, such as one called modulator of the activity of Ets (Mae). The Ets proteins are a family of polypeptides, each with a highly conserved DNA binding domain, the target of which is the consensus sequence -GGAA/T-. However, it should be noted that such transcription factors can have multiple modes of activation, phosphorylation being only one of them. Therefore, transcription factors such as Ets are a site of signalling convergence, and are of particular importance in the control of the expression of a wide range of genes, being influence by a wide range of factors.

Also under control of phosphorylation events are the STATs proteins. These are targets of the Janus kinases, JAKs, and therefore phosphorylation has a direct influence on their activity (see Chapter 6). Again, STATs function as dimers following phosphorylation.

One particular transcription factor that has attracted a lot of attention is NF-κB. It is responsible for control of the expression of well over 100 genes in humans. This transcription factor resides in the cytoplasm in a quiescent state of the cell, bound to an inhibitory peptide known as I-κB. On activation of the system, I-κB is phosphorylated, so targeting it for degradation by the ubiquitination system (see Chapter 6). This frees the NF-κB, which then moves to the nucleus and alters gene expression. Interestingly, it has also been reported that this dissociation of NF-κB and I-κB is enhanced under oxidative stress conditions, with many suggesting that NF-κB or I-κB is a target of direct oxidation, although this has been severely questioned by many others.

Dimers and palindromes

Often proteins that bind to DNA are found to do so as dimers, and the sequences of DNA to which they bind are palindromic. Palindromic DNA sequences are those that have rotational symmetry, that is, they are the same backwards as forwards.

Cyclosporin The drug cyclosporin is used to suppress immune responses and therefore can be used in many conditions in humans, for example organ transplantation. The action of cyclosporin is to inhibit the phosphatase calcineurin.

Some transcription factors are activated not by phosphorylation but by dephosphorylation. For example, the phosphatase calcineurin activates the transcription factor NF-AT by dephosphorylation.

Therefore, it can be seen that phosphorylation has a profound effect on the activities of transcription factors and the subsequent levels of gene expression. However, it should be stressed that gene expression of particular genes may be decreased. It is not always an increase in gene expression that leads to the result required by the cell.

Activation by oxidation

Under oxidative stress conditions, one of the effects seen is a change in gene expression. It has been estimated that 3% of a genome may have its level of gene expression altered by oxidative stress. However, just as with phosphorylation, oxidative stress leads to an increase in the expression of some genes, but a depression in the activation of others.

Therefore, it is no surprise to find that several transcription factors have been identified which are controlled by oxidation or reduction of amino acid side groups, as discussed in Chapter 10. One such example is the transcription factor OxyR in *Escherichia coli*, which becomes activated by the presence of cellular H_2O_2. This creates a disulphide bond between Cys199 and 208, so altering the protein's confirmation and hence activity. In the yeast *S. cerevisiae*, the transcription factor Yap1 is activated by oxidation, with reduction of the protein being mediated by thioredoxin. Oxidation of this transcription factor is mediated by another protein, glutathione peroxidase (GPx)-like enzyme, Gpx3. In the presence of H_2O_2, the thiol group of Cys36 on Gpx3 bridges to the thiol group of Cys598 on Yap1, forming an intermolecular disulphide. This subsequently results in the formation of a disulphide in Yap1 itself, and so the transcription factor is activated, altering the level of expression of the relevant genes.

Other examples of oxidation effecting transcription are also prevalent in the literature, such as the protein NPR1. This becomes reduced, altering it from an oligomer to a monomer, thus allowing it to accumulate in the nucleus and alter gene expression. It is almost certain that many more examples will be discovered in future research and that this will not be an uncommon mechanism to alter transcription factor activity.

Steroid hormone binding

As discussed in Chapter 4, some ligands are hydrophobic enough to not require the presence of receptors on the cell surface, but rather they are perceived by receptors either in the cytoplasm or nucleus. These steroid receptors are usually DNA binding proteins too, so they have transcription factor activity. The targets on DNA for binding are often palindromic sequences, and therefore these receptors bind and act as dimers, for example, the estrogen receptor that targets a sequence of DNA referred to as the estrogen-response element (ERE). However, steroid receptors do not work in isolation and many proteins called co-activators have been identified that aid steroid receptor function. More details of these transcription factors and how they are controlled were discussed in Chapter 4.

13.3   Transforming growth factor ß

Transforming growth factor ß (TGF-ß) is part of a large family of extracellular protein signals. They are involved in the control of gene expression and have profound effects in development. In humans, TGF-ß causes inhibition of cell proliferation. There are three isoforms in humans, which are each encoded for by separate genes, and these are known as TGF-ß1, TGF-ß2 and TGF-ß3. Formation of the mature proteins is different from that described for other proteins in Chapter 1, as here the protein is stored in the extracellular matrix of the cells before it is released and used. Each of the genes encode a precursor protein, which is cleaved into TGF-ß and a protein called LAP, which is derived from the pro-domain of the original protein (see Figure 13.2). Subsequent to this cleavage, a complex is formed. TGF-ß contains several cysteine residues, and this allows the formation of three intramolecular disulphide bridges, but also allows the TGF-ß peptides to dimerize through the formation of an inter-peptide disulphide. A dimer of TGF-ß then associates with a dimer of LAP, which further associates with a third protein called LTBP (latent TGF-ß-binding protein). Release of active TGF-ß then requires the presence and action of a further protein. Thrombospondin (TSP-1), which resides in the matrix, can interact with LAP, so releasing the mature and active TGF-ß dimer. Alternatively, the mature TGF-ß can be released by protease activity. Of interest is that the TGF-ß dimers can be heterodimers, so giving greater diversity and subtlety of function than just three isoforms alone would give. This is similar to that seen with the PGF as discussed in Chapter 5 (see Table 5.2).

TGF-ß signalling is through its binding to TGF-ß receptors. There are three polypeptides in humans, of 55 kDa, 85 kDa and 280 kDa, which can act as TGF-ß receptors, referred to as RI, RII and RIII (also called ß-glycan). All of these are transmembrane proteins, but RI and RII are serine/threonine kinases. Unusually, RII has constitutive kinase activity, that is, it is active all the time unless turned off in some way. The role of RIII appears to be to collect and concentrate TGF-ß so that it can activate the other TGF-ß receptor types. RI and RII are dimers in the membrane, but on binding to TGF-ß they come together to form a tetramer of two of each polypeptide. The ensuing conformational change allows RII to phosphorylate RI, causing RI itself to become active.

LTBP   The LTBP proteins are also a family of polypeptides. At least four genes encoding these proteins have been identified in humans.

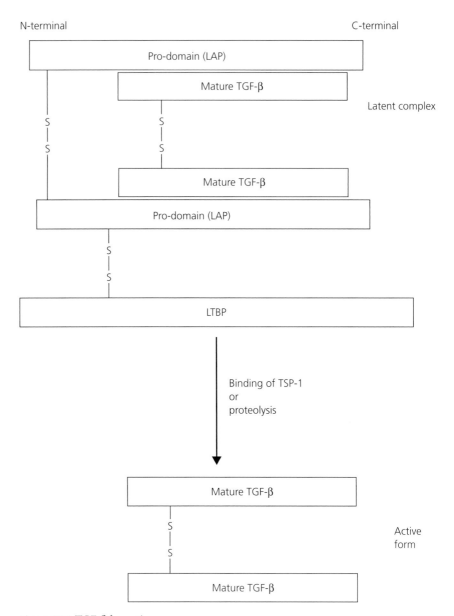

Figure 13.2 TGF-ß formation.

The substrate for the active RI TGF-ß receptor is a group of proteins called the Smads, in particular Smad-2 and Smad-3. These proteins are also a family of related polypeptides, but one group, the so-called R-Smads (receptor regulated Smads), can be phosphorylated on three serines by RI kinase activity. This phosphorylation causes a conformational change in the R-Smad protein, firstly allowing dimerization of the R-Smad protein using the phosphorylated residues and a phosphoserine binding site in each polypeptide, and, secondly, allows the association of this dimer with another Smad, also using phospho-serine binding. This second type of Smad is referred to as a co-Smad (Smad4), as it does not interact directly with the receptor. Furthermore, this complex binds another protein called importin ß, which associates to the MH1 domain of the R-Smad. The MH1 domain of the Smad contains a nucleus import signal, and this domain can bind to importin ß to allow the movement of the complex into the nucleus, where it needs to be to affect transcription. Once in the nucleus, its final effect is mediated by binding to transcription factors.

To reverse the signal, and alleviate the effects of the TGF-ß signalling, R-Smads are dephosphorylated in the nucleus and re-exported. This allows for further rounds of signalling through this system, in a similar way to that discussed for MAP kinases in **Chapter 6**.

There is also a third group of Smad proteins, the anti-Smads (otherwise known as I-Smads or inhibitory Smads), such as Smad6 and Smad7. These block Smad signalling because they inhibit phosphorylation of R-Smad and stop subsequent nuclear translocation. The expression of I-Smads is induced by TGF-ß and related proteins, and therefore there is an element of autoinhibition in Smad signalling.

As can be seen, TGF-ß signalling uses a selection of Smad proteins, but other related signalling will use a different set of Smads. One set of members of the TGF-ß superfamily which is well studied are the bone morphogenetic proteins (BMPs). These signal in a similar way to TGF-ß, but use different Smads down-stream of the receptors, for example Smads 1, 5 and 8.

The functions of Smads are regulated by other components and signalling pathways too, such as the MAP kinase pathway (see **Chapter 5**), so other signals can impinge on the final signalling. On top of this the addition of TGF-ß to some cells can activate MAP kinase pathways, perhaps leading to a different final response.

Of course, like most signalling systems, there are ramifications for disease if the mechanisms fail. Point mutations of Smad2 and Smad4 have been noted in some cancers, and mice that lack Smad3 develop colon cancer.

13.4 **Notch receptor signalling**

The Notch proteins are a group of receptors that are involved in intercellular signalling. Their role is to bind to proteins in the adjacent cell, and therefore mediate cell-to-cell signalling. Cell–cell interactions are extremely important in development, as cells need to be aware of the cells next to them, and a cell may influence the way the adjacent cell changes and differentiates. As an organ or tissue grows, different cell layers and different cell types need to be able to be generated from precursor cells, so allowing the complexity of the final tissue. Therefore, cells will communicate with adjacent cells, which may even be daughter cells. If cells are touching, signals that are released into the extracellular medium (such as shown in **Figure 1.4A**) are not necessary, and may even be detrimental. If a cell wishes to influence the development of the neighbouring cell, direct cell-to-cell communication would be a key event. The release of an extracellular signal would allow more cells to be influenced, perhaps even a long way away, and this may not be desired. Direct cell-to-cell communication would be specific and precisely targeted, and this is the type of signalling which involves the Notch proteins.

The Notch protein is made as a precursor, which is processed as it passes through the Golgi apparatus en route to the plasma membrane. The precursor is cleaved, in a process called S1 cleavage. The result is a receptor that has two subunits, an extracellular polypeptide associated with a second polypeptide that contains a single membrane spanning section and an intracellular domain.

In mammals at least four Notch proteins have been identified (Notch 1–4), with similar proteins being identified in other animals. They are characterized by having many EGF-like repeats in their domain structure. For example, Notch 1 has 36 such repeats. These stick out into the extracellular space, where binding to the appropriate protein on the other cell will take place.

The Notch proteins bind to other proteins that are accessible on the surface of other cells (**Figure 13.3**). In mammals at least five Notch ligands have been identified, which reside in the cell communicating with the Notch containing cell. These are proteins known as Delta-like (1, 3 and 4) and Jagged (1, 2). On binding, the Notch protein is cleaved again, in what is called S2 cleavage,

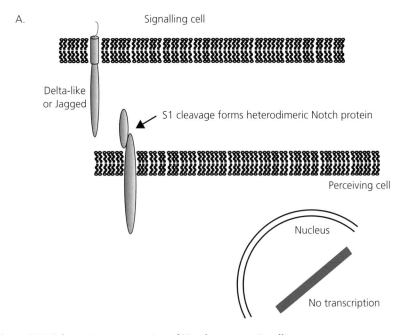

Figure 13.3 Schematic representation of Notch receptor signalling.

B.

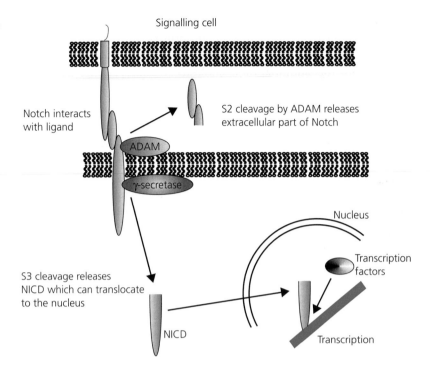

Figure 13.3 (Continued).

which releases the extracellular domain. This cleavage is catalyzed by a protein call ADAM. Following this is S3 cleavage catalyzed by an enzyme called γ-secretase. This enzyme is itself a protein complex comprising several proteins. These are presenilin, which is a transmembrane protein, nicastrin, APH-1 and PEN-2. S3 cleavage releases an intracellular domain of Notch (NICD), which can translocate to the nucleus and alter gene expression. Once there, NICD associates with a protein called CSL (CBF1, Suppressor of hairless, Lag-1) so relieving gene expression repression and activating the transcription of the genes involved.

■ Inhibitors of γ-secretase are the focus of the development of new cancer therapies.

13.5 Hedgehog signalling

Hedgehog (Hh) signalling is also very important for normal development of animals, and if the pathway is disrupted, perhaps because of a mutation, then abnormal development can result. Also, as with many of these pathways, disruption can cause the formation of tumours.

Hedgehog is a protein signal secreted from cells. The Hedgehog protein itself is produced as a precursor, but interestingly this precursor protein can cleave itself into two further sections. This cleavage takes place between a glycine (amino acid 257) and a cysteine (amino acid 258). The N-terminal section cleaved off then acts as the signal, whereas the C-terminal section, which contained the domain with the proteolytic activity, appears to be degraded. As it is being produced, the N-terminal section is also covalently modified by the addition of a cholesterol moiety to its newly formed C-terminal end, and a palmitoyl moiety to the N-terminal end. Both these added groups are by their very nature hydrophobic, and therefore the resulting signalling protein has been rendered reasonably hydrophobic, and this will encourage it to be associated with membranes and so limit its diffusion. It has been noted that the concentration of Hh perceived by a cell itself is important, as in many cases the cellular response does not just relate to the perception of the Hh protein per se, but to how much is present. Limiting the diffusion of the molecule will enable the largest concentrations, and therefore those particular responses following the perception of a high concentration, to be found near the site of generation of the original signal. In a similar way to Notch signalling, although in not quite such a stringent manner, this will mean that there is a local response, limited to the cells near by. This type of local effect, with potential for local amplification, is exactly what may be needed to guide and control the development of tissues and organs, without having an effect on other parts of the organism.

Perception of the Hedgehog protein by cells involves a 12 transmembrane spanning protein called Patched (Ptc). This protein, which resides in the plasma membrane of the cell, normally inhibits a protein called Smoothened (Smo). Smo is a protein that shows in its topology a similarity to the G protein-coupled receptors, as described in Chapter 5, that is, it has seven transmembrane spanning domains (see Figure 5.2). This latter protein is primarily found in intracellular vesicles when the system is quiescent and in the absence of the Hedgehog protein. At the same time, in the cytoplasm there is an assembled complex of proteins that is in association with the microtubules. This complex includes the proteins Costal-2 (Cos2), Fused (Fu), a protein called Suppressor of fused (SuFu) and Cubitis interruptus (Ci). Fu is a serine/threonine kinase, Ci is a transcription factor, and Cos2 is a kinesin-like protein. Together these proteins create what is known as the Hedgehog signalling complex (HSC). Cos2 can also bind to a range of kinases, including PKA, glycogen synthase kinase 3 (GSK3) and casein kinase 1 (CK1). Increased kinase activity leads to the phosphorylation of Ci, which is then cleaved. Cleavage is aided by a protein called Slimb, and the resulting polypeptide from Ci can enter the nucleus and reduce transcription of the relevant genes (see Figure 13.4). Having such a range of kinases in the HSC suggests a wide range of signalling that may influence down-stream phosphorylation. With the presence of PKA this implies an impact by intracellular cAMP levels amongst other mechanisms.

When the Hedgehog protein is present, it binds to Ptc, which then no longer inhibits the Smo protein. Smo is phosphorylated by the kinases PKA and CK1, and the proteins Fu and Cos2 are also highly phosphorylated. Cos2 associates with Smo, which has been phosphorylated, and Cos2 also acts as a

Morphogens These are signals that have different effects depending on their concentration. Therefore, Hh is often referred to as a morphogen.

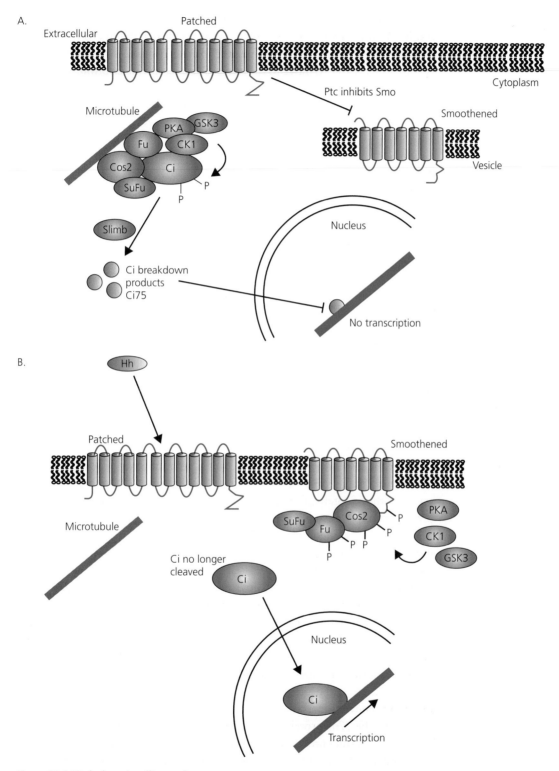

Figure 13.4 Hedgehog signalling pathway.

scaffold protein allowing Fu, SuFu and Ci to be part of the complex. The end result is that the complex is no longer associated with the microtubules, and cleavage of Ci is not longer able to take place. This means that a full length version of Ci is able to move to the nucleus and alter transcription. Therefore, there are two outcomes based on the same protein. Firstly, there are no longer Ci breakdown products to repress transcription, and, secondly, there is now full sized Ci, which can influence down-stream events.

13.6 Wnt signalling

Wnt-1 first came to the attention of researchers as it was found to be a proto-oncogene, that is, the normal version of a gene, which may cause cancer when mutated. In this case, it had been found that an alteration of the expression of the Wnt encoding gene can lead to the formation of tumours. However, the normal Wnt signalling pathway has been found to be involved in the control of developmental stages in humans and other animals, including development of the brain.

> **Wnt** Wnt is derived from wingless, as much of the early work was carried out in Drosophila, and int, which is the site for retrovirus integration.

Wnt is a family of proteins that share a conserved sequence, but have rather varied effects. They are extracellular proteins and are often referred to as growth factors, due to their influences on cell proliferation. One of the major ways in which Wnt is perceived, and hence signals, is through a pathway known as the ß-catenin-dependent pathway, otherwise known as the canonical pathway (Figure 13.5).

When the system is not stimulated, that is in the absence of Wnt, a complex is formed in the cytoplasm comprising a protein called ß-catenin, along with several other proteins. These are axin, adenomatous polyposis coli (APC) protein and a kinase, glycogen synthase kinase 3 (GSK3). GSK3 was also found in the Hedgehog signalling pathway. GSK3 phosphorylates the ß-catenin protein, and this marks it out for degradation through the ubiquitination pathway (previously discussed in Chapter 6). Therefore, the levels of ß-catenin in the cytoplasm are kept low.

The receptor for Wnt is a protein similar to the G protein-coupled receptors that were discussed in Chapter 7, as well as being mentioned above for Notch signalling. These Wnt receptors are also composed of seven α-helices that span the membrane. In this case the family of receptors are known as Frizzled (Fz). However, another protein is also involved here, known as low density lipoprotein receptor-related protein (LRP). This is a protein that only has a single pass through the membrane, but it appears that it acts as a co-receptor, both it and Fz binding to the Wnt protein. Once activated, the activated receptor complex, along with a further protein called Dishevelled (Dsh), causes an inhibition of the phosphorylating activity of the Axin/APC/GSK3 complex. However, LRP becomes phosphorylated, which causes its binding to Axin, and therefore

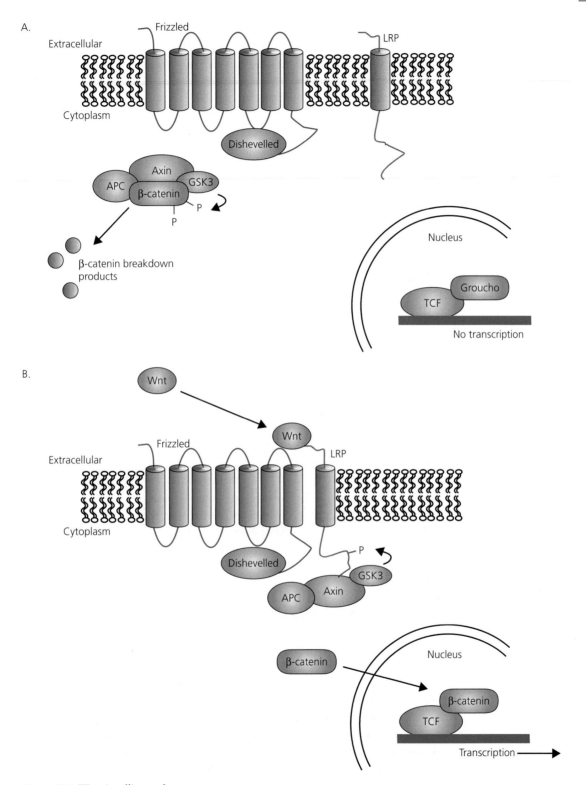

Figure 13.5 Wnt signalling pathway.

facilitates the interaction of the Axin/APC/GSK3 complex with the receptors. This phosphorylation of LRP can be carried out by at least two different kinases, including GSK3. Once the Axin/APC/GSK3 complex is inhibited and can no longer carry out its phosphorylating activity, the ß-catenin is no longer phosphorylated and broken down in the cytoplasm. It is, therefore, free to move into the nucleus, where it interacts with the transcription factors. Transcription factors of particular relevance here are those of the T-cell factor/lymphoid enhancer factor (TCF/LEF) family. Hence, gene expression is altered. TCF normally partners with another protein, for example one called Groucho, which acts as a repressor of transcription. On ß-catenin entering the nucleus, it binds to TCF in the place of Groucho, so relieving this repressor and activating transcription.

The protein Dsh is also known to be phosphorylated. This allows further subtle control of the system, through the impact on other signalling pathways.

Although Wnt signalling through ß-catenin has been extensively studied in recent years, there are other pathways through which Wnt signals. One of these involves members of the monomeric G protein family, in particular Rho, and Rac (see **Chapter 7** for more details of monomeric G proteins), and leads to alterations in the cytoskeleton of the cell. A further pathway involves the modulation of intracellular calcium ions (as discussed in **Chapter 9**), and this pathway leads to negative effects on the ß-catenin pathway also stimulated by Wnt. Therefore, it is not hard to see that the signalling initiated by the extracellular Wnt protein is quite wide ranging and complex.

Names of proteins in developmental biology

Many of the names used in this chapter sound odd, almost flippant. Many of the unusual names are derived from the systems in which they were first investigated. Several of these are from developmental phenotypes in the fly *Drosophila*, and hence names of proteins such as Frizzled, Dishevelled, etc. Loss of Hedgehog protein expression in *Drosophila* caused the embryos to develop into balls, which resembled rolled up hedgehogs.

13.7 Toll-like receptor signalling

Although not strictly used for development, the Toll-like receptors do lead to alterations of transcription and so will be discussed here. Originally Toll receptors were discovered in *Drosophila*, but homologues are also found in higher animals including humans. There are at least 10 different Toll-like receptors (TLRs) in humans. They perceive large molecules that are associated with pathogens, such as bacteria, and such molecules are usually referred to as pathogen-associated molecular patterns (PAMPs). Therefore, they are instrumental in the host recognition of invading organisms and in the control of the immune response. These receptors are commonly called pattern-recognition receptors (PRRs) because of their perception of these PAMPs. The receptors are mainly expressed in cells such as macrophages and dendritic cells, that is, those involved directly with pathogen recognition. Most are found in the plasma membrane, where receptors would normally be expected to be, but several are associated with endosomes, such as TLR3, TLR7 and TLR9.

PAMPs

Pathogen-associated molecular patterns (PAMPs) are essentially large molecules, which are found associated with pathogens but are not similar to host molecules. Examples include bacterial flagellin, lipopolysaccharide (LPS) and bacterial nucleotides. As such, these molecules are often recognized by TLRs. However, as many of these molecules are associated with microorganisms that are not pathogenic, many researchers now call them MAMPS: microorganism-associated molecular patterns.

The structure of TLRs are characterized by having a large leucine rich domain at their N-terminal end, which is on one side of the membrane, that is, an extracellular transmembrane domain, and a C-terminal domain on the other side of the membrane that can lead to down-stream signalling inside the cell. This signalling domain is referred to as the Toll-interleukin 1 receptor domain (TIR).

Down-stream signalling is quite complex, and different in detail for each of the TLRs. However, many share common components, one of which is an adaptor protein MyD88. With the exception of TLR3, this protein can associate with all the TLRs. The C-terminal end of MyD88 also contains a TIR domain, whereas at the N-terminal end there is a death domain (see **Chapter 14**). The TIR domain allows MyD88 to interact with the TLR, whereas the death domain is involved in the down-stream signalling. Down-stream of the receptor are several kinases. The MyD88 protein can interact with a kinase called IL-1 receptor associated kinase (IRAK). Other kinases further down-stream may include TGF-β-activated kinase 1 (TAK1) and IκB kinase (IKK). Once activated, IRAK proteins complex with another peptide called TRAF6 that can activate TAK1, which goes on to activate IKK. An example of such a pathway is shown in **Figure 13.6**. Phosphorylation of IκB will lead to the activation of NF-κB transcription factor, and therefore alterations in gene expression. This includes increased expression of IL-6, TNF-α and IL-1β, all of which are involved in the immune response, as would be expected from a receptor that is activated by PAMPs.

As well as activation of the NF-κB gene expression, the MyD88-mediated signalling from TLRs can also activate MAP kinases (discussed in **Chapter 6**), which can also lead to nuclear events such as altered expression of genes. Therefore, there is the opportunity for a wide divergence of the signalling leading from TLRs. Such divergence is exacerbated by the fact that, although the MyD88 pathway is now well established as a signalling mechanism initiated by TLRs, there are also MyD88-independent pathways. Such pathways can still lead to the activation of NF-κB, and indeed MAP kinases. An example here would be the association of a TLR such as TLR3 with an adaptor protein

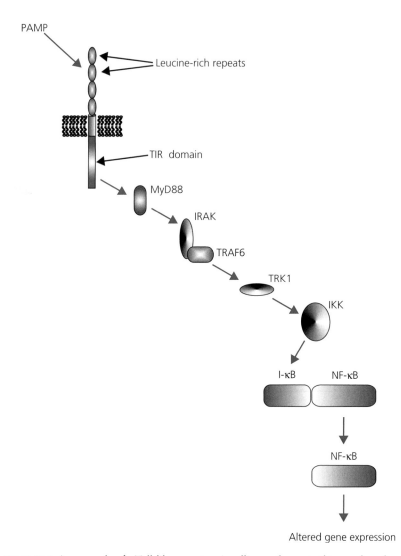

Figure 13.6 An example of a Toll-like receptor signalling pathway, in this case based on that known for TLR2 or TLR4. Activation by PAMPs leads to activation of transcription factors such as NF-κB.

called TIR-containing adaptor protein inducing IFN-β (TRIF). TRIF can lead to the activation of TRAF6 and therefore connect up with the MyD88-dependent pathway, with the same down-stream effects on NF-kB and increased cytokine production. TRIF can also interact with several other proteins, with one such pathway leading to the increased production of IFN-β.

TLRs rely on a myriad of adaptor proteins, kinases and transcription factors, with many of them interacting and influencing the functioning of others. A true representation of the down-stream signalling from all the TLRs would be a mass of interconnecting and crossing arrows; a true example of the nature of signalling was discussed in **Chapter 1** and shown in **Figure 1.8B**.

■ As TLRs are so important for the immune response, mechanisms for the manipulation of their action could be excellent targets for new drugs for allergy and autoimmune disease.

■ Summary

- Signalling is extremely important for the control of gene expression and the development of an organism.

- The use of model organisms has been instrumental to unravelling signalling in development.

- One of the final points of action of many signal transduction pathways is transcription factors in the nucleus.

- Transcription factors can be controlled in a variety of ways, including phosphorylation or oxidation.

- Some signalling molecules, such as steroid hormones, can be perceived inside the cell to act on the transcription factor machinery.

- Transforming growth factor ß is an extremely important group of signalling proteins. Their signal transduction pathways involve the Smad proteins.

- Notch signalling allows cells to have direct cell-to-cell communication.

- Hedgehog signalling is important for developmental signalling.

- Hedgehog is perceived by a system involving the proteins Patched and Smoothened, but the end result is an influence of the activity of the protein Cubitis interruptus.

- Wnt signalling can proceed through ß-catenin-dependent and ß-catenin-independent routes.

- Wnt is perceived by proteins called Frizzled and LRP, with the result being the release of ß-catenin in the cytoplasm. This ß-catenin can then move to the nucleus to act on transcription factors.

- Toll-like receptors are used in the immune response, as they perceive pathogen-derived molecules (PAMPs).

- Toll-like receptors contain a transmembrane domain and an extracellular region consisting of leucine-rich repeats.

- Down-stream signalling often leads to activation of NF-κB, to other transcription factors and alterations of transcription, including the expression of many cytokines, such as IL-6, IFN-β, TNF-α and IL-1β.

- Down-stream signalling from TLRs is complex, and involves many adaptor proteins and kinases. However, a commonly found adaptor protein used by most TLRs is MyD88, in what is called the MYD88-dependent pathway.

→ Further reading

Bryant, D.M. and Mostov, K.E. (2008) From cells to organs: building a polarised tissue. *Nature Reviews Molecular Cell Biology*, 9, 887–901.

Davidson, E.H. (2006) *The Regulatory Genome: Gene Regulatory Networks In Development And Evolution.* Academic Press, USA. ISBN 978-0-12-088563-3.

Gilbert, S.F. (2006) *Developmental Biology* (8th edn) Sinauer Associates Inc. ISBN 978-0878932504.

Lewis, J. (2008) From signals to patterns: space, time, and mathematics in developmental biology. *Science*, 322, 399–403.

Martinez Arias, A. and Stewart, A. (2002) *Molecular principles of animal development.* Oxford University Press, Oxford, UK. ISBN 0-19-879284-0 [an excellent book on development, with discussions on cell signalling].

Moody, S.A. (2007) *Principles of Developmental Genetics.* Academic Press. ISBN 978-0123695482.

Nusslein-Volhard, C. (2008) *Coming to Life: How Genes Drive Development.* Kales. ISBN 978-0979845604.

Persson, C.G.A. (2002) Mice are not a good model of human airway disease. *American Journal of Respiratory and Critical Care Medicine*, 166, 6–7.

Pound, P., Ebrahim, S., Sandercock, P. Bracken, M.B. and Roberts, I. Reviewing Animal Trials Systematically (RATS) Group. (2004) Where is the evidence that animal research benefits humans? *British Medical Journal*, 328, 514–517.

Slack, J.M.W. (2005) *Essential Developmental Biology* (2nd edn). Wiley-Blackwell. ISBN 978-1405122160.

Wall, R.J. and Shani, M. (2008) Are animal models as good as we think? *Theriogenology*, 69, 2–9.

Transcription factors

Baker, D.A., Mille-Baker, B., Wainwright, S.M., Ish-Horowicz, D. and Dibb, N.J. (2001) Mae mediates MAP kinase phosphorylation of Ets transcription factors in Drosophila. *Nature*, 411, 330–334.

Godon, C., Lagniel, G., Lee, J., Buhler, J.-M., Kieffer, S., Perrot, M., Boucherie, H., Toledano, M.B. and Labarre, J. (1998) The H_2O_2 stimulon in *Saccharomyces cerevisiae*. *Journal of Biological Chemistry*, 273, 22480–22489.

He, W., Cao, T., Smith, D.A., Myers, T.E., Wang, X.J. (2001) Smads mediate signaling of the TGF superfamily in normal keratinocytes but are lost during skin chemical carcinogenesis. *Oncogene*, 20, 471–483.

Horabin, J.I. (Ed.) (2007) *Hedgehog Signaling Protocols: Methods in Molecular Biology Series*, Vol. 397, Humana Press, ISBN 978-1-58829-692-4.

Krishna, S., Maduzia, L.L. and Padgett, R.W. (1999) Specificity of TGFß signaling is conferred by distinct type I receptors and their associated SMAD proteins in Caenorhabditis elegans. *Development*, 126, 251–260.

Latchman D.S. (2008) *Eukaryotic Transcription Factors* (5th edn) Academic Press, USA. ISBN 978-0-12-373983-4.

Logan, C.Y. and Nusse, R. (2007) The Wnt signaling pathway in development and disease. *Annual Review Cell Development Disease*, 20, 781–810.

Miyazono, K., ten Dijke, P. and Heldin, C.H. (2000) TGF-beta signaling by Smad proteins. *Advances in Immunology* 75, 115–157.

Sharrocks, A.D. (2001) The ETS-domain transcription factor family. *Nature Reviews Molecular Cell Biology*, 2, 827–837.

Yordy, J.S. and Muise-Helmericks, R.C. (2000) Signal transduction and the Ets family of transcription factors. *Oncogene*, 19, 6503–6513.

Notch signalling

Chiba, S. (2006) Concise review: Notch signalling in stem cell systems. *Stem Cells*, 24, 2437–2447.

Lai, E.C. (2004) Notch signaling: control of cell communication and cell fate. *Development* 131, 965–973.

Shih, I.-M. and Tian-Li Wang, Y.-L. (2007) Notch signaling, γ-secretase inhibitors, and cancer therapy. *Cancer Research*, 67, 1879–1882.

Hedgehog signalling

Hooper, J.E. and Scott, M.P. (2005) Communicating with Hedgehogs. *Nature Molecular Cell Biology*, 6, 306–317.

Horabin, J.I. (ed.) (2007) *Hedgehog Signaling Protocols: Methods in Molecular Biology Series*, Vol. 397, Humana Press, ISBN 978-1-58829-692-4.

Jia, J. and Jiang, J. (2006) Decoding the Hedgehog signal in animal development. *Cellular and Molecular Life Sciences*, 63, 1249–1265.

Jiang, J. and Hui, C.C. (2008) Hedgehog signaling in development and cancer. *Developmental Cell*, 15, 801–812.

Nieuwenhuis, E. and Hui, C-c. (2005) Hedgehog signaling and congenital malformations. *Clinical Genetics*, 67, 193–208.

Wnt signalling

Gordon, M.D. and Nusse, R. (2006) Wnt signaling: multiple pathways, multiple receptors, and multiple transcription factors. *Journal of Biological Chemistry*, 281, 22429–22433.

James, R.G., Conrad, W.H. and Moon, R.T. (2008) Beta-catenin-independent Wnt-pathways: signals, core proteins, and effectors. *Methods Molecular Biology*, 468, 131–144.

Logan, C.Y. and Nusse, R. (2004) The Wnt signalling pathway in development and disease. *Annual Review of Cell and Developmental Biology*, 20, 781–810.

Mazieres, J., He, B., You, L., Xu, Z. and Jablons, D.M. (2005) Wnt signaling in lung cancer. *Cancer Letters*, 222, 1–10.

Westendorf, J.J., Kahler, R. A. and Schroeder, T.M. (2004) Wnt signaling in osteoblasts and bone diseases. *Gene*, 341, 19–39.

Veeman, M.T., Axelrod, J.D. and Moon, R.T. (2003) A second canon. Functions and mechanisms of beta-catenin-independent Wnt signalling. *Developmental Cell*, 5, 367–377.

Toll-like receptor signalling

Gay, N.J. and Gangloff, M. (2007) Structure and function of Toll receptors and their ligands. *Annual Review of Biochemistry*, 76, 141–165.

Gregory W. and Konat, G.W. (2008) *Signaling by Toll-Like Receptors: Methods in Signal Transduction Series*, CRC Press, ISBN 978-1420043181.

Kaisho, T. and Akira, S. (2006) Toll-like receptor function and signalling. *Journal of Allergy and Clinical Immunology*, 117, 979–987.

Kawai, T. and Akira, S. (2005) Pathogen recognition with Toll-like receptors. *Current Opinion in Immunology*, 17, 338–344.

McCoy, C.E. and O'Neill, L.A. (2008) The role of toll-like receptors in macrophages. *Frontiers in Biosciences*, 13, 62–70.

∞ Literature link

Signalling in development and for the regulation of gene expression

Background information

The development of an organism from the moment of conception through to maturity is an extremely complex process. There are a wide variety of signalling components and pathways involved. One of the main aspects is the timing of when signalling is instigated and indeed turned off. However, underpinning the development of any organism is the production of proteins, and subsequently the action of those proteins. The presence of proteins in a cell will be determined by the expression of the genes that encoded them, and therefore to truly understand how development proceeds requires a good understanding of the regulation of gene expression, along with knowledge of the complement of proteins in a cell and how they function. A complete "picture" of which genes are being used, which proteins are present, which are quiescent or which are active, is needed before a full understanding of a cell's workings can be achieved. Therefore, many modern molecular techniques have been developed to aid the investigator, and such holistic approaches are often termed "systems biology".

Essay topic

Define the terms "transcriptome" and "proteome" in relation to cells. Discuss the main methods for studying the expression of single genes, and other methods which are available for investigations where a more holistic view of gene expression is needed. Discuss how a researcher may attempt to determine the full complement of proteins in a cell at any moment in time. What are the drawbacks to such approaches? Conversely, how will such approaches aid in the identification of diseases and development of new treatments?

Starter references

Andersen, H.S. and McArdle, H.J. (2004) How are genes measured? Examples from studies on iron metabolism in pregnancy. *Proceedings of the Nutrition Society*, 63, 481–490.

Chen, X., Jorgenson, E. and Cheung, S.T. (2009) New tools for functional genomic analysis. *Drug Discovery Today* (in press).

Churchill, G.A. (2002) Fundamentals of experimental design for cDNA microarrays. *Nature Genetics*, 32, 490–495.

Griffiths, W.J. and Wang, Y. (2009) Mass spectrometry: from proteomics to metabolomics and lipidomics. *Chemistry Society Reviews*, 38, 1882–1896.

Süss, C. and Solimena, M. (2008) Proteomic profiling of beta-cells using a classical approach - two-dimensional gel electrophoresis. *Experimental and Clinical Endocrinology and Diabetes*, 116, S13–20.

14 Life, death and apoptosis

Organisms need in many cases to develop from a single cell, and in doing so an anatomy has to be created. This process not only involves the creation of new cells, but also the removal of cells that are not needed. Fully formed organisms also need to remove old and dysfunctional cells, allowing the replacement of damaged cells. It is now recognized that molecular mechanisms are in place that lead to cells committing suicide, a process referred to as apoptosis.

Apoptosis, therefore, is a process that leads to the death of the cell, and as such needs to be very carefully controlled. It is the signalling involved in the regulation of apoptosis that will be discussed in this chapter. Many of the components involved have not been discussed previously, and therefore several new proteins with confusing names will be introduced, but the principles involved in signalling cascades are the same as discussed before. The processes are initiated with the recognition of extracellular signals or intracellular damage, which leads to cascades of events involving specific interactions, and results in a response, in this case the death of the cell involved.

This chapter highlights that it is not only the survival, adaptation, and normal functioning of the cell that are under the control of cell signalling events, but also that ultimately death of the cell may be brought about by cell signalling cascades. However, the lack of death of cells can be detrimental too, with cancer being a common result.

14.1 Introduction

Discussions in previous chapters have concentrated on the way cells respond to extracellular ligands, which in many cases enable them to adapt and survive. However, cells often need to die, at a precise time and in specific places in an organism. This death of cells results from a precise mechanism and is often referred to as programmed cell death (PCD), and in many cases is termed apoptosis.

Very few of the cells in a multicellular organism are the same age as the body as a whole. If a person says they are 36, few of their cells have been around for 36 years. Cells within an organism are dividing to create new cells continually, and cells are dying continually. Certainly not all cells divide regularly, but many do, and when cells are old and dysfunctional there is a need to get rid of them. It has been estimated that in an average adult human 50–70 billion cells undergo cell death by apoptosis every day. During development, the process of apoptosis is also critical. For example, in the nematode *Caenorhabditis elegans* development is reliant on the specific death of 131 cells. In higher organisms too, the formation of organs is dependent on selected cell death as they develop, and studies where apoptosis has been disrupted result in mis-formed and oversized organs.

Other scenarios exist where there is a need to have the death of a cell within an organism, with the good of the whole organism being the result. For example, if a plant is invaded by a pathogen, one of the responses is the hypersensitive response. Here, the plant cells surrounding the invading pathogen die, so depriving the pathogen of nutrients and depriving it of a route to invade the rest of the plant. Dead spots are seen on leaves etc., but the plant as a whole survives to fight another day. The death of cells, therefore, is essential at times, and has to be controlled carefully. This chapter will give a brief insight into some of the signalling mechanisms involved in the process of apoptosis, but the full details are still to be unravelled by the large research interest in this field.

14.2 An overview of apoptosis

The idea that cells might need to die is not a new one and was originally proposed as long ago as 1842 by Carl Vogt, who studied the development of toads. However, it was not until 1972 that the term apoptosis was actually coined. The word apoptosis is derived from a Greek word which means "to fall away from", and, amongst other things, was used to describe the fall of autumn leaves from trees. It is now a widely used term to describe the process of naturally occurring cell death, or cellular suicide. However, it is not always used correctly in the strictest sense, as apoptosis is a particular form of cell death, and a more accurate term in many cases would be programmed cell death, or PCD, meaning that an intracellular mechanism was used to bring about the death of the cell.

Apoptosis, in the true sense, is accompanied by defined morphological changes, which are distinct from the changes seen during death by necrosis.

In apoptosis, cell shrinkage and membrane ruffling (or blebbing) occur, and the cell disintegrates into small membrane bound vesicles called apoptotic bodies. Inside the cell, chromatin condensation and nuclear fragmentation occur, which are accompanied by breakdown of the DNA into regular-sized fragments, often referred to as DNA laddering (named from the pattern seen on an electrophoretic gel). On the surface of the cell, lipids are re-arranged in the bilayer of the plasma membrane with the lipid phosphatidylserine becoming exposed to the outside. Therefore, it can be seen that profound changes occur in the cell during this process of apoptosis, and the result is the death of the cell. It needs to be controlled very carefully, as the death of the wrong cells, or death of cells at the wrong time, may be catastrophic to the organism as a whole. Alternatively, the lack of death of the cells can be detrimental too, and over-proliferation of cells without a balancing death of cells can lead to cancer and the growth of tumours.

There are two main cell signalling pathways that have been identified in the control of apoptosis: the intrinsic pathway, or core pathway, which involves the mitochondria, and the extrinsic pathway, which involves cell surface receptors. These pathways, will be discussed further in **Sections 14.4** and **14.5** below.

Once cell death has been completed, the remains of the dead cell need to be removed. A multicellular organism cannot survive with a host of dead non-functioning cells as part of its main organs. Neighbouring cells and macrophages phagocytose the dead cells before they are able to release their cytosolic contents. Several genes have been identified that encode proteins involved in the two pathways of phagocytosis.

Macrophages

Macrophages are derived from monocytes, which circulate in the blood of mammals. If macrophages are required, perhaps as a result of tissue injury, monocytes leave the blood stream and differentiate and develop into macrophages. One of their major roles is to remove unwanted material, either host cells or pathogens.

14.3 Caspases

Apoptosis results in major changes to cell morphology and eventually cell death. The molecular machinery behind these events has to first control such activity, but secondly must bring about the changes seen. One central group of players in this molecular orchestra is the caspases (named from cysteinyl-aspartate-specific proteases). It is the caspases that mediate the cell shrinkage, DNA fragmentation and other changes that are seen. However, caspases are also involved in other events in cells, besides apoptosis.

Caspases are proteases, proteins that cleave and break down other proteins. In this case they are a class of proteases called the cysteine proteases, and they cleave their target proteins next to an aspartate residue. There are 14 known mammalian caspases, 11 of which have been found in humans. They can be divided into two main families. The first family are those related to caspase-1 (otherwise referred to as interleukin-1β-converting enzyme or ICE), and the second family are those that are related to CED-3, the product of a death gene from *C. elegans*. This second family can be further divided into two sub-groups, known as the caspase-2 and caspase-3 sub-families (see Table 14.1). The caspase-1 family are known to be involved in the processing of cytokines (see Chapter 4, Section 4.4), whereas the main role of the caspase-2 and caspase-3 sub-families is in cell death.

As mentioned above, caspases are cysteine proteases, which means that they contain a cysteine residue in their active site, and in fact they all contain the same conserved region around the active cysteine; that is, they all contain the peptide sequence -Glu-Ala-Cys-X-Gly- (where X is any amino acid). Caspase activity is such that they cleave the peptide bond in their target protein on the C-terminal side of an aspartate residue, but they do not all have exactly the same target preferences. Three groups of caspases can be distinguished according to their target sequence at which they will cleave. The caspases mainly involved in apoptosis are those in group II, which include caspases-3 and -7, preferring the target sequence -Asp-Glu-X-Asp-, and caspase-2, preferring -Val-Asp-Val-Ala-Asp-, and those in group III, which includes caspases-6, -8, -9, and -10, which cleave at -(Ile/Lev/Val)-Glu-X-Asp-.

The activation of the caspases is itself by proteolytic cleavage, and it has been found that one caspase can cleave and so activate another caspase, leading to activation cascades of caspases. Therefore, inactive caspases are in a non-cleaved form, referred to as the pro-caspases. The first cleavage event is an autocatalytic one, where the caspase cleaves and activates itself. This active caspase can then cleave and activate another caspase. This leads to another way of classifying caspases, either as initiator caspases or effector caspases. The former, initiator caspases, are those that can be at the start of a cascade and undergo auto-cleavage, whereas the effector caspases are down-stream

Table 14.1 The caspase families.

Caspase-1 family	CED-3 family	
	Caspase-3 sub-family	Caspase-2 sub-family
Caspase-1	Caspase-3	Caspase-2
Caspase-4	Caspase-6	Caspase-9
Caspase-5	Caspase-7	
mCaspase-11	Caspase-8	
mCaspase-12	Caspase-10	
bCaspase-13		
Caspase-14		

The prefix m denotes that it is of mouse origin, whereas prefix b denotes bovine origin.

and are activated by other caspases. Initiator caspases include caspases-2 and -9, whereas effector ones include caspases-3, -6, and -7.

Pro-caspases contain three or four domains, as seen in Figure 14.1A. Firstly, there is a pro-domain at the N-terminal end, followed by a large subunit domain. At the C-terminal end is a small-subunit domain, and between the subunit domains is often a linker region. During activation, cleavage takes place between the large subunit and small subunit domains, so releasing them as separate proteins. These then interact in a heterotetrameric fashion, that is two small subunits and two large subunits come together to form an active enzyme (see Figure 14.1A). The pro-domain may also be cleaved off and released. The active site of the functional enzyme is comprised partly of one small subunit and partly of one large subunit, so without protein–protein interaction no activity would be seen.

The structure of some caspases has been solved, and a representation of the structure of caspase-7 can be seen in Figure 14.1B. Solving of such structures really gives an insight into how these proteins work.

14.4 The intrinsic pathway

The intrinsic pathway for the control of apoptosis involves the mitochondria of the cell. This organelle is instrumental in normal cellular metabolism as the major site of ATP production, and therefore the main source of usable energy in the cell. However, alongside this important role it has become apparent that mitochondria also have a large part to play in the decision of whether a cell lives or dies.

In response to DNA damage, topoisomerase inhibition, cytoplasmic stress and many other stimuli, the mitochondria become permeabilized and release many proteins into the cytoplasm. The permeabilization involves proteins from the Bcl-2 family. Permeabilization of the mitochondria allows for the release of many proteins, but one of the most important is cytochrome *c*, a 12.5 kDa haem-containing membrane-associated (but also soluble) protein. This protein normally has a redox function, shuttling electrons between Complex III and Complex IV of the electron transport chain. Cytochrome *c* in mitochondria resides on the outer face of the inner mitochondrial membrane, where it is a peripheral protein associated with lipids. The release of cytochrome *c* involves its dissociation from these lipids, a process probably facilitated by the presence of free radicals, emanating from the mitochondria themselves. As cytochrome *c* is such an integral part of the mitochondrial respiratory chain and is used in the production of ATP for the cell, release of cytochrome *c* from the mitochondria has potentially two effects. First, it can act as a signalling molecule in the apoptosis pathway, as discussed below, and, secondly, it can no longer function in the electron transport chain and so ATP production is compromised.

A. Inactive pro-caspase

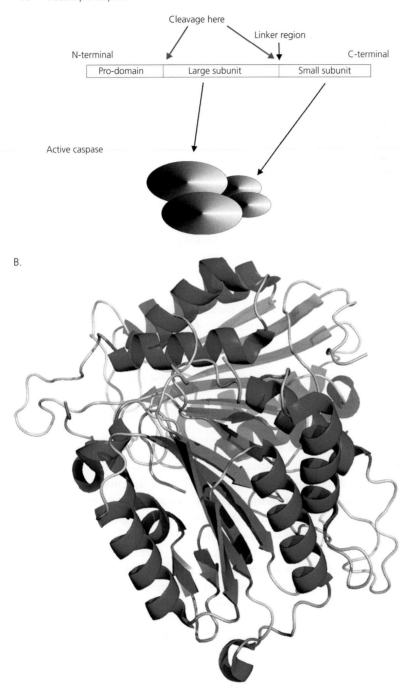

Cleavage here

Linker region

N-terminal

C-terminal

| Pro-domain | Large subunit | Small subunit |

Active caspase

B.

Figure 14.1 A. The domain structures of caspases. B. The structure of pro-caspase-7 obtained by X-ray diffraction. Structure was obtained from the RCSB Protein Data Bank (www.rcsb.org/pdb/) PDB ID: 1k86 (Chai, J., Wu, Q., Shiozaki, E., Srinivasula, S.M., Alnemri, E.S. and Shi, Y. (2001) Crystal structure of a procaspase-7 zymogen: mechanisms of activation and substrate binding. *Cell*, 107, 299–407).

Other proteins released from the mitochondria include endonuclease G, Smac/Diablo (which derives from second mitochondria-derived activator of caspases/direct IAP (inhibitor of apoptosis protein)-binding protein with low pI), and AIF (apoptosis-inducing factor), another redox protein, but this time a flavin-containing protein.

Members of the Bcl-2 family of proteins may promote apoptosis (pro-apoptotic) or inhibit apoptosis (anti-apoptotic). The major pro-apoptotic proteins here include Bax and Bak, which facilitate the release of cytochrome c from the mitochondria. Anti-apoptotic proteins include Bcl-2 and Bcl-X$_L$.

The signalling pathway that leads to apoptosis during the intrinsic route can be outlined as follows, and is shown in **Figure 14.2**. The DNA damage or cellular stress leads to the activation of pro-apoptotic Bcl-2 proteins, mitochondria become permeable and cytochrome *c* and other proteins are released. Cytochrome *c* forms a complex with a protein called Apaf-1 (apoptotic protease-activating factor 1), causing a conformational change in the structure of the Apaf-1 protein, which is dependent on the presence of ATP. The Apaf-1 forms an oligomeric complex, which then complexes with pro-caspase-9, forming a structure called an apoptosome. Caspase-9 within the apoptosome is cleaved, and activated, and so can cleave further caspases, resulting in the apoptotic effects seen. Other proteins released from the mitochondria also contribute to the morphological changes seen, as endonuclease G may migrate to the nucleus where it causes DNA degradation, and AIF may also lead to nuclear changes including condensation.

The control of apoptosis is critical and **Figure 14.2**, and the discussion above, describe a simplified version of events. There are many other proteins that can modulate this pathway. For example, a protein called p53 can neutralize the activity of anti-apoptotic Bcl-2 proteins, whereas in the cytoplasm there are inhibitors of apoptosis proteins (IAPs). These proteins are inhibitors of capases-3 and -9. During the onset of apoptosis by the intrinsic pathway, the protein Smac/Diablo released from mitochondria inhibits the activity of IAPs, so facilitating the activation of caspases. The intrinsic pathway is, therefore, a complex

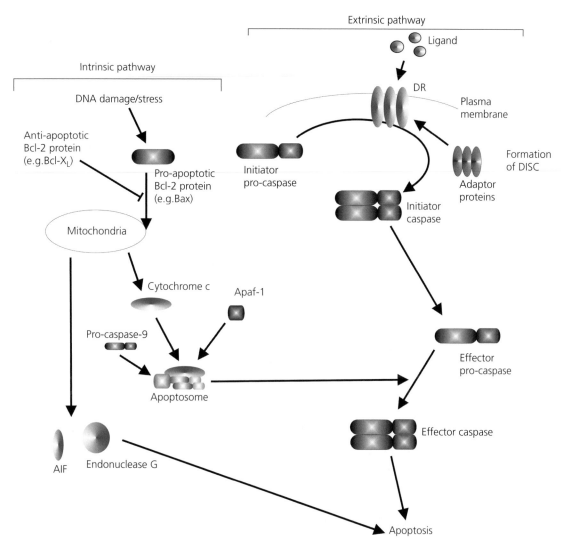

Figure 14.2 The signalling pathways of apoptosis. Two main pathways have been identified, the intrinsic and extrinsic pathways. DISC, death-inducing signalling complex; DR, death receptor.

inter-play between many proteins, some of which activate others, and some of which inhibit the activity of others, but all are focussed on the control of cellular death.

14.5 The extrinsic pathway

The extrinsic pathway is separate from the intrinsic pathway and involves the activation of protein receptors at the plasma membrane. These are the so-called death receptors (DRs). Death receptors have the role of ligand binding on the outside of the cell, and subsequent intracellular signalling leading to apoptosis.

Death receptors

A class of membrane receptors has been identified that are involved in apoptosis, and these have been termed the death receptors. They are a sub-group of the TNF receptor (TNF-R) family and include death receptor 3 (DR3), TNF-R1, and Trail-R2 (TNF-related apoptosis-inducing ligand receptor 2), among many others. They all have similar structures, having an extracellular region, a transmembrane domain and an intracellular region. The extracellular region is characterized by the presence of up to six cysteine-rich domains (CRDs), as can be seen in **Figure 14.3** for the DR3 receptor, which has four such domains. The intracellular region is characterized by the presence of a death domain, and it is this domain that is involved in intracellular signalling.

Ligands of these receptors are trimers, and often are found to be membrane proteins, although soluble forms, created either by cleavage from a membrane protein or from alternative-splicing events, are also known. Binding of the ligand to the receptor leads to trimerization of the receptors, which allows the signalling needed to take place.

Signalling from death receptors

Signalling during the extrinsic pathway starts with ligand binding to the death receptor, and is shown schematically in **Figure 14.2**. As discussed above, the ligand are trimers, and ligand binding leads to the trimerization of the receptors themselves. The death domains on the intracellular are protein–protein interaction domains (similar to those discussed in **Chapter 1, Section 1.7**). Ligand binding, trimerization and activation of the receptors

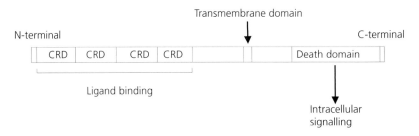

Figure 14.3 The domain structure of a typical death receptor, in this case DR3. The extracellular region has multiple cysteine-rich domains (CRDs). They all contain a transmembrane domain spanning the plasma membrane, and in the intracellular region is a death domain for protein–protein interaction and signalling. In the immature form they also contain a leader peptide at the N-terminal end for localization to the correct membrane in the cell.

allow recruitment of adaptor proteins, which bind to the death domains of the receptors. This is a similar mechanism to that seen with the receptor tyrosine kinases in **Chapter 6 (Section 6.3)**. The adaptor proteins here also contain death domains, and it is through the interaction of the death domain of the receptor and the death domain of the adaptor that the two proteins can come together. Adaptor proteins involved in apoptosis include FADD (Fas-associated protein with death domain). The adaptor protein FADD also contains a second protein interaction domain, known as the dead effector domain (DED). These interact with DED domains on initiator caspases, which leads to the formation of a complex of the receptor, the adaptor, and the caspase. This complex is known as DISC (death-inducing signalling complex; **Figure 14.4**). The complexes will activate the caspases within them, leading to the activation of the caspase cascade and resulting apoptosis.

Other signalling molecules involved in apoptosis

There are many other proteins involved in the mechanisms of apoptosis, and it is beyond the detail of the discussion in this chapter to include them all. However, there are several non-protein signals that are also involved, notably ROS and RNS such as nitric oxide. It is still unclear what their exact role is, but their production has been linked to the pathways that modulate apoptosis (see **Chapter 10** for discussion on the signalling roles of these molecules).

Other adaptor proteins, such as CRADD (caspase and RIP adaptor with death domain) contain a different protein–protein interaction domain for caspase recruitment. CRADD contains a caspase-activation and -recruitment domain (CARD), which allows CRADD to interact with pro-caspase-2.

The process of DISC formation can be modulated by the presence of other proteins which contain a DED region. Such proteins are the two isoforms of FLIP (Flice-like inhibitory proteins), $FLIP_L$ (long isoform, which has two DED domains and a caspase homology domain that is inactive) and $FLIP_S$ (short isoform, which has only two DED domains). Both contain DED regions and can compete with caspases for binding to the DISC, but as they have no caspase activity no active DISC is formed, and therefore apoptosis is not induced.

Modulation of the extrinsic pathway has also been found to occur by the presence of what has been termed decoy receptors. These are receptors that resemble the death receptors, and are able to bind to the same ligands as the death receptors, but they are unable to signal to the rest of the extrinsic pathway. Examples of decoy receptors include DcR2, which has

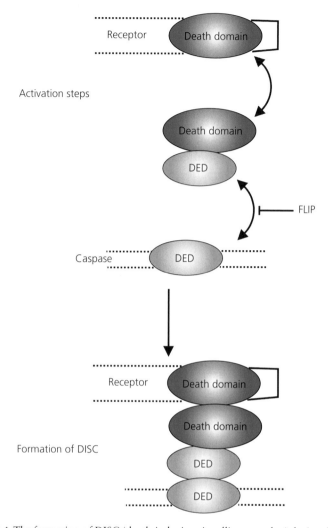

Figure 14.4 The formation of DISC (death-inducing signalling complex) during the extrinsic pathway. Protein–protein interactions between the death domains of the receptor and the adaptor protein, and the dead effector domains (DEDs) of the adaptor and the caspase, lead to the formation of a complex, termed DISC. This complex results in cleavage and activation of caspases. FLIPs (Flice-like inhibitory proteins) can disrupt caspase recruitment as they also contain DED regions.

no death domain and therefore cannot recruit the adaptor protein, and DcR1, which is a very truncated version of a death receptor and like DcR2 also has no capacity to signal into the cell. However, the presence of such receptors will compete for ligands and so stop, or reduce, the signalling that takes place through functioning receptors. In this case, DcR1 and DcR2 both bind to Trail and therefore will modulate Trail-induced apoptosis.

■ Summary

- Apoptosis is a process of cell death controlled by defined cellular pathways.

- Apoptosis is an important process during development as well as being involved in body maintenance during adult life.

- There are two signalling pathways leading to apoptosis: the intrinsic and extrinsic pathways.

- Caspases are important factors that mediate many morphological changes seen during apoptosis.

- Caspases are cysteine proteases; that is, they contain an active cysteine in their active site.

- Caspases cleave other proteins at the C-terminal side of an aspartic acid residue.

- Caspases are themselves activated by protein-cleavage events, existing in the inactive state as pro-caspases.

- Active caspases have a tetrameric structure of two large subunits and two small subunits.

- The intrinsic pathway of apoptosis involves the mitochondria, and the release of proteins from that organelle, including cytochrome c, apoptosis-inducing factor (AIF) and endonuclease G.

- Cytochrome c forms a complex with apoptotic protease-activating factor 1 (Apaf-1) and with caspase-9 to form a structure termed the apoptosome.

- Caspase-9 in apoptosomes becomes active and will cleave and activate further caspases, leading to apoptosis.

- The extrinsic pathway involves death receptors (DRs) at the plasma membrane.

- Death receptors contain multiple cysteine-rich domains and a death domain.

- Through the presence of protein–protein interaction domains, both death domains and dead effector domains (DEDs), the death receptor, adaptor proteins, and initiator caspases form a complex called the death-inducing signalling complex (DISC).

- Formation of DISC leads to caspase cleavage and results in apoptosis.

- DISC formation can be modulated by the presence of Flice-like inhibitory proteins (FLIPs) or decoy receptors.

→ Further reading

Antonsson, B. (2004) Mitochondria and the Bcl-2 family proteins in apoptosis signaling pathways. *Molecular and Cellular Biochemistry* 256, 141–155.

Conradt, B. (2001) Cell engulfment, no sooner ced than done. *Developmental Cell* 1, 445–447 [a paper on the pathways of phagocytosis, which removes cells that have undergone apoptosis].

Cotter, T.G. (ed.) (2003) Programmed cell death. *Essays in Biochemistry*, vol. 39 Portland Press, London [an excellent collection of articles on the topic, written by researchers in the field].

Diedrich, M. (ed.) (2003) Apoptosis, from signalling pathways to therapeutic tools. *Annals of the New York Academy of Sciences* 1010 [a substantial collection of papers on the topic arising from a conference in Luxembourg. It encompasses a vast amount of recent research in 799 pages].

Mor, G., Alvero, A.B. (Eds.) (2008) *Apoptosis and Cancer: Methods and Protocols: Methods in Molecular Biology Series*, Vol. 414, Humana Press, ISBN 978-1-58829-457-9.

Wang, S.L. and El-Deiry, W.S. (2003) TRAIL and apoptosis induction by TNF-family death receptors. *Oncogene*, 22, 8628–8633.

⌬ Literature link

Background information

The term apoptosis, as a description of cell death, was first used in 1972. Specifically, the term is used to describe a process that is highly regulated, but leads to the death of the cell as a desired outcome, and this is often termed "cell suicide". Since the term was first used there has been a vast amount of research to understand the molecular processes underpinning apoptosis and how the death of cells impinges on the physiology of organisms. The process is not confined to animals as in plants cell death is also important, with programmed cell death (PCD) after a pathogen attack often showing as lesions on leaves and other tissues.

Apoptosis was originally distinguished from necrosis, but it has more recently been realized that there are common features of both, and that the exact mechanisms which control cell death are complex and not easy to classify as simply apoptosis or necrosis. However, the death of cells is important for the continued survival of organisms, both during their development and during maturity. If apoptosis fails, or cell death programmes dysfunction, then this can lead to disease and death of the whole organism.

Essay topic

Discuss the role of apoptosis, or programmed death, in the disease of organisms. How is the death of cells measured?

How will understanding apoptosis better lead to improved disease identification and treatments in the future?

Starter references

Bignon, J., Bénéchie, M., Herlem, D., Liu, J.M., Pinault, A., Khuong-Huu, F. and Wdzieczak-Bakala, J. (2009) A novel iodomethylene-dimethyl-dihydropyranone induces G2/M arrest and apoptosis in human cancer cells. *Anticancer Research*, 29, 1963–1969.

Brown, V., Elborn, J.S., Bradley, J. and Ennis, M. (2009) Dysregulated apoptosis and NF kappaB expression in COPD subjects. *Respiration Research*, 10, 24.

Cotter, T.G. (2009) Apoptosis and cancer: the genesis of a research field. *Nature Reviews Cancer*, 9, 501–507.

Formigli, L., Conti, A., Lippi, D. (2004) "Falling leaves": a survey of the history of apoptosis. *Minerva Med*, 95, 159–164.

Schattenberg, J.M. and Schuchmann, M. (2009) Diabetes and apoptosis: liver. *Apoptosis*. (in press).

Suomeng, D., Zhengguang, Z., Xiaobo, Z. and Yuanchao, W. (2008) Mammalian pro-apoptotic bax gene enhances tobacco resistance to pathogens. *Plant Cell Reports*, 27, 1559–1569.

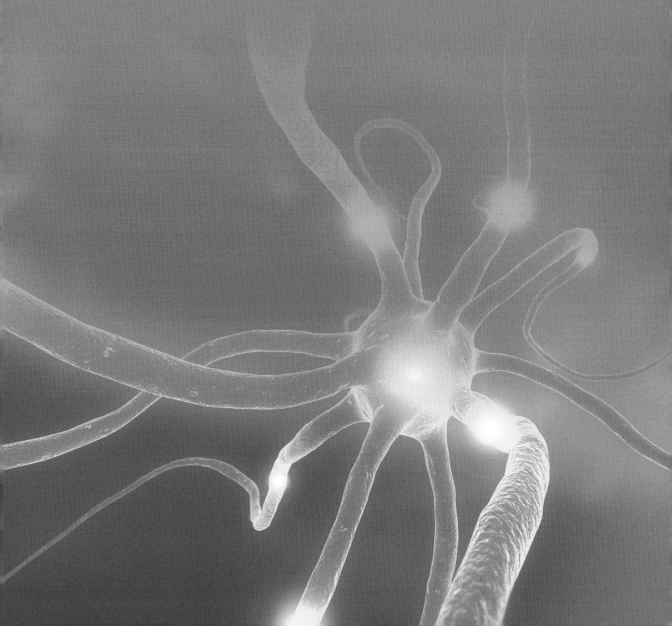

Part 4
Final thoughts

Cell signalling: importance, complexity and the future

15

On reading this book, or even just dipping into it, it is hoped that the reader will have gained an insight into the mechanisms used by cells to control their activities. However, it is also hoped that many questions may have been raised. In this final chapter, some such questions are discussed briefly, leading to a look to the future of cell signalling research.

15.1 **Introduction**

This book started with the remark that "*Cell signalling has arguably become one of the most important aspects of modern biochemistry and cell biology*", and it is hoped that the chapters which have preceded this one have emphasized that point. Organisms from their very moment of existence, either from a relatively simple cell division event, or from the instant that the DNA from a sperm enters the egg, rely on functional cell signalling for their very survival. Development, of their limbs, organs and form rely on cells to signal to each other and for cells in the neighbourhood to respond appropriately. When fully formed, the obtainment of a food source, the avoidance of predators, and adaptation to environmental stresses, whether that is heat, cold, too much salt, or lack of light, all rely on cells to sense and respond, using cell signalling mechanisms. Therefore, there is no part of modern biology, from environmental sciences to medicine, which is not impacted by the field of cell signalling. Dysfunctional cell signalling, after all, is responsible for a plethora of diseases and conditions.

Cells respond not only to external influences of course, but also to their own internal signals, and the inner workings of both plants and animals are awash with hormones and extracellular signals, such as cytokines. All these have to be received at the right time and by the right cells, and the perception has to be followed by appropriate action, the activation or inhibition of intracellular signal transduction, all of which comes under the umbrella of cell signalling.

15.2 Specificity, subtlety and crosstalk

It is the uniqueness of the signals used, and the specific perception of the signals that gives the cell the range of responses it needs to survive at any given moment, or event. Molecules that are structurally very similar can have very different effects, such as seen with cAMP and cGMP, and it is the specific nature of the interactions between these signals that is crucial. Often, a cell signalling pathway will start with the specific interaction between a receptor and its ligand, and only then will the cell recognize the presence of the ligand and respond. This perception has to be specific, and of course the complement of receptors possessed by a cell determines whether the cell is capable of responding to an external signal or not.

It must be remembered, however, that at any given moment in time, a cell is not simply sitting there responding to a single signal. It is not like a telephone operator talking to one caller at a time. It is more like the train operator on the platform when the train does not arrive, being surrounded by dozens of people, some asking the same question (Where is my train?), others asking a different but related question (When is the next one due?), and some just shouting nonsense, all in an atmosphere dominated by other unrelated noises. The cell has to respond as efficiently and quickly as possible, and ignore the "noise" that does not apply to it. By having specific receptors, specific protein interactions and specific effectors all ready to act, the cell can filter and take action.

However, many conundrums exist. A glance through the chapters above will reveal that the same components exist in many different pathways. The demands on a given component to signal might come from many sources at the same time, with different amplitudes, and perhaps even opposite messages. For example, if both hormone X and hormone Y lead to an increase in the concentration of signal Z inside the cell, when signal Z increases, how does the cell know that it is due to the perception of X or Y? It clearly does, but how? The answer must come from the complexity of the events that are taking place. It is clear from the literature that few signals prompt a single response from cells, and often several events are initiated when a component is activated. Phospholipase C, for example, generates $InsP_3$ and DAG, both potent signals. Therefore, signalling is not a series of parallel signal transduction pathways as depicted in many books, but rather it is a complex web of events, and this has been argued by some in the literature. However, it cannot be the case that every component can interact with every other component.

At any moment in time, there are many signals being initiated, reduced, or created, many of which will impact on others. The term crosstalk is often used to indicate that one pathway is interfering with, or at least having an influence on, another. Perhaps a good analogy would be if we consider our perception of music, a signalling event in itself. If one plays a D major chord it sounds harmonic and rather happy. We perhaps respond in a certain way to music in the major keys, they sound joyful and make us pleased. However, change a single note, an F sharp down to an F natural, and all of a sudden we have a minor key, and it sounds very different – it sounds depressing and sad. We respond differently. A little change, a mere semitone amongst three notes and we have a different response. Perhaps the cells are the same. It is the suite of signals that are responded to, but the singularity, and subtle changes within, can make large differences. Perhaps future computer modelling will reveal the answer.

Furthermore, as discussed in Chapter 1, the cell architecture as we now view it appears to be contrary to the findings that we sometimes see. For example, if the cytoplasm of the cell is a homogeneous fluid entity as described by many, why does calcium only appear in "hot-spots" within it when cells are activated? Why is it that reactive oxygen and nitrogen species also appear in "hot-spots"? Why don't they quickly diffuse. Why do apparently soluble proteins not interact with each other? What is the cellular architecture that we do not understand? Some receptors, which in theory are perceiving extra-cellular signals, appear to exist in the endoplasmic reticulum, but how does the ligand ever get perceived? What is it that we simply do not understand?

Clearly, these and many other questions raised show that we are a long way from a full understanding of cell signalling in any single organism, but future research will no doubt reveal the answers.

15.3 Longer term effects

Cell signalling often refers in many papers to the here and now, and most of the text above falls into this trap. But we must not neglect the longer term. A hint as to what might be happening in the long term may come from the studies using microarrays. In such studies, it is often found that in response to the perception of an extracellular ligand there is a profound change in gene expression. In these papers, the authors usually try to classify the types of genes that are expressed according to the type of protein they are thought to encode. It is no surprise to find that a significant proportion of such genes encode products that are involved in signalling, such as receptors, kinases, G proteins, etc. Therefore, the perception of the ligand has not just led to the immediate response, but has led to the alteration of the complement of signal-ling components within the cell, and therefore has altered the way the cell might respond a second time. Such events no doubt enable a cell to adapt to its new environment, or allow specialist cells to adapt for the good of the whole organism. Sometimes the genes that have their expression raised will even encode the proteins which produced the original signal, perhaps a way of the cells amplifying and emphasizing the original message, making it last longer with a more profound effect, until signalled to do otherwise by the perception of another different ligand.

But it does not always stop there. A careful look at the type of signalling proteins encoded by the genes for which expression is raised, or lowered, in microarray studies often reveals that a significant number are transcription factors, which themselves will alter the expression of yet more genes. It is an assumption without the studies to back it up, but it is almost certain that these transcription factors will alter the expression of more components of signal transduction pathways. Therefore, a ligand will put in place a cascade of events, leading to gene expression, leading to production of new proteins, which lead to altered gene expression, which will lead to altered signalling, some of which might involve even more transcription factors, and so the cascade might continue. Therefore, the long-term ramifications of signalling need to be considered further in many cases. Only further external and internal influences will modify such on-going events. Clearly, therefore, a cell's functioning and activity is controlled by signalling that has taken place in the recent past, as well as the signalling taking place in the present, all of which will be influenced by the signalling of the future. A complex, dynamic and fluid system, which we are only just starting to understand.

Mechanisms used in cell signalling also exist that are not discussed in detail above. Mechanical stresses and pressures have been shown to have profound effects on cells, for example used in auditory perception. Pressure pulses appear to be sensed by plants too, and others have used mechanical shock to study gene expression in cells from both plants and animals.

15.4 The future

The above discussion highlights the complexity of signalling and perhaps paints a negative view of the future, but recent technological advances mean that studies in cell signalling, along with all aspects of molecular biology, are at an exciting point. Microarrays, and similar technologies, now allow the study of expression of all the genes in a genome at the same time, whereas genome sequencing studies are expanding the availability of such arrays, across a wide range of species, from prokaryotes to humans. These sequencing studies also allow for the comparison of protein function across species, so protein domains with new functions will no doubt come to light. Proteomics, with mass spectroscopy, is also opening the door to holistic studies of cells and promises to yield a wealth of information.

RNA interference, and the on-going study of diseases and mutants, allows signalling to be studied where one or more components have been removed, or at least their activity ablated. Therefore, the ramifications of the lack of a function to the cell as a whole, and to the web of signalling involved, can be studied. At the moment, often such studies are hard to interpret, as the lack of a protein might do very little, suggesting that cells have degeneracy in their signalling, and removal of one signal allows another simply to take over. Or, removal of one component has a profound effect, and authors of papers claim that they have found the key component, only for other researchers to remove another protein and claim that their protein is in fact the key one. How can they both be right? Perhaps they are, and it is the inter-play and subtlety of signalling that is important.

Perhaps a final thought should consider what is perceived as one of the world's greatest challenges, that is global climate change, often referred to as global warming. If the temperature rises, and a plant or animal needs to adapt to this change, it will be the cell signalling that allows it to achieve this. The signalling components, including many of those discussed above will ensure that the cells, tissues or organisms survive (if possible). Furthermore, the signalling will change the complement of transcription factors, and so the suite of gene expression, which will alter the cellular components and so the possible future responses the cells or organisms can mount. Future evolution will alter the proteins, and their expression, to tailor the cellular conditions and responses in line with the new environment. Problems may arise if the temperature changes so rapidly, or to such an extreme, that nature, and the signalling that controls it, cannot keep up.

With such challenges ahead, along with the plethora of diseases that cell signalling can cause, or indeed be part of a cure for, much future work in this area will be carried out. There is no doubt that individual studies are making great advances into our understanding of cell signalling, but a full understanding is some way off. Assimilation of data, with authors taking a holistic view of the workings of the cell, will no doubt reveal how signalling truly works and allow manipulation of signalling events, which will lead to the creation of better crops and curing of many diseases, or even allow us and the plants and animals around us to survive into future millennia as the planet changes.

→ Further reading

Hancock, J.T. (2008) Cell signalling is the music of life. *British Journal of Biomedical Sciences*, 65, 205–208.

Hetherington, A.M. and Woodward, I. (2003) The role of stomata in sensing and driving environmental change. *Nature*, 424, 901–908.

Latchman, D.S. (2003) *Eukaryotic Transcription Factors*. 4th edn. Academic Press. ISBN 0124371787 [the topic of transcription factors has not been covered in detail above, so excellent texts such as Latchman are recommended].

Pawson, T. (2004) Specificity in signal transduction: from phosphotyrosine-SH2 domain interaction to complex cellular systems. *Cell*, 116, 191–203.

Wan, X., Steudle, E. and Hartung, W. (2004) Gating of water channels (aquaporins) in cortical cells of young corn roots by mechanical stimuli (pressure pulses): effects of ABA and of $HgCl_2$. *Journal of Experimental Botany*, 55, 411–422.

■ INDEX

A

14-3-3 proteins 25
A23187: 215
abscisic acid (ABA) 64
accessory proteins 87, 88
acetylcholine 41, 73
    receptor 85–86
Achlya 75
ADAM 303
adaptins 101
adaptor proteins 23, 35, 86, 128, 133, 178,
        31, 323
adenosine 73
adenosine monophosphate (AMP) 168
adenylate cylase, *see* adenylyl cyclase
adenylation 141
adenylyl cyclase 17, 40, 152, 154–156
    control by G proteins 156–161
    soluble 155
ADP-ribosylation factor (ARF) 159,
        180, 203
ADP-ribosyl cyclase 236, 237
adrenaline, *also see* epinephrine 31
adrenergic receptor 82, 124, 160
    desensitization 97
adrenergic receptor kinase 97, 123, 288
Aequorea forskalae 238
aequorin 238
ageing 258
agonists, defintion of 92
Agranoff 41
AHLs 75
AICAR 125
AIF 320
allene oxide cyclase 206
Allomyces 75
α-aminobutyric acid (GABA) 73, 82
amylase 277
β-amyloid 194
alternative splicing 72, 86, 118, 123, 137,
        161, 176
Alzheimer's disease 86, 194, 210, 258
8-amino-cADPR 236
amoeba 12
AMP-activated protein kinase 125, 270
AMPK kinase 125
amplification of signals 6, 16–17, 32, 108,
        133, 153, 173, 286
anabolic 269
androgens 60, 90

angina 245
annexin 202, 221
antagonist, definition of 92
Anthopleura elegantissima 75
anthopleurine 75
AP-1: 254
Apaf-1: 320
APC 306
APH-1: 303
apoptosis 138, 255, 315–324
    extrinsic pathway 321–323
    intrinsic pathway 318–321
apoptosome 320
apoptotic bodies 316
β-ARK 84
April 66
Arabidopsis 25, 48, 136, 251, 255, 296
arabinose operon 154
arachidonic acid (AA) 59, 178,
        199, 232
arachidonoyl-CoA synthase 202
arachidonoyl-lysophopholipid
        transferase 202
arteriosclerosis 72
Armadillo protein 13
arrestin 84, 97, 98, 124, 288
Ashman 165
aspartate phosphorylation 136
aspirin 60
AtGORK 221
ATP, as signal 74
AtSKOR 221
AtTPC1: 221
autism 86
autocrine 12, 66
autophosphorylation 127, 128, 133,
        175, 273
auxin 40, 62, 63
Axel 290
axin 19, 306

B

bacteriorhodosin 82
Bak 320
Baldridge 252
barbiturates 86
bathorhodosin 284
Bax 320
Bcl-2: 318, 320

betacellulin 72
binding measurements 93–94
bioinformatics 21, 22, 49,
        108, 213
Bitensky 41, 285
blue-light receptors 289
BMPs 301
bombykol 75
Bombyx mori 75
Bourne 41
bradykinin 248
brassinosteroids 65
Bredt 246
bride of sevenless (Boss) 13
Brody's disease 210
Buck 290
Burgoyne 39
Butenandt 75

C

C2 domain 213–214
Ca^{2+}- ATPase 120, 219, 222
Ca^{2+}-calmodulin-dependent protein
        kinase 121–123
cachectin 70
Caenorhabditis elegans 89, 315
caffeine 225, 226
Ca^{2+}-influx factors (CIF) 221
Ca^{2+}-ions 185, 210–239
    endoplasmic reticulum 221–227
    mitochondria 228–229
Ca^{2+}/Na^+ exchanger, mitochondrial 228
Ca^{2+}-uniporter 228
Calbindin-D_{28K} 215
calcineurin 138
calciosomes 222
calcitonin 82
calcitrol 60
calcium buffering 214, 215
calcium gradients 229
calcium ionophores 215
calcium oscillations 229–230
calcium pumps 219–220
calcium re-entry 220
calcium signalling, history 40
calcium spikes 220
calcium waves 229–230
calmodulin 171, 185, 212,
        216–219